"This book is both a strong reference manual for practitioners in public finance and operations fields and a good textbook for the training of future public financial and operations managers. The material will stand the test of time as an essential resource for practitioners while remaining accessible for students. The extended examples and figures demonstrate thoroughly the principles and techniques explained in the text."

Kenneth A. Kriz, *Wichita State University, Kansas, USA*

Cost and Optimization in Government

The careful management of costs and operations are two of the most essential elements of operating any successful organization, public or private. While the private sector is driven by profit-maximizing incentives to keep costs to a minimum, the public sector's mission and goals are guided by a different set of objectives: to provide a wide range of essential goods and services to maintain social order, improve public health, revitalize the economy, and, most importantly, to improve the quality of life for its citizens. Although the objectives are different, it is just as important for public decision makers to make the best use of available resources by keeping the cost of operation to a minimum. This book demonstrates that with a careful emphasis on cost accounting, operations management, and quality control, all organizations and governments can increase efficiency, improve performance, and prepare to weather hard times.

This book is divided into three parts: Part I offers thorough coverage of cost fundamentals, with an emphasis on basic cost concepts, cost behavior, cost analysis, cost accounting, and cost control. Part II examines optimization in costs and operations in government including traditional or classical optimization with applications in inventory management and queuing, followed by mathematical programming and network analysis. Finally, Part III explores special topics in cost and optimization, in particular those related to games and decisions, productivity measurement, and quality control. Simple, accessible language and explanations are integrated throughout, and examples have been drawn from government so that readers can easily relate to them. *Cost and Optimization in Government* is required reading for practicing public managers and students of public administration in need of a clear, concise guide to maximizing public resource efficiency.

Aman Khan is Professor of Political Science and Public Administration at Texas Tech University, USA.

PUBLIC ADMINISTRATION AND PUBLIC POLICY
A Comprehensive Publication Program

EDITOR-IN-CHIEF
DAVID H. ROSENBLOOM
Distinguished Professor of Public Administration
American University, Washington, DC

Founding Editor
JACK RABIN

RECENTLY PUBLISHED BOOKS

Cost and Optimization in Government: An Introduction to Cost Accounting, Operations Management, and Quality Control, Second Edition, by Aman Khan

The Constitutional School of American Public Administration, edited by Stephanie P. Newbold and David H. Rosenbloom

Contracting for Services in State and Local Government Agencies, Second Edition, William Sims Curry

Democracy and Civil Society in a Global Era, Scott Nicholas Romaniuk and Marguerite Marlin

Development and the Politics of Human Rights, Scott Nicholas Romaniuk and Marguerite Marlin

Public Administration and Policy in the Caribbean, Indianna D. Minto-Coy and Evan Berman

The Economic Survival of America's Isolated Small Towns, Gerald L. Gordon

Public Administration and Policy in the Caribbean, Indianna D. Minto-Coy and Evan Berman

Sustainable Development and Human Security in Africa: Governance as the Missing Link, Louis A. Picard, Terry F. Buss, Taylor B. Seybolt, and Macrina C. Lelei

Information and Communication Technologies in Public Administration: Innovations from Developed Countries, Christopher G. Reddick and Leonidas Anthopoulos

Creating Public Value in Practice: Advancing the Common Good in a Multi-Sector, Shared-Power, No-One-Wholly-in-Charge World, edited by John M. Bryson, Barbara C. Crosby, and Laura Bloomberg

Digital Divides: The New Challenges and Opportunities of e-Inclusion, Kim Andreasson

Living Legends and Full Agency: Implications of Repealing the Combat Exclusion Policy, G.L.A. Harris

Politics of Preference: India, United States, and South Africa, Krishna K. Tummala

Cost and Optimization in Government

An Introduction to Cost Accounting, Operations Management, and Quality Control

Second Edition

Aman Khan

NEW YORK AND LONDON

Second edition published 2017
by Routledge
711 Third Avenue, New York, NY 10017

and by Routledge
2 Park Square, Milton Park, Abingdon, Oxon, OX14 4RN

Routledge is an imprint of the Taylor & Francis Group, an informa business

First edition published by Quorum Books (imprint of Greenwood Publishing Group, Inc.) 2000

Library of Congress Cataloging in Publication Data
Names: Khan, Aman, author.
Title: Cost and optimization in government : an introduction to cost accounting, operations management and quality control / by Aman Khan.
Description: Second Edition. | New York : Routledge, 2017. | Series: Public Administration and Public Policy | Previous edition: 2000. | Includes bibliographical references and index.
Identifiers: LCCN 2016052234| ISBN 9781420067217 (hardback : alk. paper) | ISBN 9781315207674 (ebook)
Subjects: LCSH: Government productivity. | Public administration–Cost control. | Public administration–Cost effectiveness. | Mathematical optimization.
Classification: LCC JF1525.P67 K48 2017 | DDC 352.4/3–dc23
LC record available at https://lccn.loc.gov/2016052234

ISBN: 978-1-4200-6721-7 (hbk)
ISBN: 978-1-315-20767-4 (ebk)

Typeset in Times New Roman
by Wearset Ltd, Boldon, Tyne and Wear

To my family – Terri, Junaid, Jasmine, and Tyler

Contents

About the Author

Aman Khan is Professor of Political Science and Public Administration at Texas Tech University in Lubbock, Texas, where he teaches public budgeting, financial management, and quantitative methods in public administration. He is the author, coauthor, editor, and coeditor of numerous professional publications, including *Cost and Optimization in Government* (first edition), *Case Studies in Public Budgeting and Financial Management* (with W. Bartley Hildreth), *Budget Theory in the Public Sector* (with W. Bartley Hildreth), and *Financial Management Theory in the Public Sector* (with W. Bartley Hildreth). He previously served as the Director of the Graduate Program in Public Administration. Trained as an economist and planner, he has an MS in Urban & Regional Planning, an MA in Economics, and a PhD in Public Administration, with specialization in public budgeting and finance from the University of Pittsburgh in Pennsylvania.

Preface

This book is about cost and optimization, with an emphasis on cost accounting, operations management, and quality control. Managing costs and operations are two of the most important components essential for successful operation of an organization. Whereas cost has always been the focal point of business organizations, it has not received the same level of attention in government until recently. There is a simple reason for this: the private sector is driven by profit-maximizing incentives and cost is an integral part of it. Without knowing how to manage cost, it will be impossible for any firm or business to survive, let alone make profits. In fact, there is an inverse relationship between cost and profit: for a firm to maximize profit, it must keep its costs to a minimum. This plain and unassumingly simple principle that lies at the heart of all business operations explains why firms and businesses must remain efficient at all times.

In recent years there has been a shift in focus from profit maximization, especially in the short run, to revenue maximization in the long run. This apparent shift in business practice has been necessary to ensure that firms build up a clientele that remains loyal to it for a long time. Since consumer satisfaction is essential for the long-term viability of a firm, most firms are willing to sacrifice their short-term profit maximization objective for long-run revenue maximization goals. In contrast, a government does not operate to maximize profit either in the short run or the long run. Therefore, it does not necessarily have the same incentive to undertake measures that will increase efficiency in the same way the private sector does. Additionally, a government provides a wide array of goods and services, the net gain of which to society cannot be measured in precise monetary terms yet they are essential to maintain social order, improve public health, revitalize the economy, and raise the level of living for its citizens. In other words, they are necessary to improve the welfare of society, assuming one is able to measure "welfare" in some meaningful ways. In reality, it is almost impossible to measure welfare that would be consistent and sufficiently reflective of the social, economic, and political differences that exist in a democratic society. This makes it even more difficult for the decision makers to determine how efficiently the resources of society should be allocated to achieve a desired level of welfare.

From a theoretical point of view, the question of efficiency in resource allocation should not pose any major problems as long as the resources of a government are sufficient to meet the needs of its citizens. From a practical point of view, however, it is only when a government does not have sufficient resources to address the needs of the public or when public expectations exceed the ability of the government to meet its needs, which is often the case, that it becomes a major concern of the public and the government. Even without resource limitations or public expectations, it is still necessary, from the point of view of good management, for a government to function efficiently and effectively. Inefficiency in operation can not only contribute to poor performance, but also cause serious economic and financial problems, leading eventually to financial bankruptcy and default. Whereas the latter may not be so prevalent in government as in the private sector, there are instances such as those experienced by the cities of Cleveland, New York, Bridgeport, and Orange County, among others, where governments literally went bankrupt. In each instance, the respective governments were forced to take extraordinary measures, including hiring freezes, reduction in essential services, and raising taxes in some cases to regain control of their fiscal health. What this means is that to remain financially viable, all organizations, including government, must be willing to undertake measures and adopt methods and approaches that will increase their efficiency, improve performance, and enable them to deal with complex problems with relative ease. This book is a small attempt toward that objective. It is hoped that the approach taken in this book will generate interest among those who are constantly searching for better ways to deal with their everyday problems. To that end, if it helps even a single individual or organization, it will be considered worthwhile.

The book is divided into three parts. Part I deals with cost basics, with an emphasis on basic cost concepts, cost behavior, cost analysis, cost accounting, and cost control. Part II examines optimization in costs and operations in government. Included in this category are traditional or classical optimization with applications in inventory management and queuing, followed by mathematical programming, and network analysis. Finally, Part III looks at several special topics in cost and optimization; in particular, those related to games and decisions, productivity measurement, and quality control. While many of the topics discussed in the book appear to be analytically demanding, every effort has been made to keep them simple. Furthermore, examples have been drawn primarily from government so that readers can relate to them without much difficulty.

In putting together the chapters, the book has drawn heavily from a number of disciplines, including economics, accounting, statistics, operations research, industrial engineering, and public management. Each of these disciplines plays an important role in cost studies – how they are defined, measured, analyzed, and evaluated. For instance, there is a direct relationship between cost and economics; in particular, microeconomics, which studies the economic behavior of individuals, firms, and businesses (i.e., individual consumers and individual firms). Cost plays a critical role in the determination of that behavior. To give an example, when economists study the behavior of individual consumers and firms, they measure costs in terms of resource scarcity. What this means is that when resources are used for one purpose they cannot be used for anything else at the same time. Therefore, the cost of resource use is the cost of alternatives forgone. In other words, it is the sacrifice individuals, as well as firms and businesses, make by forgoing the opportunities they could have availed.

On the other hand, accounting, which has long claimed an exclusive interest in costs, measures costs in absolute terms. From an accounting point of view, costs are regarded as historical outlays rather than as opportunities forgone. This has an intuitive appeal since no one quite knows all the opportunities that are available to a decision maker at any given time. This does not mean that economists' notion of cost is irrelevant. What this means is that both perspectives are useful to study the cost behavior of an organization, similar to the cost behavior of individual consumers and firms. For instance, historical costs are important where legal, operational, and financial requirements of an organization are concerned, whereas opportunity costs are important where predictions have to be made on future costs, based on alternative courses of action. In fact, to be able to make good and reliable forecasts, one must have good historical data. Therefore, the two views on costs are complementary, not conflicting.

As an applied branch of mathematics, statistics is concerned with two things: providing summary information and drawing inferences about the characteristics of a population based on a sample drawn from the same population. The former is called descriptive statistics, and the latter inferential statistics. Both descriptive and inferential statistics are useful in cost studies, especially in measuring the cost behavior of an organization, such as measuring the characteristics of cost and related variables, analyzing their relationship, setting up confidence intervals, testing the significance of the estimated cost parameters, and so forth. Knowledge of statistics is also important in dealing with cost variables whose behavior is difficult to predict with certainty.

Like statistics, operations research can provide valuable information on cost and related variables that are important in decision making in government, but operations research is much broader in scope than statistics. It provides a comprehensive approach to a decision problem by constructing a model, collecting relevant data, finding a solution to the problem by evaluating the outcomes of alternative courses of action, and selecting the one that would best address the problem. By constructing models, which are at the heart of operations research, one is better able to deal with a complex array of relationships for a large number of variables that define the problem, which otherwise would not have been possible. However, the real advantage of using operations research in cost studies lies in its ability to experiment with solutions by sensitizing them to alternative scenarios without altering the basic structure of the model.

Industrial engineering, which uses some of the same approaches as operations research, was developed initially to study and analyze the manufacturing processes. Over time, as the field grew in importance, it began to shift its focus from manufacturing processes to the overall operation of an organization. This shift to the broader consideration of an organization and its operation makes industrial engineering much more useful in dealing with a multitude of cost problems such as scheduling, inventory control, setting cost standards, planning facilities location, and determining estimates of resource consumption, to name just a few.

Finally, a good management system must be in place to appreciate the critical need for integrating diverse knowledge into a coherent whole. Without this integration, these goals will not produce the kinds of results one would expect from their use. It is essential that the management understands this critical need and is supportive of those who are directly responsible for utilizing them to solve everyday problems. Proper communication between those who set up organizational policies and those who carry them out is critical to the success of an organization; particularly in government, where acceptance of new ideas and innovations are more gradual, such openness can provide a real and welcome change.

Acknowledgments

Several of my colleagues here and elsewhere took their time to read parts of the original manuscript, including those that appeared in the earlier edition. My sincere thanks and gratitude to each one of them; in particular, to Professors Jim Jonash, Robert McComb, and Klaus Becker of the Department of Economics; Professor Surya Limon of the Department of Industrial Engineering in the College of Engineering; Professors Omar Saatchioglu and James Burns of the Department of Information Systems and Quantitative Analysis in the College of Business; and Professor Olga Murova of the Department of Applied Economics in the College of Agriculture at Texas Tech University. Thanks are also due to Professor Irwin Morris at the University of Maryland, College Park; late Professor Hector Correa of the Graduate School of Public and International Affairs, and Professor Luis Vargas of the Department of Operations, Decision Science and Artificial Intelligence in the Graduate School of Business at the University of Pittsburgh; Professor Vic Valcarcel of the Department of Economics in the School of Economic, Political and Policy Sciences at the University of Texas at Dallas. Several of my former students, Drs. Christine Farias, David Yarskowitz, and Ken Hansen, took the time to read parts of the original manuscript; I very much appreciate their time. Again, my sincere thanks to each one of these individuals who took time from their busy schedules to read parts of the earlier and the current edition, but I alone bear the responsibility for any error that the book may contain.

In the same vein, I want to extend a special thanks to several individuals at Taylor & Francis without whose unfailing help, patience, and support this publication would not have been possible – Mr. Richard O'Hanley, publisher, CRC Press, who has been there right from the beginning, in particular for his kind and unfailing support throughout the entire process; Ms. Lara Zoble, editor at CRC Press, who had the patience to bear with me all this time and never failing to extend her support while I was working on the revised edition; Ms. Laura Stearns, publisher, Routledge Press, for her patience, help, and support in the true sense of the words; and Ms. Misha Kydd, assistant; Ms. Francesca Hearn, production editor; Ms. Ashleigh Phillips, production manager, and the entire production team for making sure that everything was in order during the production process. I owe each one of them my heartfelt thanks and gratitude.

Finally, this work would never have been complete without the love and support of my family – my wife Terri, my children Junaid and Jasmine, and a new member of the family, Tyler. I am fortunate to have them all in my life.

Part I

Cost Basics

1 Basic Cost Concepts

Cost is central to effective decision making in all organizations – public, private, quasi-public,[1] and nonprofit. It is the cement that holds together the different elements of an organization into a viable operating structure. For government in particular, it has become the principal guiding force that underlies a majority of decisions it makes, especially those related to the provision of public goods and services. To put it simply, without some knowledge of costs it will be difficult for a government to know what kinds of goods and services it can provide, the levels of their provision, and the length of time for which they will be provided. Even for those goods and services that are considered meritorious, such as education and healthcare, cost remains the single most overriding concern.

This chapter introduces several key concepts that are useful in cost studies; in particular, it defines and distinguishes between different types of costs, illustrates the contrast between cost, expense, and expenditure, highlights several commonly used methods of depreciation in cost studies, explains the relationship between cost and price, and discusses how to adjust costs for inflation. The chapter concludes with some basic guidelines for cost commitments.

What is Cost?

The term cost ordinarily means the value of economic resources used in the production or delivery of a good or service. The use of the word "value" has a special meaning in cost calculations: it indicates the actual dollar amount spent on materials, labor, and other factors of production. Since a government does not produce goods for direct public consumption, with some exceptions such as electricity, water, and a few others, the costs a government incurs in providing goods and services include the costs of acquisition, maintenance, and delivery.

Costs are generally expressed in cost units. A cost unit is a unit of a good or service in relation to which costs are ascertained. It is the basic unit of measurement used in all cost calculations. Examples of cost units in government will be the number of accidents prevented, gallons of water supplied, tons of garbage collected, and so on. However, costs can also be expressed in terms other than cost units, such as costs incurred by an agency, a department, a function, a program, a product (i.e., good or service). These individual units of a government and their functions, programs, and services to which costs are frequently charged are called cost objects. Theoretically a cost object could be anything, as long as it is possible to attribute costs to a responsible body unit.

When costs are charged to a cost object, it entails only those costs that belong to it and not to any other entity. On occasion, however, one may find two or more cost objects sharing a service responsibility and, therefore, the costs associated with it. These types of costs are generally known as joint costs. A good example of joint costs in government will be a public safety program that is jointly provided by a city police department and a county sheriff's office, where both organizations share the responsibility as well as the cost. Cost sharing for joint provision is an old concept, but is becoming increasingly common, especially for communities that are financially strapped.

Cost Classification

Classification generally means organizing activities with the same attributes into distinct groups or classes. In cost studies, classification means maintaining effective record keeping and establishing proper linkage to cost objects. Interestingly, there is no single particular type of cost that can uniformly apply to all government activities; different types and variations of costs exist, depending on the purpose they serve. This section presents several different types of costs that are used in varying degrees in government and other organizations: they are (1) direct vs. indirect; (2) fixed vs. variable; (3) mixed vs. step; (4) current vs. historical; (5) sunk vs. future; (6) incremental vs. total; (7) implicit vs. explicit; (8) product vs. period; (9) standard vs. actual; (10) relevant vs. irrelevant; (11) controllable vs. uncontrollable; and (12) one time vs. life cycle.

Direct vs. Indirect Cost. A cost is said to be direct if it can be unequivocally traced to a cost object. Since, by definition, direct costs must be unequivocally traced to a cost object, they do not involve any cost sharing between cost centers. This makes it possible to calculate them with relative ease compared to other types of costs. Examples of direct costs include personnel costs such as wages, salaries, and fringe benefits, as well as other non-personnel costs such as materials and supplies. In contrast, an indirect cost, also known as an overhead, cannot be unequivocally attributed to a specific cost object because they often involve multiple cost objects, although the cost may be necessary for the completion of an activity. Examples of indirect costs are indirect labor, supplies, insurance, depreciation, utilities, repair, and maintenance.

Fixed vs. Variable Cost. A cost is considered fixed if it does not change in relation to the quantity produced or consumed. This condition for fixed costs is true only in the short run (i.e., throughout the relevant range) because in the long run all costs are variable, meaning they do not remain fixed over a long period of time. Examples of fixed costs include salary, rent, interest payments on debt, and depreciation on structures and equipment. A variable cost, on the other hand, changes in direct proportion to changes in the quantity produced or consumed. It is fixed per unit of output, but varies in total as output level changes (Figure 1.1).

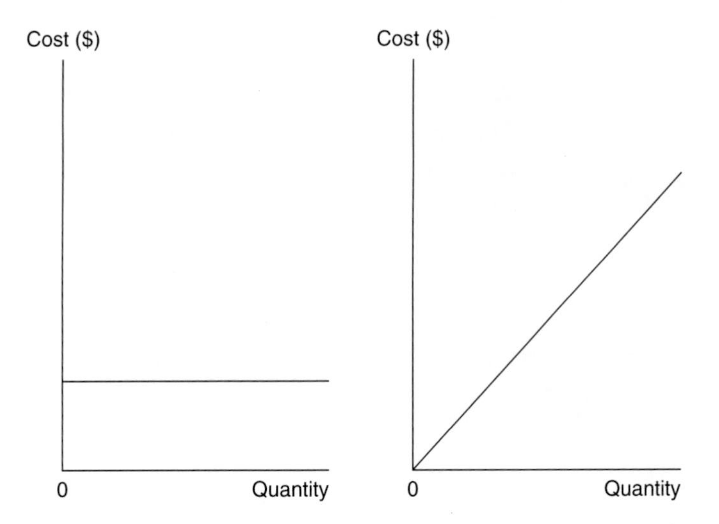

Figure 1.1a Fixed Cost

Figure 1.1b Variable Cost

For instance, if demand for a good consumed by a government increases, say, by 5 percent from 100 units last year to 105 units this year, where it sells for $25 per unit, the variable cost, which constitutes, say, 75 percent of the total cost, will also increase by 5 percent from $1,875 [100 × $25 × 0.75 = $1,875.00] to $1,968.75 [105 × $25 × 0.75 = $1,968.75]; that is, ($1,968.75 − $1,875)/$1,875 = 0.05. Wages, utilities, materials, and supplies are typical examples of variable costs.

Mixed vs. Step Cost. There are circumstances, however, where a cost is neither entirely fixed nor completely variable. These types of costs are called mixed costs. Mixed costs, also known as semi-variable costs, are a common occurrence in government. For instance, a public utility's billing rate may contain a flat (fixed) charge per month that is independent of consumption, plus a rate per unit in excess of the fixed monthly rate. A step cost, on the other hand, is a cost that does not change continuously as the quantity increases, but rather increases at discrete points, as in a staircase (Figure 1.2). For instance, the construction cost of a building that remains steady until the next level is added when it increases to a higher level as new costs are incurred, then becoming steady again until the next level is added, and so forth.

Current vs. Historical Cost. A current cost represents the value of a good or service consumed at the moment. The term current means that costs incurred in the past have no bearing on consumption in the present. For instance, a government that purchased, say, a desktop computer for $1,500 a year ago, which, if purchased today, would cost $1,200. The current cost of the item will be $1,200, even though the government paid $1,500 for it. Historical costs, on the other hand, represent the costs incurred or the amount paid at the time of purchase. Thus, the $1,500 the government paid a year ago for the computer will be the historical cost, regardless of how much it costs today.

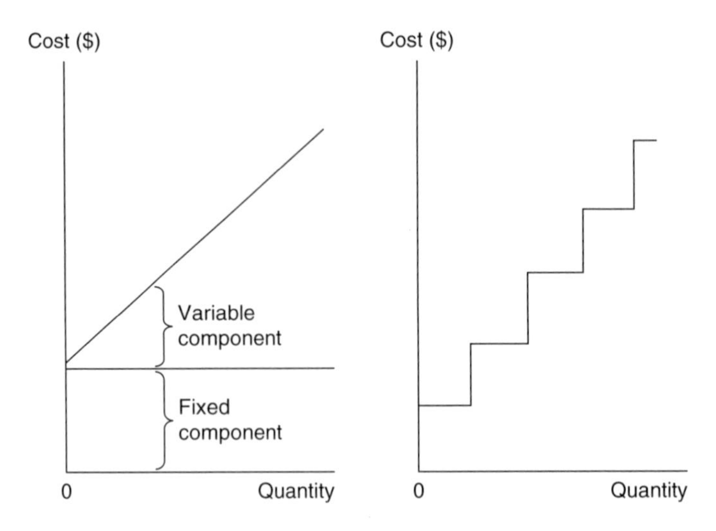

Figure 1.2a Mixed Cost

Figure 1.2b Step Cost

Sunk vs. Future Cost. A special case of historical cost, a sunk cost can be defined as the cost incurred in the past that has no resale or salvage value and, as such, it cannot be recovered. Therefore, if the computer in our previous example has no resale value (assuming it has become obsolete) but it cost the government $1,500 when first purchased, the sunk cost will be the entire amount of $1,500. A future cost, as opposed to a sunk or historical cost, will be incurred at certain points in time in the future. For instance, if the government decides to purchase a computer six months from now, the price it will pay for it at that time will be the future cost.

Incremental vs. Total Cost. An incremental cost, also called a marginal cost, is the additional cost one incurs as a result of a change in the course of an action. It is the cost of producing an additional unit of a good or service. For instance, if it costs $11.50 to produce 10 gallons of water and $12.25 to produce 11 gallons, the marginal cost or cost of producing the 11th gallon would be $0.75 [$12.25−$11.50=$0.75]. On the other hand, total cost is the sum of all costs involved in the production and delivery of a good or service. For instance, if it costs a government $500 in fixed cost and $1,500 in variable costs to undertake an activity, the total cost of the activity will be $2,000 [$500+$1,500=$2,000].

Implicit vs. Explicit Cost. Implicit cost, also known as imputed or implied cost, is the cost of what one must give up in order to consume or produce a good or service. The term is often used synonymously with opportunity cost. It is not really the cost one incurs in the true sense of the term, but is essential for making economic decisions. Since these are not actual costs, they do not show up in accounting records. For instance, if a government uses internally generated funds for which it receives no interests, but had the funds been invested they would have earned an interest income of, say, $1,500; the loss of interest income will be the implicit cost, which is the cost of opportunity forgone. Opportunity cost, therefore, is the cost one incurs by not selecting the alternative; in fact, all costs in some sense are opportunity costs. Explicit cost, on the other hand, is the cost one incurs in carrying out an activity and is recognized as such. Thus, if a government pays $500 in rent for a piece of equipment it uses to carry out an activity, it will be an explicit cost. Other examples of explicit costs will be salaries and wages, payments for materials and supplies, interest payments on borrowing, and so forth.

Product vs. Period Cost. Product costs are costs of direct labor, materials, and overheads that can be directly attributed to the production of a good or service. They are quite useful in inventory management, where they are often treated as inventory and do not appear on a financial statement unless they are used to produce a good. Period costs, on the other hand, include all costs other than product costs. They are usually not associated with the production or delivery of a good or service; as such, they cannot be assigned to the cost of production or delivery. Marketing costs, office depreciation, and costs of research and development are good examples of period costs.

Standard vs. Actual Cost. Standard costs are predetermined costs based on industry standards or on estimated value of labor, materials, and overheads. They are usually established on a unit cost basis. Actual costs, on the other hand, are costs an organization incurs in carrying out an activity. They are often compared against standard costs to determine the variation in costs of providing a good or service. For instance, suppose that it cost a government $1,500 to purchase 10 units of an item at a cost of $150 per unit, while the standard cost for the items is $1,250, based on, say, the current market price of $125 per unit, the difference will indicate an overpayment of $250 [$1,500−$1,250=$250] by the government.

Relevant vs. Irrelevant Cost. A cost is considered relevant if its value changes with the changes in the course of an action. Relevant costs are decision-specific, such as acquiring an asset or outsourcing a service; as such, it has a direct bearing on the cost of operation. An irrelevant cost, on the other hand, does not change with the changes in the course of an action. Sunk costs are good examples of irrelevant costs. This does not mean that the cost is no longer necessary, only that it does not have any quantitative effect on a decision. There may be a tendency to confuse between fixed and irrelevant costs since both remain unaffected by changes, but the two are not the same. In a fixed cost, cost remains the same regardless of the quantity produced or consumed, whereas for an irrelevant cost it remains the same regardless of the changes in the course of an action.

Controllable vs. Uncontrollable Cost. A cost is considered controllable if there exists a significant degree of control on its incurrence. In general, for a cost to be controllable, an organization must have sufficient discretion over it. Wages, salaries, fringe benefits, and various non-personnel costs are good examples of controllable costs. However, controllable does not mean 100 percent controllable, only that the organization has sufficient influence over it. Uncontrollable costs, on the other hand, are costs over which an organization does not have any discretion. Interest payment on debt, contract obligations, and entitlements are among the best-known examples of uncontrollable costs. They are called uncontrollable because when an organization, in particular a government, makes a commitment to purchase an item or provide a certain service guaranteed by legislative actions, the government is obligated to pay for the item or provide the service. In other words, it cannot avoid the cost of purchase or the provision of the service without violating any legal or contractual obligations.

One-Time vs. Life-Time Cost. A cost can be one time, or it can extend over the entire useful life of an asset, service, or program. One-time costs are those costs that do not involve any additional commitment of resources once a good or service has been provided. In other words, there are no subsequent obligations, including repair and maintenance, except for the initial cost of procurement. Seasonal contracts, with no direct or indirect benefits, are good examples of one-time costs. In contrast, a life-cycle cost includes all costs that are associated with the ownership of an asset, including acquisition, operation, and maintenance over its useful life. Costs associated with buildings, equipment, and service vehicles are typical examples of life-cycle costs.

Although they appear distinct, costs are not necessarily mutually exclusive. For instance, a cost can be direct as well as fixed, or it can be current as well as uncontrollable. When dealing with cost items with multiple characteristics, one must pay attention to the innate differences that exist in their characteristics and treat them in a manner that will avoid duplication in cost calculations.

Cost, Expense, and Expenditure

It is often necessary to make a distinction between three very important terms that frequently appear in cost calculations: cost, expense, and expenditure. As noted earlier, a cost is the value of economic resources used in the production or delivery of a good or service. An expense, on the other hand, is an expired cost resulting from the productive use of an asset. An asset is defined as anything of value acquired or owned. Since a majority of public services are provided for direct and immediate consumption by the public, they do not have any acquired value and, consequently, are not regarded as assets. When assets are expended in the course of providing a good or service such as preventing a crime or providing healthcare or education, they benefit the community a government serves; as such, they serve a productive purpose that is necessary for the welfare of society. Consequently, they are considered an expense. If such provisions do not result in any benefit – direct or indirect – they are considered a waste or loss. Finally, expenditure is a payment, or disbursement for a good or service an organization provides. From an accounting point of view, it can be defined as an unadjusted cost in that it does not reflect any changes in the consumption of an asset in the course of providing the good or service.

Reconciling the Difference

There is a fundamental reason why it is necessary to understand this distinction between cost, expense, and expenditure: when a government acquires an asset or purchases a cost item, it does not always consume it in the same period or it may consume it at different rates over time. This creates a lag in consumption, but also affects the way in which costs and expenditures are calculated (Cory & Rosenberg, 1984). From an accounting point of view, the lag indicates the amount of adjustment that is necessary for cost calculations. If there is no time lag (i.e., if goods purchased in one period are consumed in the same period), then expenditures and costs will be the same, but when they are not consumed in the same period, an adjustment is necessary to convert them to costs in order to maintain consistency in cost calculations.

Let us look at a simple example to illustrate this. Suppose that the public works department of a government spent $1,250,000 last year on street cleaning, including $84,000 on a new vehicle, $26,000 on materials and supplies, of which $16,000 worth of materials and supplies remain unused, and the rest went to wages, salaries, insurance, and miscellaneous other costs. Since most vehicles have a life expectancy of several years, let us assume that the vehicle purchased has a life of seven years in that it can be used next year, the year after, and so forth, although it may lose some of its use value each year due to wear and tear. By the same token, we can assume that not all of the materials and supplies will be fully consumed in the current year, such that a portion of the materials and supplies will be available for consumption next year or the year after. Assume that $16,000 worth of materials and supplies remain unused. Therefore, to determine the cost of street cleaning to the government we will need to make some adjustments to the total expenditure by taking into account these changes in consumption.

We can do this in two steps. First, subtract the costs of items such as the cost of the vehicle as well as the cost of materials and supplies from total expenditure for the year when they were purchased. The rationale for this is that since these items were not fully consumed in the year in which they were purchased, a portion of them will be available for consumption in the future. Therefore, subtraction would allow for the adjustments to be made for the portion of resources that remain unused during the accounting period. Second, add the portion of materials and supplies as well as the vehicle used for the year. For the vehicle, it is important to depreciate it to account for wear and tear. The cost of street cleaning, therefore, is the residue. In other words, it is the balance after all the adjustments have been made to the total expenditure. This is shown in Table 1.1.

As the table shows, the total cost of street cleaning after all the adjustments have been made came out to be $1,162,000, which is $88,000 less than the total expenditure for the year. The $88,000 is the unused portion of the resources ($72,000 for the vehicle plus $16,000 for materials and supplies) that will be available for future use.

Table 1.1 Cost and Expenditure Calculation

Street Cleaning: FY 2XXX ($)		
Total Expenditure:		*$1,250,000.00*
(–) Vehicle Purchase	84,000	
(–) Materials & Supplies	26,000	(110,000.00)
		$1,140,000.00
(+) Depreciation*	12,000	
(+) Materials & Supplies*	10,000	22,000.00
Total Cost:		$1,162,000.00

Note
*Consumed during the year.

To see if our calculation is correct, we can add $88,000 to $1,162,000, which will produce the total amount of expenditure for the year [$1,162,000 + $88,000 = $1,250,000]. Thus, we can formally define the cost of operation (i.e., street cleaning, in our example) as the total expenditure for any given year minus the value of the materials, equipment, and supplies purchased in that year plus the value of the materials, equipment, and supplies used up during the year. However, one must be careful in using this simplistic approach because it can complicate the conversion process when too many cost items are involved.

Cost and Depreciation

In the example on street cleaning we just presented, recall that we adjusted the cost of the service by depreciating the value of the vehicle for the year during which it was used. Without this adjustment, we would have had a different cost figure than the one we obtained. Depreciation, therefore, plays a critical role in all cost calculations, especially where physical assets are involved with several years of useful life. As noted earlier, depreciation occurs because of wear and tear as well as technological obsolescence. When an asset depreciates, it loses some or all of its use value because it is wearing out and, as such, fails to perform to its fullest capacity. On the other hand, an asset may be slow in wearing out, but may still lose its usefulness and in the process its value because of obsolescence such as changes in technology. Thus, the loss of value either because of wear and tear or because of technological obsolescence must be taken into account when calculating the depreciation of an asset.

There are different ways of depreciating an asset. This section discusses four frequently used depreciation methods: straight-line, declining balance, sum-of-the-years'-digits, and units of production. Of these, declining balance and the sum-of-the-years'-digits belong to a category of depreciation, known as accelerated depreciation.

Straight-Line Depreciation

The depreciation method we used in the street-cleaning example is the classic straight-line depreciation. An important characteristic of straight-line depreciation is that it is uniformly distributed over an asset's useful life, which means that depreciation will be the same for each year of the useful life of the asset, regardless of the nature of wear and tear. Determining the depreciation for a straight line is quite simple: all we need to do is divide the initial value of an asset by the number of years of its useful life (i.e., V_0/T, where V_0 is the initial value of an asset and T is the number of years in the useful life of the asset). However, in practice it is obtained by multiplying the beginning book value of an asset by the depreciation rate. The book value is the difference between the value of an asset and its accumulated depreciation, whereas the depreciation rate for the straight-line method is obtained by taking the reciprocal of the number of years in the useful life of an asset (i.e., $1/n$, where n is the number of years).

We can formally express depreciation under the straight-line method as

$$d_t = BV_{Beg} \times DR \quad [1.1]$$

where d_t is the depreciation at time t, BV_{Beg} is the beginning book value, and DR is the depreciation

To illustrate the method, let us return to the street-cleaning example. We know that the vehicle has a useful life of seven years. Let us assume that the vehicle has no salvage value; the depreciation for the vehicle for the first year will thus be \$12,000 [$\$84{,}000 \times 0.1429 = \$12{,}000$ (after rounding off)]. Note that \$84,000 is the initial value of the vehicle, also known as the beginning book value, and 0.1429 is the depreciation rate, obtained by taking the reciprocal of the useful life of the vehicle; that is, $1/7 = 0.1429$. The net book value at the end of the year, which is the ending book value, will be \$72,000 [$\$84{,}000 - \$12{,}000 = \$72{,}000$]. Since the vehicle depreciates at the same rate for each of the seven years of its useful life, the depreciation will be the same for each year; that is, $d_1 = d_2 = \ldots = d_7$. For T years, it will be $d_1 = d_2 = \ldots = d_T$ and the sum of the depreciations will the equal to its purchase price; that is, Σd_i (for $i = 1, 2, \ldots, 7) = \$12{,}000_1 + \$12{,}000_2 + \$12{,}000_3 + \$12{,}0000_4 + \$12{,}000_5 + \$12{,}000_6 + \$12{,}000_7 = \$84{,}000$. This is shown in Table 1.2.

We made an assumption early in the example that there is no salvage value at the end of the useful life of the vehicle, but what if there is a salvage value? In that case, we will need to adjust the book value by subtracting the salvage value from the initial value. Assume that the vehicle has a salvage value of \$14,000, the adjusted book value will thus be \$70,000 [$\$84{,}000 - \$14{,}000 = \$70{,}000$], which is our initial or beginning book value; the depreciation for the first year will be \$10,000, obtained the same way as before; that is, $\$70{,}000 \times 0.1429 = \$10{,}000$ (after rounding off). This is the amount by which we would have depreciated the vehicle to obtain the cost of street cleaning instead of the \$12,000 we used, had we taken the salvage value into consideration. This is shown in Table 1.3.

Table 1.2 Straight-Line Depreciation (without the salvage value)*

Year	*Annual Depreciation ($)***	*Accumulated Depreciation ($)*	*Net Book Value ($)*
1	84,000×0.1429=12,000	12,000	72,000
2	84,000×0.1429=12,000	24,000	60,000
3	84,000×0.1429=12,000	36,000	48,000
4	84,000×0.1429=12,000	48,000	36,000
5	84,000×0.1429=12,000	60,000	24,000
6	84,000×0.1429=12,000	72,000	12,000
7	84,000×0.1429=12,000	84,000	0

Notes
*Rounded-off to the nearest integer thousand.
**Beginning Book Value × Depreciation Rate.

Table 1.3 Straight-Line Depreciation (with the salvage value)*

Year	*Annual Depreciation ($)***	*Accumulated Depreciation ($)*	*Net Book Value ($)*
1	70,000×0.1429=10,000	10,000	60,000
2	70,000×0.1429=10,000	20,000	50,000
3	70,000×0.1429=10,000	30,000	40,000
4	70,000×0.1429=10,000	40,000	30,000
5	70,000×0.1429=10,000	50,000	20,000
6	70,000×0.1429=10,000	60,000	10,000
7	70,000×0.1429=10,000	70,000	0

Notes
*Rounded-off to the nearest integer thousand.
**Beginning Book Value × Depreciation Rate.

The greatest advantage of using a straight-line method is that it is simple and easy to use. Also, forecasting the future operation of an asset is much easier when it depreciates at a constant rate. On the other hand, it may not accurately reflect how an asset depreciates, as some assets may depreciate faster than others. Additionally, since it depreciates at the same rate, it is difficult to know when an asset needs to be replaced. Finally, it may not be useful for assets that go through periodic additions or other changes such as land, buildings, and physical plant.

Declining Balance Depreciation

In a declining balance, the depreciation is usually high during the early years of an asset's useful life. The rationale for this is that it tends to reflect the use pattern of assets, indicating that they are highly productive during the early years of their useful lives and their productivity declines gradually with time. There is also a tax advantage of using accelerated depreciation: because assets are highly productive when they are new, they are more heavily used during the early years, which allows businesses to write off the assets before the end of their useful lives or before they become obsolete. Consequently, it is mostly used by firms and businesses rather than by government. Like the straight-line method, it is also easy to understand and simple to use.

In general, the depreciation in a declining balance is treated as a fixed percentage of the book value of an asset. As noted earlier, the book value of an asset is the difference between the initial value of an asset and the accumulated depreciation over time. To give an example, suppose that we have an asset whose acquired value at the beginning of the year was $1,850 and the accumulated depreciation at the end of the year was $350. The book value of the asset (at the end of the year) will be $1,500. We can formally illustrate this with the help of a simple expression. Let FP_{BV} be a fixed percentage of the book value of an asset at the end of the first year of its useful life. The depreciation for the first year can thus be calculated as

$$d_1 = BV_{Beg} \times FP_{BV} \qquad [1.2]$$

where d_1 is the depreciation in Year 1, and BV_{Beg} is the beginning book value (i.e., the depreciable value for Year 1).

The depreciation for the second year is calculated the same way – multiply the beginning book value of the asset in Year 1, which is the depreciable value for Year 2, by the fixed percentage of the book value. That is,

$$d_2 = BV_{Bge1} \times FP_{BV} \qquad [1.3]$$

where d_2 is the depreciation in Year 2, BV_{Beg1} is the ending or net book value of the asset in Year 1, and FP_{BV} is the fixed percentage of the book value, as before.

The depreciation for the third year is calculated similarly, as

$$d_3 = BV_{Beg2} \times FP_V \qquad [1.4]$$

and so forth. The process usually stops when the salvage value is reached. The reason for this is that it is not worth pursuing if the net book value reaches the depreciation below the salvage value.

Double Declining Balance Depreciation. A special case of declining balance depreciation is the double declining balance depreciation, where the FP_{BV} is obtained by doubling the rate used for straight-line depreciation; hence, the name double declining depreciation. Although the rate is doubled, the total amount of depreciation will be the same as in straight-line depreciation. What this means is that depreciation will be high during the early years and decline gradually in the later years of the useful life of an asset.

To illustrate the method, we can apply it to the vehicle purchased by the government in our street-cleaning example. Assume that it depreciates at twice the rate of the straight-line depreciation, which will be 28.57 percent [$0.1429 \times 2 = 0.2857$]. Given this, we can easily obtain the schedule of annual depreciation and the book value for each of the seven years, as shown in Table 1.4. As the table shows, the depreciation is high in the early years of the vehicle and begins to decrease as time progresses.

Table 1.4 Declining Balance Depreciation (without the salvage value)*

Year	*Annual Depreciation ($)***	*Accumulated Depreciation ($)*	*Net Book Value ($)*
1	84,000×0.2857=23,999	23,999	60,001
2	60,001×0.2857=17,142	41,141	42,859
3	42,859×0.2857=12,245	53,386	30,614
4	30,614×0.2857=8,746	62,132	21,868
5	21,868×0.2857=6,248	68,380	15,620
6	15,620×0.2857=4,463	72,843	11,157
7	11,157×0.2857=3,188	76,031	7,969

Notes
*Rounded-off to the nearest integer thousand.
**Beginning Book Value × Depreciation Rate.

Table 1.5 shows the double declining balance depreciation with salvage value. Interestingly, salvage value is not generally used for the declining balance depreciation calculation. Part of the reason for this is that depreciation does not technically reduce the net book value to 0, as can be seen from the table (although it will eventually be, but will take much longer than the useful life of the asset). Since, by definition, the net book value is supposed to be 0 at the end of the useful life of an asset, as in straight-line depreciation, it does not make much sense to subtract it from the original value under declining balance because of the amount of time it will take to produce zero net book value.

A couple of points are worth noting here: whereas the double declining balance is useful in situations in which an asset is consumed at a higher rate during the early years of its useful life, it may not be suitable if it is consumed at a more or less consistent rate. An alternative would be to use moderate rates such as 1.5, which may better reflect the rate of consumption. From the perspective of business organizations, depreciating an asset at a faster pace can distort investment decisions if the depreciation fails to accurately reflect profitability. Consequently, most businesses tend to use a variation of accelerated depreciation called modified accelerated cost recovery system (MACRS).[2]

Table 1.5 Declining Balance Depreciation (with the salvage value)*

Year	*Annual Depreciation ($)***	*Accumulated Depreciation ($)*	*Net Book Value ($)*
1	70,000×0.2857=19,999	19,999	50,001
2	50,001×0.2857=14,285	34,284	35,716
3	35,716×0.2857=10,204	44,488	25,512
4	25,512×0.2857=7,289	51,777	18,223
5	18,223×0.2857=5,206	56,983	13,017
6	13,017×0.2857=3,719	60,702	9,298
7	9,298×0.2857=2,656	63,358	6,642

Notes
*Rounded-off to the nearest integer thousand.
**Beginning Book Value × Depreciation Rate.

Sum-of-the-Years'-Digits Depreciation

The sum-of-the-years'-digits depreciation, on the other hand, is a much slower version of an accelerated depreciation. The difference between this and the declining balance depreciation is that it depreciates against the full values rather than against the depreciated value of an asset. The result produces a rate of depreciation that is much slower than the double declining balance depreciation, but still higher than the straight-line depreciation. The way it is calculated is that each year a fraction, called fractional depreciation, is multiplied by the initial value of an asset to obtain that year's depreciation. That is,

$$d_{SYD}=BV_{Beg}\times d_f \quad [1.5]$$

where d_{SYD} is the sum-of-the-years'-digits depreciation, BV_{Beg} is the beginning book value, and d_f is the fractional depreciation.

The fractional depreciation, in turn, is obtained by dividing the remaining years in the useful life of an asset by the sum-of-the-years'-digits, as shown below:

$$d_f=t_R/SYD \quad [1.6]$$

where d_f is the fractional depreciation, t_R is the year remaining in the useful life of an asset, and SYD is the sum-of-the-years'-digits.

In general, the denomination of the fractional depreciation remains constant, while the numerator changes from year to year, representing the number of years remaining in the useful life of an asset, and is obtained by simply reversing the order of the years; that is, $t_{R_1}=T$, $t_{R_2}=-1$, $t_{R_3}=T-2,\ldots$, and so forth. The denominator, on the other hand, is obtained by adding the total number of years in the asset's useful life; that is, $SYD=1+2+3+\ldots+T$.

We can use the same street-cleaning example to illustrate the method. Since the vehicle in our example has a useful life of seven years, the denominator of the fraction, which is the sum of the digits, will be 28 (1+2+3+4+5+6+7=28). From this, we can write the fractional depreciation for each year, as follows: Year 1: $t_{g_1}/SYD=7/28=0.250$; Year 2: $t_{g_2}/SYD=6/28=0.214$; Year 3: $t_{g_3}/SYD=5/28=0.179$; Year 4: $t_{g_4}/SYD=4/28=0.143$; Year 5: $t_{g_5}/SYD=3/28=0.107$; Year 6: $t_{g_6}/SYD=2/28=0.071$; and Year 7: $t_{g_7}/SYD=1/28=0.036$. Next, we multiply the depreciable value of the vehicle by its corresponding fractional depreciation to obtain the depreciation for that year. This is shown in Table 1.6.

Table 1.7 shows the sum-of-the-years'-digits depreciation with the salvage value. As before, the net book value becomes 0 at the end of the seventh year of the vehicle's life. To see if our calculation is correct, we can add the depreciation for every single year of the asset; if the sum of these depreciations equals the initial cost of the asset, the result should be considered correct. The results presented in the table clearly show that the sum of the depreciation for all seven years adds up to $70,000, which is the initial depreciable value of the vehicle.

Table 1.6 Sum-of-the-Years'-Digits Depreciation (without the salvage value)*

Year	*Annual Depreciation ($)***	*Accumulated Depreciation ($)*	*Net Book Value ($)*
1	84,000×0.250=21,000	21,000	63,000
2	84,000×0.214=17,976	38,976	45,024
3	84,000×0.179=15,036	54,012	29,988
4	84,000×0.143=12,012	66,024	17,976
5	84,000×0.107=8,988	75,012	8,988
6	84,000×0.071=5,964	80,976	3,024
7	84,000×0.036=3,024	84,000	0

Notes
*Rounded-off to the nearest integer thousand.
**Beginning Book Value × Depreciation Rate.

Table 1.7 Sum-of-the-Years'-Digits Depreciation (with the salvage value)*

Year	*Annual Depreciation ($)***	*Accumulated Depreciation ($)*	*Net Book Value ($)*
1	70,000×0.250=17,500	17,500	52,500
2	70,000×0.214=14,980	32,480	37,520
3	70,000×0.179=12,530	45,010	24,990
4	70,000×0.143=10,010	55,020	14,980
5	70,000×0.107=7,490	62,510	7,490
6	70,000×0.071=4,970	67,480	2,520
7	70,000×0.036=2,520	70,000	0

Notes
*Rounded-off to the nearest integer thousand.
**Beginning Book Value × Depreciation Rate.

Although it has some of the same attributes of the declining balance depreciation, sum-of-the-years'-digits can be confusing and harder to use than the straight-line or even the double declining balance depreciation. Consequently, it has lost some of its appeals in recent years; still, it is used by a wide range of organizations, including banking and regulated industries.

Units of Production Depreciation

The final method, which is occasionally used in depreciation calculations in the private sector but has potential for use in government, especially for utilities, is the units of production depreciation. According to this method, the depreciation is based on the output of an asset rather than on its useful life. In order to calculate the method, we first divide the book value of an asset by the output the asset produces to obtain the depreciation cost per unit of production, and then multiply it by the (estimated) output to obtain the annual depreciation, as given below:

$$DC_U=(V_0-V_S)/O \quad [1.7]$$

$$d_U=O\times C_U \quad [1.8]$$

where DC_U is the depreciation cost per unit of production, V_0 is the initial value of an asset, V_S is the salvage value, O is the output, d_U is the unit of production depreciation, and C_U is the cost per unit of production.

As before, we use the street-cleaning example to illustrate the use of the method, but to do so we first need to know the total amount of output the vehicle would produce over its useful life. Let us say that it is given by the amount of vehicle miles to be covered over a seven-year period. It makes sense to use vehicle miles as a substitute for output, since the depreciation of the vehicle is directly related to the miles driven during its useful life. Assume that the vehicle would be driven an average of 100,000 miles each year for a total of 700,000 miles over the seven-year period. The depreciation cost per unit of production will therefore be

$$DC_U = (V_0 - V_S)/O = (\$84{,}000 - \$14{,}000)/700{,}000 = \$0.10$$

which is 10 cents per mile. Next, we multiply the vehicle miles for a given year by the cost per unit of production to obtain the depreciation for that year, which will be \$10,000 [$100{,}000 \times \$0.10 = \$10{,}000$] for the first year and repeat the process for the remaining years.

If we assume the miles to remain constant for each of the seven years, as we have in the current example, the depreciation will remain the same for all seven years, in which case it will give the appearance of a straight-line depreciation. This is where the method has an inherent weakness in that in order to obtain the correct depreciation one must be able to accurately estimate the output; that is, the estimated units of production over the useful life of the asset. If not, an average has to be used, as we did here.

Other Methods of Depreciation

In addition to the declining balance, sum-of-the-years digits, and units of production methods of depreciation, there are other methods that one can use to depreciate an asset. A good example of this will be sinking fund depreciation. A sinking fund is a fund in which money is deposited at regular intervals to pay for the principal and interests on a debt. According to this method, the depreciation for any year is the increase for that year, which is the deposit for the year plus any interest accrued for that year. The method produces a yearly depreciation schedule, which increases systematically with time and hence is viewed as the opposite of accelerated depreciation.

Cost and Inflation

Inflation is an increase in the general price level. Inflation affects both the consumer and the producer in the same way by lowering the value of money they spend on goods and services. From a cost point of view, inflation is a factor in how costs are measured, evaluated, and compared. Variations in costs over time reflect the changes in inflation, assuming everything else remains the same. Therefore, to get an accurate measure of costs, unaffected by inflation and called the real cost, one must take out the effects of inflation from costs.

The simplest way to remove the effect of inflation from costs is to deflate it by a price index. A price index compares the average price of a group of commodities in one period with the average price for the same group in another period. As a general rule, prices are determined first for a base period, and the prices of all subsequent periods are then measured in relation to the base year price. The price index that is most commonly used in cost adjustment is the consumer price index (CPI), which is based on the price of a group of basic items required by an average family. The Bureau of Labor Statistics (BLS), which calculates the CPI, uses a formula based on an index called the Laspeyres Index.[3]

Let us look at the street-cleaning example again to illustrate the effect of inflation on the cost of providing the service. As shown in Table 1.8, it cost the government $945,000 in Year 1 to provide the service, $961,000 in Year 2, $982,000 in Year 3, and so on. Note that Year 6 happens to be the year when the government bought the vehicle. Let us assume that we have the price indices for each of these years from BLS. As the table shows, the adjusted costs (real values) are much lower than the unadjusted costs (nominal values), as expected, obtained by dividing the unadjusted costs by their corresponding index values, as can be seen in the expression below:

$$RV=(CPI_0/CPI_t)\times NV \qquad [1.9]$$

where RV is the real value, NV is the nominal value, CPI_0 is the base year CPI, CPI_t is the CPI at time t, and the ratio CPI_0/CPI_t is the index value.

Now, applying the expression in Equation 1.9 to the street cleaning problem we get the following adjusted or real costs for street cleaning using Year 6, which is the most recent year, as the base year; that is, $1.3818\times\$945{,}000=\$1{,}305{,}801$ for Year 1, $1.3217\times\$961{,}000=\$1{,}270{,}154$ for Year 2, and so forth (Table 1.8).[4] What this means is that the cost of street cleaning in Year 1 in terms of Year 6, which is the most recent year, would be $1,305,801. Put differently, it means that had there been no inflation, the real

Table 1.8 Cost of Street Cleaning (Adjusted)*

Year	*Nominal Cost** ($)*	*CPI*	*Index Value*	*Real Cost ($)*
1	945,000	110	152/110=1.3818	1,305,801
2	961,000	115	152/115=1.3217	1,270,154
3	982,000	122	152/122=1.2459	1,223,474
4	1,117,000	130	152/130=1.1692	1,305,964
5	1,135,000	140	152/140=1.0857	1,232,270
6	1,162,000	152	152/152=1.0000	1,162,000

Notes
*Rounded-off to the nearest integer; Year 6=Current (Base) Year; ** Unadjusted.

cost and the initial cost of service provision would have been the same. In other words, the difference between the current cost of operation and its real cost is the cost of inflation.

In the event that the CPI is not available, we can determine the index using a simple procedure: divide the base year value, which is usually the most recent year, by the value for the year under consideration. Next, multiply the index value by the corresponding cost for a given year (the historical cost) to obtain the real cost for that year; that is, index value × historical cost = real cost.

To give an example, suppose that the water rate for a government for residential property is currently \$5 per 1,000 gallons of water, which was \$3.50 ten years ago ($t-10$). The index value for the water rate for Year 10 will be \$5.00/\$3.50 = 1.4286 (i.e., base year rate/rate for Year 10). For the current year, it will be 1.0 [\$5/\$5 = 1.0000]. Let us say that the cost of supplying residential water for Year 10 was \$2 million; in current year terms, it will be \$2,857,200 [\$2,000,000 × 1.4286 = \$2,857,200], which is the real cost of water. We can do the same for Year 9. Assume that the rate for Year 9 ($t-9$) was \$3.65 with a corresponding cost for the year of \$2.25 million; the cost of water for Year 9 in current year terms (i.e., the real cost) will thus be \$3,082,275 [(\$5.00/\$3.65) × \$2,250,000 = 1.3699 × \$2,250,000 = \$3,082,275], and so forth.

Besides the CPI, governments, in particular the federal government, have been using the GNP Implicit Price Deflator instead of the traditional CPI for deflating costs. The GNP Implicit Price Deflator is an index that takes into account not only the changes in prices, but also some changes in the quality of goods and services that individuals consume. Another alternative would be to use the GDP deflator, which is calculated the same way as the GNP deflator.[5]

Cost and Price

A term that frequently appears alongside cost is price. A price is simply a measure of value. It is the premium we pay for the goods and services we consume, but determining price for public goods is far more complex than it is for private goods. In the private sector, price is generally determined by calculations that reflect the profit-maximizing interest of a firm or business. These calculations typically include factors such as the target return on an investment, the market share the firm must maintain, the price it must charge to match competition, the need to maintain a stable price to avoid fluctuations in demand, and so forth (Lanzillotti, Dirlam, & Kaplan, 1958). Unfortunately, the same calculations do not readily apply to public goods because of congestion costs, the free-rider problem, and a variety of other factors that separate them from most private goods (Davis and Whinston, 1967; Samuelson, 1954). Nevertheless, it is important for government to have a basic understanding of the methods it can use to determine the price it must charge the consumers, vis-à-vis the public.

A Note on Natural Monopoly

Before discussing the pricing methods, it is useful to briefly introduce a term frequently used in microeconomics, called natural monopoly, since for the vast majority of goods and services a government provides, its behavior often reflects that of a monopolist; in particular, a natural monopolist. Monopoly is a market condition in which a single firm dominates the market. However, unlike a typical monopoly, which often controls the market by controlling the resource base and preventing other firms from entering the market, a natural monopoly does not prevent others from entering the market. Additionally, it can produce goods at a cheaper price because of economies of scale resulting from technology and specialization, which keeps the long-run average cost of production low. Contrary to conventional wisdom, natural monopolies are more efficient than most small firms, which are often constrained by a limited market and can be easily priced out.

Although the natural monopolies deviate from the conditions of free-market competition, they are not necessarily undesirable from society's point of view, as long as the benefits to society of lower prices from economies of scale outweigh the higher prices generally associated with traditional monopolies. Under these circumstances, society may be better off allowing some monopolies to exist, or it can assume the responsibility by producing the goods itself, or use regulation to control the price a monopoly firm may charge for the provision of the goods. Goods such as electricity, water, sewerage, and other capital-intensive goods are classic examples of activities that have built-in economies of scale that can be produced at a lower cost without the complications of higher prices. Also, the initial set-up costs of these activities are quite high, making them prohibitive for small firms to enter the market to provide the good. This may explain why a government often allows large private firms to undertake these activities or, in some cases, assumes the ownership of these activities itself, thereby serving the role of a natural monopolist.

Pricing Methods

In spite of the complexity of how price is determined, a government needs to have some knowledge of the methods it can use to determine the price it must charge the public, especially for those goods that are provided in a business-like manner. One such approach that has considerable appeal among the practitioners is an accounting approach, where costs serve as the threshold. Since no organization, public or private, would like to produce a good at a price that is less than its cost of production, such that it would incur a loss, an accounting approach can serve as a useful guide for determining price. A number of pricing methods have been developed over the years that have come to rely on this approach. Although most of these methods are designed for private goods, they can also be applied to public goods, especially those that are provided in a business-like manner.

This section focuses on several pricing methods that have been widely used in the private sector, but have considerable merits in government and not-for-profit organizations. They are (1) marginal cost pricing, (2) average cost pricing,

(3) total-cost pricing, (4) cost-plus pricing, (5) levelized pricing, and (6) transfer pricing. All of these methods, with the exception of marginal cost pricing, are non-marginal in that they do not base their analysis on incremental changes.

Marginal Cost Pricing

One of the most widely used concepts in price behavior is marginal cost pricing. Marginal cost pricing is based on a simple microeconomic rule that resources are best allocated if the price of goods and services is set equal to their marginal cost (i.e., the cost of producing an additional unit, MC); that is, $P=MC$. For instance, if it costs a firm $100 to produce 10 units of a good and $109 to produce the eleventh unit of the same good, the marginal cost (i.e., the cost of the eleventh unit) will be $\$109-\$100=\$9$. The $9 will then serve as the price at which the good will be sold in the market. If this were a public good, then the marginal cost would measure the cost of resources to society for producing an additional unit of the good and the price would measure the value of the additional unit to the public (Tyndall, 1951). However, for marginal cost pricing to work, the goods produced must be divisible so that they can be measured in discrete units such as bushels of wheat produced, tons of garbage collected, number of meals served, and so on.

The economic rationale for setting the price equal to marginal cost is that at that level the profit will be maximum, but for the profit to be maximum the rule further suggests that marginal cost (MC) must be equal to marginal revenue (MR) (i.e., the revenue earned from an additional unit of a good sold or service provided); that is, $MC=MR$. The rationale for this is quite simple: as long as $MR>MC$, the firm should continue to produce because there is a net gain from additional production until the marginal revenue becomes equal to marginal cost; that is, $MR=MC$. If, on the other hand, $MR<MC$, it should continue to lower production because there is a net loss from additional production until the marginal revenue becomes equal to marginal cost; that is, $MR=MC$. By extension, then, if $MR=MC$ and $P=MC$, then $P=MR$. Alternatively, if $MR=MC$ and $P=MR$, then $P=MC$. To give an example, suppose that a privately run outpatient surgical clinic can treat a maximum of 25 patients per day. The marginal cost of running the facility at that level is $2,500, which corresponds to its marginal revenue and produces a maximum profit for the clinic of, say, $7,500 per day. The price of surgery should then be set at $2,500, based on the MC rule. However, the clinic could see additional patients but it will not be as profitable (i.e., will not be maximum) since its marginal revenue will be below the marginal cost.

Assume now that it is a government-run clinic, in which case the price could be set at a level where it will be equal to average cost ($AC=TC/Q$), where TC is the total cost of running the clinic and Q is the quantity (i.e., number of patients treated). In general, when price is set equal to average cost, it is usually lower than the marginal cost and the price is known as the break-even price, because at that level the profit will be zero but the quantity produced (i.e., the number of additional surgeries, in our case) would be higher than the number under marginal cost pricing. This makes good sense, since for most government activities the goal is not to maximize profit, as long as they break even.

Although as a concept marginal cost pricing makes good sense when applied to firms and businesses, it does not necessarily work so well in government; in particular, where public goods are concerned. Part of the reason for this is that most public goods, except for those provided in a business-like manner, are not divisible; as such, they cannot be measured in discrete units. As a result, the marginal concept does not readily apply to them. Additionally, marginal cost pricing works well in situations where there is perfect competition because perfect competition forces firms to be efficient in their resource use. In reality, the market is neither perfect nor fully competitive, which means that inefficiency in resource use can get easily translated into the price the consumers pay for the goods and services they consume.

Average Cost Pricing

Unlike the private sector, a government is not in the business of making profit (although there are activities where it would prefer a net positive return). As noted earlier, as long as the government is able to break even, it will have achieved its basic economic and financial goal. The pricing method that is most suitable where break-even is concerned is average cost pricing, which is determined by setting the price of a good or service at a level where it is equal to the average cost; that is, $P=AC$.

To give an example, suppose that it costs a state-run four-year college with 10,000 students \$45 million annually to operate, 50 percent of which comes from the state, 40 percent from tuition fees, and the remaining 10 percent from a variety of other sources such as charitable contributions. Students take on average 30 credit hours during an academic year, including summer. Therefore, the price per credit hour for a student will be \$150, which is the average cost for a credit hour, obtained in the following way:

$$\begin{aligned} AC &= TC/Q \\ &= \text{Total cost}/N \text{ of credit hours} \\ &= \$45{,}000{,}000/(10{,}000 \times 30) \\ &= \$45{,}000{,}000/300{,}000 \\ &= \$150 \\ &= P \end{aligned}$$

At a price of \$150 per credit hour, the total revenue ($TR=P\times Q$) for the college will be \$45 million [$P\times Q=(\$150)(30\times 10{,}000)=\$150\times 300{,}000=\$45{,}000{,}000$], the same as the total cost, $TC=TR$; in other words, it will break even. However, given that 40 percent of the cost is covered by tuition fees and charges, the actual price the student will pay for each credit hour of college education will be \$60 [$\$150\times 0.40=\$60$], excluding books, supplies, transportation, rooms, meals, and so on. This means that 40 percent of \$45 million, or \$18 million

[$45,000,000 × 0.40 = $18 million] will come from tuition fees and the rest from other sources, although theoretically the college could charge the entire $150 for each credit hour, which is the theoretical break-even price.

Since average cost pricing is primarily used in situations where the goal is to break even, it will not produce maximum profit like marginal cost pricing. Also, because the price is set equal to average cost, it means that some would pay more and some less, depending on the level of consumption, but on average the price of the good will be the same for all consumers. Going back to the tuition fee example, because the college needs 300,000 credit hours to generate the break-even revenue, those taking more hours are essentially subsidizing those taking fewer hours for the college to be able to charge $150 per credit hour. This is known as cross subsidization. Cross subsidization is a common practice in government, such as in the postal service, water, sewerage, electricity, and so forth, where revenue generated from a particular service area or a group of consumers is used to compensate the loss in other areas or other consumers.

It is worth noting that average cost pricing is also used by government to regulate natural monopolies providing private goods. Since the monopolies have a tendency to produce less than the optimal quantity, pushing the prices up, government can use average cost pricing as a tool to regulate prices that the monopolies otherwise would charge. Put simply, average cost pricing can force a monopoly firm to reduce the price to a point where the firm's average cost equals the market demand. Alternatively, the government can provide the good itself, as long as it breaks even.

Total-Cost Pricing

Total-cost pricing is the price consumers must pay or an organization must charge consumers to recover the full cost of a good or service. It is called total-cost pricing because the focus is on the recovery of the full cost of a good rather than a fraction of it. A typical example of total-cost pricing in government would be the cost of infrastructure such as roads, bridges, highways, and other physical assets that are produced in lump sums. The fact that these goods cannot be produced in discrete units and, once produced, cannot exclude individuals from their consumption creates a free-rider problem and makes it difficult for government to determine the exact price it must charge the consumers – that is, the public – for their use. Consequently, governments often use tax as a means to provide them, although there are exceptions such as toll roads and other high-end public goods. Therefore, taxes can be construed as a form of price the public must pay for the consumption of the goods.

To illustrate the point above, let us suppose that a government wants to build a road at a cost of $100 million and the money needed to construct it must come from gasoline tax, earmarked for road construction and maintenance. Theoretically, then, the revenue from the gasoline tax used for the construction of the road would be the price the public must pay for its consumption, which will be $100 million (i.e., the total cost of construction).

Cost-Plus Pricing

Cost-plus pricing is basically an extension of total-cost pricing and is obtained by adding a predetermined margin to the total cost of a good, called markup. The rationale for using cost-plus pricing is that a government, like most private firms and businesses, often needs a cushion to provide safeguards against uncertainty, especially where fixed costs are concerned. Cost-plus pricing provides that cushion, known commonly as a safety margin, to ensure that these costs can be eventually recovered. The margin is essentially a predetermined "markup" on an organization's estimated per unit cost of production at the time of setting the price.

To determine the cost-plus price, one must first obtain the per unit variable cost, then add to it a per unit allocation for fixed cost which would produce the allocated per unit cost of production. The information is then used to determine the price the organization will charge with a markup as

$$P_{\text{Cost-Plus}} = TC(1+k) \qquad [1.10]$$

where $P_{\text{Cost-Plus}}$ is the cost-plus price, TC is total cost, and k is the percentage markup.

As the total cost is the sum of fixed and variable costs, Equation 1.10 can be written as

$$P_{\text{Cost-Plus}} = [FC + VC(Q)](1+k) \qquad [1.11]$$

where FC is the fixed cost, VC is the variable cost per unit, and Q is the quantity.

To give an example, suppose that a government wants to determine the price of garbage collection next year, $t+1$. The variable cost of garbage collection is $5.25 per ton and the fixed cost is $750,000 per year. Assume that the government has estimated the volume of collection for next year to be 1.25 million tons. Assume further that the government has decided to use a 5 percent markup as a safety margin to estimate the cost of collection. The total cost of garbage collection for next year will then be

$$\begin{aligned} P_{\text{Cost-Plus},\, t+1} &= [FC + VC(Q)](1+k) \\ &= [\$750{,}000 + \$5.25(1{,}250{,}000)](1+0.05) \\ &= [\$7{,}312{,}500](1.05)] \\ &= \$7{,}678{,}125 \end{aligned}$$

which includes the markup. Therefore, the cost-plus price for each ton of garbage will be $6.14 [$7,678,125 / 1,250,000 = $6.14], which is the price the consumers – the public – must pay for each ton of garbage collection.

As a pricing method, cost-plus pricing is the surest way to recover the cost of a good, but it raises an interesting question in how one determines the markup price. Ideally, it should be determined in such a way as not to have any negative effect on demand. This is possible in situations where demand is price inelastic, meaning that any increase in price will not have any effect on consumption, which is possible only if there are no substitutes for the good and the consumers have no alternative but to consume the good. However, if there are substitutes with lower prices, the consumers, in all likelihood, will change their preference for goods to those with the lower prices, known as the substitution effect. Alternatively, they may lower their consumption. In either case, a higher markup or safety margin will have a dampening effect on consumption, which, in turn, will have a negative effect on net return for government. Interestingly, there are ways in which one can determine this markup, such as by inflation rate, service usage, or tailoring it to the financial needs of a government.

Levelized Pricing

Levelized pricing is used mostly for public utilities such as electricity. It is unique in some sense in that unlike most other pricing methods the price the public utilities charge is often regulated by the regulatory commission. Consequently, utility companies, as well as others who are responsible for these services, are careful about what they include in setting these prices. Since investments in utilities can cost large sums of money, utility companies are allowed a return on their investments. In most instances, they are allowed to treat the interest cost as part of investment, as well as other costs such as depreciation as expenses, and include them in the price consumers will pay for these services. Therefore, the price consumers pay reflects the cost they must absorb to compensate the producers for interest and other costs.

To give an example, consider a case in which a government expects to earn $1.525 million in net revenue from electric utility next year, based on an estimated production of 35 million kilowatt-hours (kWh) of electricity. The levelized price per kilowatt of electricity will be a little over four cents, obtained by dividing the estimated revenue by the estimated output; that is, $1.525 million/35 million kWh = $0.04. This is the amount of cost the consumers must absorb for each unit of electricity they will consume.

It should be pointed out that the calculation of levelized pricing is somewhat more involved than most other pricing methods because it takes into consideration a variety of factors that are directly associated with the asset used in the production of a good over its entire lifespan. This would include initial investment, operations and maintenance, cost of fuel, cost of capital such as debt, and so on. Since the calculations run over the life-time of an asset, it requires some basic knowledge of the time value of money.

Transfer Pricing

A pricing method that applies to both public and private goods is transfer pricing. Transfer pricing is usually applied to a situation in which there is a transfer of services between different units within the same organization. It can be defined as the price a unit within an organization charges other units for rendering a service. Since the service provided is internal to the organization and no cash transfer is involved, it is also known as internal service pricing. To give an example, suppose that the Highway Department of a government buys $10,000 worth of legal services from the Public Attorney's office, which it records as an internal service transfer. The government as a whole did not spend any cash for the transaction but, from an accounting point of view, it needs to move the amount from the Highway Fund to the Legal Services Fund, where the transaction (payment) will be recorded. This will allow the government to keep track of the cost it incurred in providing the service, similar to other service transactions it had during the fiscal year, except that it was provided internally.

There are several ways in which one can determine the transfer price. The simplest way to determine it would be to use one's professional judgment, since there are no actual monetary transactions involved in these transfers, and administer it from the top. However, the problem with any judgmental or qualitative price determination is that it could lead to inefficiency in service provision since the price would not reflect the real value of the service. It could also lead to accounting problems for the organization if the price is artificially inflated or deflated, which would distort the actual cost of the service. An ideal alternative would be to use the market price; that is, the price the service would have charged if it were sold in a competitive market (assuming there is an actual demand for the service outside of the organization). Another alternative would be for the units to negotiate among themselves and settle on a price that is realistic and will be satisfactory to all parties concerned. This is known as the negotiated price. The rationale for using negotiation as a tool to establish a price is that it can stimulate competition and also encourage noncompetitive units to be more competitive.

Other Pricing Methods

In addition to the methods discussed above, there are other types of pricing methods a government can use, depending on the nature of the goods and the circumstances in which the methods are applied. Examples of some of these prices will be: strategic price (short-run price which may or may not be used in the long run); penalty price (the price an individual or a firm pays for delinquent behavior such as pollution or delinquent taxes); target price (the price set equal to a predetermined target such as the return on an investment); value-based price (the price determined by the value a good or service will produce for the public); and price subsidy (the price support a firm or business receives from government to compensate for the difference between a prevailing price in the market and what they would have earned under normal conditions).

As noted previously, there is no one particular method that can uniformly apply to all goods and services a government provides, since they vary in number and characteristics. This explains why it is not uncommon to find a government using not one, but several different pricing methods for different goods.

A Note on Price Discrimination

It is customary to discuss price discrimination when discussing pricing of private goods, but not so much for public goods, since price discrimination is not a common occurrence in government. Price discrimination is the practice of charging different prices to different consumers for an identical good at the same time that cannot be traced to cost differentials. In other words, if price differences are not due to cost differentials, they are discriminatory. In a perfectly competitive market with perfect knowledge and no transaction costs (i.e., costs other than the price one pays resulting in part from market imperfections), there should not be any price discrimination. However, in an imperfect market with monopoly, oligopoly, and consumers with inelastic demand, discrimination is not uncommon, since the firms can artificially determine the price. Price discrimination may also occur if it costs a firm more to serve one consumer than it does another, but the firm charges the same price for both, in which case it penalizes the customer that costs less to serve (Baye, 2010). Interestingly, many pricing practices that appear discriminatory may not really be discriminatory, while those that appear non-discriminatory may actually be discriminatory, depending on the circumstances.

In the case of government, it cannot discriminate by law against its own citizens, but there are situations in which one may find some variations of price discrimination in government. A good example would be the price a government charges its consumers for the consumption of utility services such as gas or electricity. Since the consumption of these goods takes place in block, the price tends to go down with increased consumption, known as declining-block pricing. In a declining-block price, the consumers pay at a higher rate or a flat fee for the first few kilowatt-hours of electricity and successively lower rates for additional blocks of consumption. The lowest rate, called the tail rate, generally applies to all consumption over a specific maximum.

Cost Reporting and Audit

Historically the information generated by cost and managerial accounting has been used for internal consumption, but in recent years there have been some interests in external reporting of cost information, especially where contractual obligations are concerned. However, the problem with cost reporting is that unlike the financial statements, which are prepared using formal guidelines such as the generally accepted accounting principles (GAAPs), there are no such principles that provide guidelines for reporting costs. As such, cost reports produced by financial organizations are often considered as ad hoc, but as approaches to cost and managerial accounting become established and universally recognized it should be possible to balance the information needed for cost reporting for internal as well as external consumption.[6]

Once a cost report has been prepared, it may be necessary for an organization to carry out a cost audit. The purpose of a cost audit would be to help the organization verify that the cost records and accounts used in a study are consistent with the established standards and that they have followed prescribed cost accounting procedures. Understandably, because of the ad hoc nature of the cost reports, there is no such requirement for a formal cost audit at this time, similar to the financial audit most organizations, including government and nonprofit organizations, are either required or expected to complete at the end of the financial year. As such, there are no formal guidelines as to how the auditing should be prepared. However, the fairness and compliance requirement that serves as the basis for financial audits could easily apply to a cost audit.

To illustrate the point above, financial audits are based on two essential components: fairness and compliance. The former requires that audited entities have presented their basic financial statements fairly, in conformity with GAAPs, and that they are in compliance with the stated laws, regulations, and contractual obligations. In addition to financial audits, organizations often conduct a performance audit to assess the efficiency and economy of the audited entities' operations and to determine the extent to which they have achieved their stated goals and objectives. Since the GAAPs do not extend to cost accounting, a cost audit can focus on accuracy of cost records (fairness), statutory requirements (compliance), and efficiency and economy (performance) using standard or prescribed accounting procedures.

In sum, the benefits of a cost audit, even an informal audit, are enormous. If properly conducted, a cost audit can easily help improve an organization's internal control, increase efficiency of its operation, and create cost consciousness among its employees. Additionally, a formal cost audit can help ensure maximum utilization of available resources and, more important, provide decision makers with information for effective financial decision making.

Rules of Cost Commitment

When costs are incurred by an organization, it is due to one of two things: either due to a commitment made by the organization or owing to a third-party effect. A majority of costs are due to the commitments an organization makes and, as such, they cannot be avoided in most instances. This is particularly true for government, where costs result from policy decisions, often with long-term consequences. Therefore, it is important that these decisions are made on a sound understanding of the costs that underlie them.

The following provides some basic rules of cost commitment:

There should be a direct relationship between a cost and the purpose it serves. This is the first rule of cost commitment. There must be a clearly defined objective for an activity for which costs will be ascertained. Setting clear-cut objectives can not only help an organization identify the costs it will incur for specific activities, but also facilitate its resource allocation, improve organizational performance, and increase accountability in the long run.

Costs should be charged only when they are incurred. Although most organizations are careful about it, occasional glitches can occur in an accounting system where a cost unit could be charged before a cost has been actually incurred. This can happen especially where programs or activities are recurring. Regular monitoring of accounting and related activities can easily correct this problem.

Costs should be properly segregated to avoid the problems of multiple counting. Serious problems can arise if costs are not properly traced to their sources of origin, especially in government, where revenues and expenditures are required by law to be segregated by funds. Therefore, any mix-up in costs can affect fund balance, which, in turn, can affect future financial management decisions of an organization. With good record-keeping practices, this can be easily avoided.

Past costs should be avoided in cost calculations. It is an axiom in cost studies that only those costs that are relevant to a specific activity should be included in cost calculations. Past costs, while they may be significant, are not considered relevant except when they are used for forecasting purposes since they have already been incurred. However, if a cost has been incurred, it can be included in cost calculations provided that it still has some realizable values left that can be retained or replaced.

All cost transactions should be recorded on a double-entry accounting system where possible. The double-entry accounting system has become a standard practice for most organizations, including government. However, for cost accounting, a lot of organizations still use a single-entry accounting system. In a single-entry system, cost ledgers are prepared differently from the financial ledgers (as in fund accounting), yet the two use the same basic data. This creates a discrepancy in cost analysis, but it can be corrected with a double-entry accounting system.

Abnormal cost data should be treated separately from normal data. This is more of a statistical than an accounting problem, but it can easily influence the outcome for the latter. From time to time, organizations may encounter costs that are clearly outside the realm of normal data. In statistics, they are called outliers. Their presence in normal data can distort cost calculations and provide misleading information to decision makers. They should not be totally thrown out of cost calculations but, like most outliers, should be treated separately.

Data used in cost studies should be verified periodically for validity and reliability. Like any financial data, cost data should be verified on a regular basis to see whether they are correct and follow proper cost accounting procedures. This is known as cost auditing. Cost auditing ensures that the data used in accounting are correct and that cost units, cost centers, and cost accounts have been appropriately charged.

Cost statements should be presented fairly and accurately. Finally, all financial organizations, including government, are required to provide fair and full disclosure of all their transactions and changes in financial position, which, as noted earlier, can also be extended to include cost statements. Failure to accurately measure costs and present them in a consistent manner can have serious impacts on organizational performance both in current and in future years. Inaccuracy can enter into cost statements when, for instance, contingencies are added to cost calculations. If contingencies are necessary, they should be maintained in a separate account and not merged with accounts for regular and normal activities.

Ethics in Cost Accounting

It is almost impossible to discuss any subject these days without having any reference to ethics. This is more so for public and nonprofit organizations where ethical norms and behavior remain a central focus of how employees in these organizations should be conducting themselves. Ethical conduct generally includes behavior that one would consider "right" or "proper," but it is not quite as simple to define what is right or proper. Nevertheless, there are some commonalities or common values that bind individuals within an organization, which could provide the basis for designing ethical norms or standards.

To ensure that public organizations carry out their responsibilities ethically and professionally, the American Society for Public Administration (ASPA) has developed several core values for organizational ethics that would apply to all professional activities, including cost accounting (Svara, 2013). They are: (1) advance the public interest; (2) uphold the constitution and the law; (3) promote democratic participation; (4) strengthen social equity; (5) fully inform and advise; (6) demonstrate commitment to duty, principle, and personal integrity; (7) promote ethical organizations; and (8) strive for professional excellence. Two of these codes (5 and 6) have a direct bearing on accounting practices as they focus on efficiency and standards of behavior. Code 5, for instance, emphasizes accurate, honest, comprehensive, and timely information; Code 6 focuses on the highest standards of conduct to inspire public confidence and trust in public service. Collectively all of these values are important in public organizations as they focus on striving for and maintaining the highest ethical and professional standards at all times.

Chapter Summary

Cost lies at the heart of all financial activities of an organization. Without some knowledge of costs, cost behavior, and accounting it will be difficult for an organization to carry out its normal activities. Understanding the basic cost concepts is an important first step toward that. This chapter has presented an overview of a number of key cost concepts that are frequently used in government and nonprofit organizations. In presenting these concepts, the chapter made a distinction between terms that frequently appear alongside cost, such as expense, expenditure, depreciation, inflation, and price. Some of these terms, in particular cost, expense, and expenditure, are often used interchangeably, which can create confusion in cost calculations. As such, the chapter provided a brief explanation of what contributes to their differences and how to reconcile them. The chapter also provided a brief explanation of why it is difficult to determine the price of public goods and, more important, the need for using different pricing methods in government. Finally, the chapter concluded with a brief discussion of some basic principles of cost commitments that are useful in cost studies.

Notes

1 These are organizations that are neither completely public nor completely private. In reality, these are mostly private organizations that may have been a part of government at one time, but are no longer a part of it and have government mandates to provide specific services. A good example would be the Federal National Mortgage Association.

2 Following the passage of the 1986 Tax Reform Act, the government began to allow greater accelerated depreciation over a longer time period, ranging from three to fifty years, called the Modified Accelerate Cost Recovery System (MACRS). According to this method, assets are divided into classes by type of asset or by the type of business in which the asset is used. The idea is to create greater incentives for firms and businesses to refurbish their capital to remain competitive.

3 The index is based on three sets of information: the average quantity of an item consumed in the base period; the price in the base period; and the price in the current period. It can be expressed as

$$L = \Sigma P_{jt} Q_{j0} / \Sigma P_{j0} Q_{j0}$$

where L stands for the Laspeyres Index, P_{jt} is the price of the jth item at time t, Q_{j0} is the quantity of the jth item at time 0 (which is the base year), and P_{j0} is the price of the jth item at time 0. The Bureau updates the index on a regular basis, which can be easily accessed from its website.

4 Interestingly, we could have obtained approximately the same result using a more direct expression:

$$Cost_{Adjusted} = (Cost_t / CPI_t) \times CPI_0$$

where $Cost_t$ is the cost at time t, CPI_t the consumer price index at time t, and CPI_0 is the base year price index. To illustrate, we can apply the expression to the street cleaning cost for Year 2: ($961,000/115) × 152 = $1,270,191, which is close to the adjusted cost we obtained.

5 It is obtained by dividing the nominal GNP by the real GNP, and multiplying the product by 100. For instance, if nominal GNP is $175,000 and real GNP is $125,000, the deflator will be 140 [($175,000/$125,000) × 100 = 140]. If GDP is used instead of GNP, it will be the GDP deflator.

6 In fact, in 1970, Congress established the Cost Accounting Standards Board under Public Law 91-379, as an agency of Congress to: (1) promulgate cost accounting standards to achieve uniformity and consistency in cost accounting principles to be used primarily by defense contractors and subcontractors for any contract exceeding

$100,000; and (2) establish regulations requiring the contractors and subcontractors, as a condition of contracting, to disclose in writing their cost accounting practices. It started with 19 standards, which have changed over the years under successive administrations (Rayburn, 1996).

References

Baye, M. R. (2010). *Managerial economics and business strategy*. New York: McGraw Hill/Irwin.

Cory, R. C., & Rosenberg, P. (1984). Costing municipal services. In J. Metzer (Ed.), *Practical financial management* (pp. 50–62). Washington, DC: International City Management Association (ICMA).

Davis, O. A. and Whinston, A. B. (1967). On the distinction between public and private goods. *American Economic Review, 57*(2): 360–373.

Lanzillotti, R., Dirlam, J. B., & Kaplan, A. D. H. (1958). *Pricing in big business*. Washington, DC: Brookings Institution.

Rayburn, F. R. (1996). Cost Accounting Standards Board (1970–1980; 1988–). In M. Chatfield & R. Vangermeersch (Eds.), *History of accounting: An international encyclopedia* (pp. 170–180). New York: Garland Publishing.

Samuelson, P. A. (1954) The pure theory of public expenditure. *The Review of Economics and Statistics, 36*(4): 387–389.

Svara, J. H. (2013). *Public administrator's code of ethics*. Phoenix, AZ: Lincoln Center for Applied Ethics, Arizona State University.

Tyndall, D. G. (1951). The relative merits of average cost pricing, marginal cost pricing, and price discrimination. *The Quarterly Journal of Economics, 65*(3), 342–372.

2 Cost Behavior

A fundamental principle that guides the cost behavior of an organization is that there exists a direct relationship between cost and output. Although variables other than output, such as the price of factors of production, the structure of an organization, and the latter's ability to efficiently utilize the input resources may have a bearing on cost, they are usually held constant in most cost analyses, especially in the short run. In the long run, all costs and related factors of production are variable. Since organizations at different levels of production operate differently, it is difficult to determine a priori the time period for the short run, although the rule of thumb is that it should be a year or less. However, the time period for the long run can be determined by the relationship between inputs and the production process. In general, the more capital intensive an operation (i.e., the more capital an organization uses in relation to labor), the longer it takes for the production process to change.

This chapter discusses three basic elements of cost behavior that are important for understanding the nature of costs and how they affect the functioning of an organization: time frame of costs, general properties of cost functions, and cost estimation. Of these, cost estimation is particularly important for determining the nature of current as well as future costs of an organization.

Time Frame of Costs

Although it may be difficult to determine the exact time period that separates a short-run from a long-run cost, the concepts nevertheless have important implications for decision making in an organization. For instance, the short run is an operating concept. As an operating concept, it deals with the day-to-day operations of an organization. Thus, when an organization wants to deal with its routine activities such as collecting garbage, providing water and sewerage, or fixing traffic lights, it relies on short-run cost functions. On the other hand, the long run, which includes many short runs, is a planning concept and, as a planning concept, it deals with nonroutine activities of an organization. These activities usually take a longer time to complete and often have impacts that extend over a long period of time. Thus, when an organization plans to make major changes in its operation, such as improving traffic congestion on major highways or upgrading the delivery system for healthcare, it relies on its long-run cost functions.

Because in the long run all factors of production are variable, it is possible to make adjustments in the factors more easily in the long run than in the short run. To a large measure, this is due to the fact that once a production process has been set in motion it is not possible for an organization to easily change the usage of its input factors. Some inputs such as plant and equipment may have already been committed to the process, thus making it difficult for the organization to change these factors without affecting the efficiency of operation.

Short-Run Costs

As noted previously, the short run is a period during which some of the input factors such as machines, tools, and equipment remain fixed. Since they remain fixed, they produce a cost to the organization that is independent of the level of output. Two factors determine the cost behavior in the short run: the price of variable inputs such as labor and capital, and the production function that underlies a production process. A production function shows the relationship between input and output based on the state of technology. If technology remains constant in the short run, it is likely that the production function will also remain constant in the short run.

To understand the cost structure of an organization and see how it affects decision making in the short run, it is necessary to reintroduce some of the cost terms we introduced earlier; in particular, average cost (AC), fixed cost (FC), variable cost (VC), marginal cost (MC), and total cost (TC). In reality, all of these costs can be directly obtained from observations on total cost. For instance, since the total cost is expressed as the sum of fixed and variable costs using the standard expression $TC=FC+VC$, we can obtain from it the average cost by dividing the total cost by the quantity, Q; that is, $AC=TC/Q$. It is worth noting that since the variable cost depends on quantity, it is expressed as a function of quantity, so that the total cost can be written as $TC=FC+VC(Q)$.

Furthermore, if we assume Q to be discrete, then we can obtain the marginal cost by taking the ratio of changes in total cost to changes in output; that is, $MC=\Delta TC/\Delta Q$. On the other hand, if it is continuous, then $MC=dTC/dQ$, which is the first derivative of the cost function. Table 2.1 presents a cost schedule showing the relationship between fixed, variable, total, average, and marginal costs based on cost data obtained for the water department of a local government.

According to the table, it costs the department \$20 in fixed cost and \$3.50 in variable cost to produce the first gallon of purified water (to mix with recycled water to improve water quality). While the fixed cost remains the same for any amount of water produced (supplied), the variable cost increases as the amount of water supplied increases. For instance, it costs almost five times as much to produce 7 gallons of water, or 13 times as much to produce 9 gallons of water as it does to produce 10 gallons of water. Since the fixed cost remains constant, the only factor that affects the total cost of production is the variable cost.

Table 2.1 Cost Schedule for Water Department

Quantity (Gallons)	*Fixed Cost ($)*	*Variable Cost ($)*	*Total Cost ($)*	*Marginal Cost ($)*	*Average Cost ($)*
1	20	3.50	23.50	–	23.50
2	20	5.00	25.00	1.50	12.50
3	20	6.00	26.00	1.00	8.67
4	20	7.00	27.00	1.00	6.75
5	20	8.75	28.75	1.75	5.75
6	20	11.96	31.96	3.21	5.33
7	20	17.29	37.29	5.33	5.33
8	20	26.69	46.69	9.40	5.84
9	20	45.00	65.00	18.31	7.22
10	20	73.20	93.20	28.20	9.32

Elasticity of Cost

Let us introduce a term that frequently appears in discussions on cost behavior: elasticity. Elasticity measures the responsiveness of a variable (dependent) as a result of a change in another variable (independent), usually expressed in percentage terms. The responsiveness is measured by the coefficient, ϵ, called the elasticity coefficient, and is determined by two interrelated factors: its sign and magnitude. The sign reflects the direction of movement between two variables, and the magnitude indicates the degree of responsiveness of the dependent variable to a change in the independent variable.

Elasticity can be greater than, equal to, or less than 1. If it is less than 1, $|\epsilon|<1$, it is inelastic, meaning that a 1 percent change in the independent variable (say, output) will lead to a less than 1 percent change in the dependent variable (say, cost). Expressing it in cost terms, it means that cost increases at a slower rate than output; in other words, there are economies of scale indicating a decreasing average cost of production. If it is equal to 1, $|\epsilon|=1$, it is unit elastic, meaning that a 1 percent change in the independent variable will lead to a proportionate change in the dependent variable. In cost terms, it means that cost and output change proportionately – by exactly the same percentage. In other words, there are no economies or diseconomies of scale. If it is greater than 1, $|\epsilon|>1$ it is elastic, meaning that a 1 percent change in the independent variable will lead to a more than 1 percent change in the dependent variable. In cost terms, it means that cost increases faster than output. In other words, there are diseconomies of scale indicating increasing average cost.

The following presents a simple expression of cost elasticity to measure the responsiveness of cost to a change in output:

$$\epsilon_C = \frac{\Delta TC}{TC} / \frac{\Delta Q}{Q} \qquad [2.1]$$

where ϵ is the elasticity of cost, ΔTC is the change in total cost, and ΔQ is the change in quantity, TC and Q are total cost and quantity, respectively.

Equation 2.1 can also be expressed as $\epsilon_C=(\Delta TC/\Delta Q)\times(Q/TC)$, where the first term represents the marginal cost and the second term the cost per unit of output. Thus, if we have information on marginal cost and the cost per unit of a good or service, we can easily obtain the elasticity of cost by taking the product of two. To illustrate the point, let us look at the water production problem again. For instance, at $Q=6$, the elasticity of cost is 0.6; that is, $e,=MC/AC=3.21/5.3$ $3=0.6$. What this means is that for a 1 percent change in output, total cost will change by 0.6 percent, indicating that cost is inelastic at $Q=6$. By the same token, cost is elastic at higher levels of output. For instance, at $Q=9$, the elasticity is 2.54 [$18.31/7.22=2.54$], meaning that it changes by a more than proportionate change in output. In common sense terms, it means that it will cost the department more at a higher level of output, especially when it exceeds 8 gallons of water, than it will when the output level is low. Put differently, when the output reaches a level where the production cost increases at a faster rate, it becomes necessary to make adjustments in the production process to bring the costs down, which then becomes a long-run cost problem.

Point Versus Arc Elasticity

The elasticity concept discussed above is known as point elasticity because it represents incremental movements from point to point along a cost curve, but there are situations where we may need to measure elasticity over a large segment of the curve. In that case, we need to use a term called arc elasticity. The arc elasticity for a total cost function can be written as

$$\epsilon'_C = \left[\frac{TC_1 - TC_0}{(TC_1 + TC_0)/2}\right] / \left[\frac{Q_1 - Q_0}{(Q_1 + Q_0)/2}\right] \qquad [2.2]$$

where ϵ', is the arc elasticity, and the subscripts 1 and 0 are the new and the initial costs and quantities, respectively. The purpose of dividing the denominator by 2 is that it measures the average distance or the distance halfway between the designated end points of the arc.

To give an example, let us take two output quantities from the water supply problem, $Q_0=5{,}000$ and $Q_1=8{,}000$, and two total costs corresponding to these outputs, $TC_0=\$28.75$ and $TC_1=\$46.69$, then substitute these values into Equation 2.2:

$$\begin{aligned}
\epsilon'_C &= \left[\frac{TC_1 - TC_0}{(TC_1 + TC_0)/2}\right] / \left[\frac{Q_1 - Q_0}{(Q_1 + Q_0)/2}\right] \\
&= \frac{(TC_1 - TC_0)(Q_1 + Q_0)}{(TC_1 + TC_0)(Q_1 - Q_0)} \\
&= \frac{(46.69 - 28.75)(8+5)}{(46.69 + 28.75)(8-5)} \\
&= \frac{(17.94)(13)}{(75.44)(3)} \\
&= 1.03
\end{aligned}$$

which comes out to be 1.03.

The result indicates that the total cost is elastic, although marginally, over the range of the cost curve designated by the end points of the arc (TC_1, Q_1) and (TC_0, Q_0). In other words, it represents the average elasticity for the range of output between Q_1 and Q_0 (i.e., between 8,000 and 5,000 gallons of water). Interestingly, both point and arc elasticities serve as good examples of how cost responds to a change in the independent variable. Although our example has been limited to cost behavior in the short run, it can be used in any situation, including long-run costs, as long as there is a logical justification for it.

It is worth noting that when elasticity is used for long-run cost functions, the results may be less than precise because so many other factors may intervene in the relationship between a dependent and an independent variable. Since elasticity measures a one-to-one relationship, it cannot account for any intervention that may affect the responsiveness of the variable and the elasticity being measured.

Long-Run Costs

As we noted earlier, the short-run cost is an operating concept, which means that an organization has no control over the fixed costs it incurs for a given level of operation in the short run. Therefore, any effort to increase its efficiency must come from a reduction in its variable costs. In contrast, an organization dealing with long-run costs does not have any such problem. Since all factors of production are variable in the long run, it is possible for the organization to select the most efficient (i.e., least cost) combination of factors at which its operation will be optimal. An important implication of this behavior is that in the long run it is possible for organizations to have greater economies of scale. Economies of scale occur in an organization when the average cost of production decreases as a result of improvements in the production process, but economies of scale cannot be maintained throughout the process. Eventually, costs will rise. The eventual rise in costs is due to what is known as diseconomies of scale, which occur once the level of operation becomes too large, making it difficult for the decision makers to efficiently coordinate all the activities of an organization.

Properties of Cost Functions

All cost functions have one characteristic in common: they all contain two basic parameters – one representing the fixed cost and the other the variable cost. Fixed costs, as we know, are costs that do not change in the short run over a given range of output, whereas variable costs change in direct proportion to changes in output. Note that the notion of range is important in cost functions because it sets the limits within which one's definition of cost remains valid. Since fixed costs are constants, they can be assumed away in a cost function in that if one would remove the parameter representing the fixed cost from a cost function, it would leave the function with variable costs only, meaning that fixed costs do not have any effect on the function. In other words, the fixed cost will be zero.

Linear Cost Function

We can explain the above relationship between fixed, variable, and total cost, and output with the help of a simple linear equation, as shown below:

$$TC = a + bQ \qquad [2.3]$$

where a is the parameter representing fixed cost, and b is the variable cost per unit of output, Q.

In conventional statistical terms, TC will be the dependent variable and Q the independent variable. If, according to this relationship, the fixed cost does not have any effect on total cost, TC, Equation 2.3 will become

$$TC = bQ \qquad [2.4]$$

This means that if we can draw a line corresponding to this equation on a two-dimensional plane, one representing total cost and the other quantity, it will pass through the points of origin of the two axes (as shown in Figure 1.1b in Chapter 1).

Nonlinear Cost Functions

Although linear relationship is common in most cost studies, it is not uncommon to have a nonlinear relationship such as quadratic, cubic, or a higher degree polynomial. While a linear cost function shows a straight-line relationship between cost and one or more independent variables, as in Equation 2.4, a nonlinear relationship shows changes in relation to the behavior of one or more independent variables such as the rate of change in output, change in direct labor, change in direct materials, and so on.

For instance, a typical quadratic function describing the relationship between cost and quantity can be expressed as

$$TC = a + bQ - cQ^2 \qquad [2.5]$$

where TC is the total cost, Q is the quantity, and a, b, and c are the parameters representing fixed and variable costs, respectively.

Similarly, a typical cubic function describing the cost behavior of an organization can be expressed as

$$TC = a + bQ - cQ^2 + dQ^3 \qquad [2.6]$$

where the terms of the expression are the same as before. We can easily extend Equation 2.6 to show higher degree polynomial cost functions.

Nonlinear cost functions, by their very nature, are more complex than a linear cost function. As such, the cost terms one derives from these functions, in particular average and marginal costs, are also more complex. In general, the average cost of a quadratic function (Equation 2.5) is linear, $(a/Q + b - cQ)$, but is quadratic for a cubic function (Equation 2.6), $(a/Q + b - cQ + dQ^2)$. On the other hand, the marginal cost is constant for a linear function, b, whereas it is linear for a quadratic function, $b - 2cQ$, and quadratic for a cubic function, $b - 2cQ + 3dQ^2$. An advantage of using these functions is that once we know the average and marginal costs, we can use this information to estimate the elasticity of costs from these functions without having to go through the formalities of Equation 2.1. However, it should be worth noting that although most cost functions in the real world are nonlinear, there is an advantage of using a linear cost function: it is easier to estimate costs empirically with a linear function than with a nonlinear function. Nonlinear functions such as cubic functions often require theoretical generalizations that make it difficult to fit the curves to empirical data.

Cost Estimation

Cost functions and their properties serve two important objectives in all cost studies: first, they help us understand the patterns of cost behavior over time; and, second, they provide us with the information necessary to estimate current, as well as future, costs of operation for an organization. Cost estimation is an important element of cost behavior. Without good estimates of cost, it will be difficult for an organization to effectively plan its current, as well as future activities. There is a wide range of methods with varying degrees of complexity that can be used for cost estimation. This section discusses three such methods, selected primarily for their simplicity and ease of use: the graphical method, the high–low method, and the regression method.

The Graphical Method

The graphical method is used to show the relationship between two variables (a dependent and an independent) by fitting a visual line into a set of data on a two-dimensional plane. Usually the cost data are presented on the vertical axis and the service data (quantity) on the horizontal axis, based on the conventional wisdom that cost depends of the quantity of service provided. For instance, if we assume that there is a linear relationship between total cost (*TC*) and the level of service activity Q such that $TC=a+bQ$, then using the graphical method we can estimate the parameters a (called intercept) and b (called slope), of this relationship by visually fitting a line to the plotted observations (coordinates). The intercept represents the value the dependent variable (*TC*, in our case) will assume regardless of the value of the independent variable (Q, in our case), while the slope represents the amount by which the dependent variable (*TC*) will change for a unit change in the independent variable (Q).

We can use a simple example to illustrate this. Suppose that a local school district wants to know the cost of running a computer literacy program for disadvantaged children; in particular, fixed and variable costs that include, among others, spending several hours a month after school. Let us say that the district has collected data for the number of hours the students have spent (Q) and its corresponding cost (*TC*) over a 12-month period. Our objective is to find the values of a and b, given the cost and quantity information we have on the two variables. Table 2.2 presents these data.

To estimate the values of a and b, we begin first by plotting the data on a graph with Q on the horizontal axis and *TC* on the vertical axis, as shown in

Table 2.2 Cost of Running the Computer Literacy Program

Month	*Number of Hours*	*Total Cost ($)*
January	1,500	50,000
February	1,300	45,000
March	1,200	42,500
April	1,000	37,500
May	1,200	40,000
June	1,600	47,500
July	2,000	55,000
August	1,800	52,000
September	1,700	50,000
October	1,800	50,000
November	2,000	57,500
December	2,200	60,000

Figure 2.1. Next, we visually construct a line that would pass through the data points in such a way that some of the observations will lie above the line and some below it, thus, giving it the appearance of a good fit. As the figure shows, the line intercepts the cost axis roughly at $15,000. This is the fixed cost (*FC*), *a*, of running the program per month. The variable cost (*VC*), *b*, is given by the

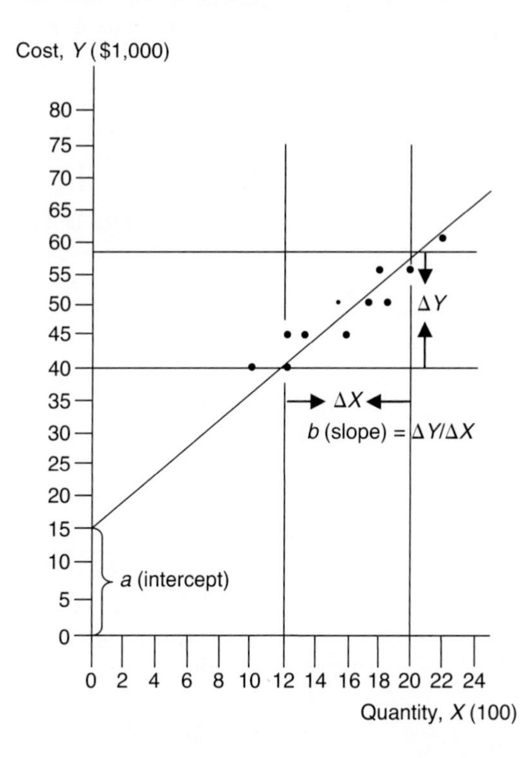

Figure 2.1 Cost and Output Data for the Computer Literacy Program

slope of the line. We can obtain this slope by taking the ratio of two changes, changes in total cost (ΔTC) to the changes in the number of hours the student met (ΔQ) during a month; that is, $\Delta TC / \Delta Q$, as shown below:

$$b = \frac{\Delta TC}{\Delta Q}$$

$$= \frac{\$57,500 - \$40,000}{2,000 - 1,200}$$

$$= \frac{\$17,500}{800}$$

$$= \$21.875 \approx \$21.88$$

The result produces a value of $21.88 (after rounding off). This is the variable cost, *VC* (i.e., cost of running the program) per hour.

The High–Low Method

Unlike the graphical method, where one tries to visually fit a line to a set of sample observations, the high–low method uses two observations, one high and one low, to estimate the slope, b, and the vertical intercept, a. The advantage of using this method is that it utilizes a minimal amount of information to estimate the parameters compared to most methods, including the graphical method. To illustrate how the method is used in practice, let us return to the computer literacy program. As before, we assume that the relationship between total cost (TC) and the number of hours attended (Q) per month is linear; that is, $TC = a + bQ$. In fact, the high–low method always assumes a linear relationship between a dependent and an independent variable. Our objective is to find the values of a and b, vis-à-vis the intercept and slope. We use the following procedure to determine these values: let Y represent the cost and X the number of hours spent, Y_H and Y_L the high and low cost, and X_H and X_L the high and low quantity of hours spent, respectively.

Since we assumed a linear relationship between the two variables, we can set up the problem in terms of two linear equations, given by

$$Y_H = a + bX_H \qquad [2.7]$$

$$Y_L = a + bX_L \qquad [2.8]$$

where a is the intercept and b is the slope. The equations indicate that at both levels, Y_H and Y_L, the total cost consists of a fixed component, a, and a variable component equal to the rate, b, times the level of activity such that $TC = FC + VC(Q)$.

We begin with the slope, the variable component of the total cost function, because we need to have this information in order to compute the intercept. To obtain the slope, we simply subtract Equation 2.8 from Equation 2.7, and solve for b. That is,

$$Y_H - Y_L = bX_H - bX_L$$

$$\therefore b = \frac{Y_H - Y_L}{X_H - X_L}$$

$$= \frac{\$60,000 - \$37,500}{2,200 - 1,000}$$

$$= \frac{\$22,500}{1,200}$$

$$= \$18.75 \qquad [2.9]$$

The result produces a slope of \$18.75. This is how much it cost the school district per hour to run the program, which is about three dollars less than what we obtained for the graphical method.

The calculation of the intercept term, a, is relatively straightforward once we know the value of the slope, b. We can use either of Equations 2.7 or 2.8 for this purpose. Equations 2.10 and 2.11, which are obtained from Equations 2.7 and 2.8, show below how this intercept can be estimated:

$$Y_H = a + bX_H$$

$$\therefore a = Y_H - bX_H \qquad [2.10]$$

Alternatively,

$$Y_L = a + bX_L$$

$$\therefore a = Y_L - bX_L \qquad [2.11]$$

Suppose now that we want to use Equation 2.10. Thus, substituting the respective values of the terms b, Y_H, and X_L into Equation 2.10, we obtain the intercept (i.e., the estimate of a). That is,

$$a = Y_H - bX_H$$

$$= \$60,000 - (\$18.75)(2,200)$$

$$= \$60,000 - \$41.250$$

$$= \$18,750$$

which produces a value of \$18,750. This is the fixed cost of running the program, which turns out to be higher than the value we obtained under the graphical method.

Alternatively, we could have used Equation 2.11 to obtain the same result, since they are identical equations:

$$\begin{aligned} a &= Y_L - bX_L \\ &= \$37{,}500 - (\$18.75)(1{,}000) \\ &= \$37{,}000 - \$18{,}750 \\ &= \$18{,}750 \end{aligned}$$

We can now put all of this information together and present the estimated equation as

$$Y = \$18{,}750 + \$18.75X$$

The result indicates that the fixed cost of operating the program, regardless of the number of hours spent, is \$18,750. The variable cost, on the other hand, is \$18.75, which indicates that for each additional hour spent, the total cost of running the program will increase \$18.75 (i.e., by an amount equal to the value of the slope).

What appears to be an obvious advantage for the high–low method can also be a disadvantage. Because the method does not utilize all the information in a dataset, the estimated equation may not be an accurate representation of the exact relationship between the variables under study. In other words, it may not be the perfect line that can be fitted through the data points (coordinates). This is where the regression method becomes useful in that it makes full use of all the available information and, as such, provides a more accurate measure of the estimates than the high–low method.

The Regression Method

The regression method is by far the best of the three methods discussed here for cost estimation. Besides being able to fully utilize the available information in a dataset, there are a number of other advantages that make the regression method considerably superior to either the graphical or the high–low method. For instance, it is computationally simple and does not have excessive data requirements. Another advantage of the method is that its parameters have certain optimal properties that make it possible to apply various tests of significance. These tests are primarily used to determine the reliability of the estimated parameters, as well as of the estimated equation. Furthermore, it is comprehensive in analytical detail and frequently serves as the foundation for other more advanced methods used in cost studies.

This section focuses on a simple regression model involving two variables: one dependent and one independent, as shown below:

$$Y = a + bX \qquad [2.12]$$

where Y is the dependent and X is the independent variable, and a and b are the respective parameters (i.e., intercept and slope) of the model. As before, we assume a linear relationship between the two variables.

Steps in a Regression Method

The basic idea behind the regression method is quite simple. As with the graphical and the high–low method, the objective is to construct a line that would best fit a set of sample observations. However, unlike the first two methods, the regression method is based on a number of mathematical conditions that ensure that the line is the best that could be drawn for a dataset. Important among these conditions are:

1 *Specification of the model.* The regression model must be specified in terms of the relationship between a dependent and one or more independent (explanatory) variables. For instance, for a two-variable case, the model can be specified as $Y=f(X)$, where Y is the dependent and X is the independent variable. For an n-variable case, it will be $Y=f(X_1,X_2,X_3,\ldots,X_n)$.
2 *Expression of precise mathematical relationship between the variables.* The simplest way to achieve this is to assume a linear relationship between the variables in a model so that one can write it in the familiar form $Y=a+bX$, for a two-variable case, and $Y=a+b_1X_1+b_1X_2+\ldots+b_nX_n$, for an n-variable case. Where the relationship is nonlinear, one can always linearize it by using measures such as taking logarithms on both sides of the equation, called log transformation, unless the relationship is inherently nonlinear, in which case it must be solved without the benefit of transformation.
3 *Estimation of the model.* One must be able to estimate the values of the parameters, along with all the relevant statistics, and set up the estimated equation once the model has been estimated.
4 *Tests for the significance of the estimated parameters and the model as a whole.* The model must be tested for statistical significance of the estimated parameters to ensure that it is a good fit and that the estimated parameters did not occur by chance.

As a matter of convention, the signs of the parameters in a regression model (i.e., the intercept and the slope) are always presented in positive terms although, in reality, the values of the parameters can be positive, negative, or zero. In general, a positive slope indicates a positive relationship between X and Y, meaning that a change in X in a certain direction will lead to a change in Y in the same direction; that is, if X goes up, so will Y, and vice versa. A negative slope, on the other hand, indicates an inverse relationship between X and Y, meaning that a change in X in a certain direction will lead to a change in Y in the opposite direction; that is, if X goes up, Y will go down, and vice versa. The slope, however, cannot be zero because it will mean a total redundancy of the independent variable, in which case it should not be included in the first place.

The intercept term, on the other hand, can be positive, negative, or zero. When it is positive, it means Y is positive, when X is zero. When it is zero, it means Y is zero, when X is zero. In other words, there is no intercept; that is, the line passes through the points of origin, where X and Y intersect at 0. Finally, when the intercept is negative, it means that Y has a value that is less than zero, when X is zero, in which case it depends on the researcher whether to keep or ignore the negative portions of the equation.

The Error Term

When a linear equation is expressed as $Y = a + bX$, it implies that the relationship between X and Y is exact; that is, all the variation in Y is due entirely to the variation in X. If this were true, then all the data points on a two-dimensional plane would fall on a straight line. However, when observations come from the real world and are plotted on a graph, as we saw in our garbage collection example, one will most certainly observe that they will not fall on a straight line. In other words, there will be some deviations from the line, however small they may be.

There are several explanations for why these deviations occur; in particular, the following;

1. *Omission of variables.* We may have excluded from our regression equation those explanatory variables that would have played a significant role in explaining the variations in the dependent variable.
2. *Random behavior of human beings.* The scatter points around the regression line may be due to the unpredictable elements in human behavior.
3. *Imperfect specification of the mathematical form of the model.* We may have linearized a nonlinear relationship or have left out some equations from the model that should have been included. This happens, especially when one does not have a full understanding of the factors affecting the model relationship.
4. *Errors in aggregation.* When specific data are not available, one is occasionally forced to use aggregate data. When this happens, it produces errors in aggregation, meaning that when data are aggregated they often add magnitudes referring to individuals whose behaviors are dissimilar.
5. *Errors in measurement.* This refers to the deviations of observations from the line. These types of deviations occur often due to the methods employed in collecting and processing empirical data. This is a common problem in social science, where data are mostly generated from primary sources.

Of the five sources of errors mentioned here, the first four are called the errors of omission and the fifth is called the error of measurement. In order to correct for these errors, the convention is to introduce an error term into the standard regression equation (Equation 2.12), so that the new equation will now appear as

$$Y = a + bX + e \qquad [2.13]$$

where e is the error term. Therefore, the true relationship that connects the variables in this equation consists of two components – one represented by the line, $a + bX$, and the other by the error term, e.

The Least Squares Criterion

The linear equation we have just introduced, $Y=a+bX+e$, holds true for situations in which one is dealing with population data but, in reality, we seldom deal with population data. Instead, we collect a sample of observations on X and Y, specify the distribution of the error term, and try to obtain a satisfactory estimate of the true value of the parameters of the relationship. As noted earlier, we do so by fitting a regression line through the observations of a sample, assuming that it would be a good approximation of the true line. Therefore, the true relationship between X and Y can be expressed as $Y=a+bX+e$, and the true regression line as $Y=a+bX$. On the other hand, the estimated relationship can be expressed as

$$Y = \hat{a} + \hat{b}X + \hat{e} \qquad [2.14]$$

and the estimated regression line as

$$\hat{Y} = \hat{a} + \hat{b}X \qquad [2.15]$$

where $\hat{Y}$ is the estimate of Y, given a specific value of X, $\hat{a}$ is the estimate of the true intercept, a, $\hat{b}$ is the estimate of the true slope, b, and $\hat{e}$ is the estimate of the true value of the error term, e. Note that the symbol ^ (hat) represents the estimated value of the parameters, as well as of the regressed (dependent) variable in the equation.

The weakness of this procedure is that we can obtain an infinite number of regression lines by assigning different values to the parameters a and b that will fit the data, but our objective in regression analysis is to choose a line that will best approximate the true line. Therefore, we need to use some guidelines or criteria by which we can determine this line. The criterion we generally use for this purpose is the least squares criterion, also known as the ordinary least squares (OLS) method. According to this criterion, the line selected must be such that it will produce the smallest possible error for the observations from the line (i.e., it will be optimal). In other words, select the line for which the sum of the squares of the deviations of the observations from the line is the minimum.

Estimating the Parameters

The least squares criterion serves a very important purpose in regression analysis in that it allows us to estimate the parameters directly from it. The only requirement is that they must be estimated in such a way as to ensure they are efficient (i.e., they have the minimum variance [i.e., errors]). To illustrate this, let us recall Equation 2.15 for the two-variable case, X and Y, and the equation $Y=\hat{a}+\hat{b}X+\hat{e}$ which, with slight reorganization, can be written as

$$\begin{aligned}\hat{e} &= Y - \hat{Y} \\ &= Y - (\hat{a} + \hat{b}X) \\ &= Y - \hat{a} - \hat{b}X\end{aligned} \qquad [2.16]$$

with respect to $\hat{a}$ and $\hat{b}$.

Our objective is to find the estimates of a and b in such a way that the sum of the squares of the deviations, $\Sigma(\hat{e}^2)$, will be minimum. That is,

$$\text{Minimize} \sum \hat{e}^2 = \sum (Y - \hat{a} - \hat{b}X)^2 \quad [2.17]$$

with respect to $\hat{a}$ and $\hat{b}$.

Now, solving the expression in Equation 2.17 would produce two equations called normal equations:[1]

$$\sum Y = n\hat{a} + \hat{b}\sum X \quad [2.18]$$

$$\sum XY = \hat{a}\sum X + \hat{b}\sum X^2 \quad [2.19]$$

From Equations 2.18 and 2.19 we can obtain the values of $\hat{a}$ and $\hat{b}$ by solving them simultaneously, as shown below:

$$\hat{a} = \frac{\sum Y}{n} - \frac{\hat{b}\sum X}{n} \quad [2.20]$$

$$\hat{b} = \frac{n\sum XY - \sum X \sum Y}{n\sum X^2 - (\sum X)^2} \quad [2.21]$$

Note that the values of $\hat{a}$ and $\hat{b}$ in Equations 2.20 and 2.21 correspond to the points where the first derivatives of the squared errors taken with respect to $\hat{a}$ and $\hat{b}$ (Equation 2.17) are zero; therefore, guaranteeing that the sum of the squared deviations is minimum.[1] Since it requires minimum amount of information to compute the estimates of intercept and slope, it is often called the computational method.

We can now apply the above expressions to the data in Table 2.2 to obtain the intercept and slope for the computer literacy program (Table 2.3). To be consistent with the model, we will call the total cost of the program our dependent variable, Y, and the number of hours spent by the students for a given month our independent variable, X, as shown below:

$$\hat{a} = \frac{\sum Y}{n} - \frac{\hat{b}\sum X}{n}$$

$$= \frac{587{,}000}{12} - \frac{(17.757)(19{,}300)}{12}$$

$$= 48{,}916.667 - 28{,}559{,}175$$
$$= 20{,}357.492 \approx \$20{,}357.49$$

for $\hat{a}$. Similarly,

$$\hat{b} = \frac{n\sum XY - \sum X \sum Y}{n\sum X^2 - (\sum X)^2}$$

$$= \frac{11,659,200,000 - 11,329,100,000}{391,080,000 - 372,490,000}$$

$$= \frac{330,100,000}{18,590,000}$$

$$= 17.757 \approx \$17.76$$

for $\hat{b}$.

However, it is possible to show that these coefficients can also be obtained by the following equations.[2] That is,

$$\hat{a} = \bar{Y} - \hat{b}\bar{X} \quad [2.24]$$

$$\hat{b} = \frac{\sum (X - \bar{X})(Y - \bar{Y})}{\sum (X - \bar{X})^2} \quad [2.25]$$

where the equations are expressed in terms of the deviations from the means of the variables X and Y.

Table 2.3 Calculation of Summary Statistics for Intercept and Slope

Month	*Number of Hours,* X	*Total Cost,* Y *($)*	X^2	XY
January	1,500	50,000	2,250,000	75,000,000
February	1,300	45,000	1,690,000	58,500,000
March	1,200	42,500	1,440,000	51,000,000
April	1,000	37,500	1,000,000	37,500,000
May	1,200	40,000	1,440,000	48,000,000
June	1,600	47,500	2,560,000	76,000,000
July	2,000	55,000	4,000,000	110,000,000
August	1,800	52,000	3,240,000	93,600,000
September	1,700	50,000	2,890,000	85,000,000
October	1,800	50,000	3,240,000	90,000,000
November	2,000	57,500	4,000,000	115,000,000
December	2,200	60,000	4,840,000	132,000,000
	$\Sigma X = 19,000$	$\Sigma Y = \$587,000$	$\Sigma X^2 = 32,590,000$	$\Sigma XY = 971,600,000$

Notes

$n\Sigma XY = (12)(971,600,000) = 11,659,200,000.$

$\Sigma X \Sigma Y = (19,300)(587,000) = 11,329,100,000.$

$n\Sigma X^2 = (12)(32,590,000) = 391,080,000.$

$(\Sigma X)^2 = (19,300)^2 = 372,490,000.$

As before, we can apply the above equations to the computer literacy program to obtain the estimated values of intercept and slope. Table 2.4 presents the summary measures that were computed for the data to help us estimate the parameters of the model and other relevant statistics.

Using the appropriate summary data from the table and substituting them into the respective terms in Equations 2.24 and 2.25, we now obtain the estimates of a and b, as shown below:

$$\hat{b} = \frac{\sum(X-\bar{X})(Y-\bar{Y})}{\sum(X-\bar{X})^2}$$

$$= \frac{27,508,333.33}{1,549,166.68}$$

$$= 17.757 \approx \$17.76$$

and

$$\hat{a} = \bar{Y} - \hat{b}\bar{X}$$

$$= 48,916.67 - (17.578)(1,608.33)$$

$$= 20,357.11 \approx \$20,357.11$$

As expected, the results appear to be the same as those obtained from the earlier expressions, except for the rounding-off errors. We can now put all of this information together and present the estimated equation of the regression line as

$$\hat{Y} = 20,357.11 + 17.757X \quad [2.26]$$

What the result in Equation 2.26 tells us is that, given the estimated values of the intercept and slope ($\hat{a}$ and $\hat{b}$), it will cost the government \$17.76 in variable costs to provide each additional hour of the program and \$20,357.11 in fixed

Table 2.4 Summary Statistics for Intercept and Slope

Summary Data	*Estimated Parameters*
$\bar{X}$ = 1,608.33; $\bar{Y}$ = 48,916.67	$\hat{a}$ = 20.357.11
$\Sigma(X^2)$ = 32,590,000.00	$\hat{b}$ = 17.757
$\Sigma(X-\bar{X})^2$ = 1,549,166.67	
$\Sigma(Y-\bar{Y})^2$ = 514,905,038.20	
$\Sigma(X-\bar{X})(Y-\bar{Y})$ = 27,508,333.33	
$\Sigma(Y-\hat{Y})^2$ = 26,455,087.80	
$\Sigma(\hat{Y}-\bar{Y})^2$ = 488,449,950.40	
σ_e^2 = 2,645,508.80	

costs per month, regardless of the amount of hours spent. Interestingly, the results compare well with those we obtained earlier for the high–low method.

Significance Tests for the Estimated Parameters

So far, our discussion has concentrated on the procedure for estimating the numerical values of the parameters of the regression equation, but we do not know how good these estimates are. Therefore, to establish the goodness (i.e., reliability) of these estimates we need to conduct a statistical test called the *t* test (used mostly for small samples involving 30 or fewer cases). The purpose of this test is to determine whether the estimates of *a* and *b* are significantly different from zero. In other words, whether the sample from which the data have been drawn and the parameters of the regression equation estimated could have come from a population whose true parameters are zero and, if so, what are the chances of observing that. What this means is that if the *t* values could show with considerable degrees of certainty that they did not come from the population whose true parameters are zero, one should be able to consider the estimates statistically significant (reliable), provided that everything else remains the same.

In general, the *t* values are computed by taking the ratio of the estimated parameters to their corresponding standard errors. That is,

$$t_{\hat{a}} = \frac{\hat{a}}{S_{\hat{a}}} \qquad [2.27]$$

$$t_{\hat{b}} = \frac{\hat{b}}{S_{\hat{b}}} \qquad [2.28]$$

where *S* is the standard error and the rest of the terms are the same, as before.

The standard errors are essentially the standard deviations of the estimated parameters and are given by the square roots of their variances, as shown below:

$$S_{\hat{a}} = \sqrt{VAR(\hat{a})} = \sqrt{\frac{[(\sigma_E^2)(\sum x^2)]}{n\sum(X-\bar{X})^2}} \qquad [2.29]$$

for $S(\hat{a})$, and

$$S_{\hat{b}} = \sqrt{VAR(\hat{b})} = \sqrt{\frac{\sigma_e^2}{\sum(X-\bar{X})^2}} \qquad [2.30]$$

for $S(\hat{b})$, where $\sigma^2 = \Sigma\hat{e}^2/(n-k)$, *k* is the number of parameters in the model (i.e., the regression equation) and *n* is the sample size (i.e., the number of observations in a dataset). The $\hat{e}^2$ is called the error variance and is given by the expression $(Y-\hat{Y})^2$; that is, $\hat{e}^2=[Y-(\hat{a}+\hat{b}X)]^2=(Y-\hat{Y})^2$.

To obtain the t values for our estimated parameters, we simply substitute the values from Table 2.3 into the respective terms in Equations 2.29 and 2.30, so that

$$t_{\hat{a}} = \frac{\hat{a}}{\sqrt{\frac{[(\sigma_e^2)(\sum X^2)]}{n\sum(X-\bar{X})^2}}}$$

$$= \frac{20,357.11}{\sqrt{\frac{(2,645,509.80)(32,590,000.00)}{(12)(1,549,166.67)}}}$$

$$= \frac{20,357.11}{2,153.56}$$

$$= 9.4528$$

is the observed t corresponding to $\hat{a}$, and

$$t_{\hat{b}} = \frac{\hat{b}}{\sqrt{\frac{\sigma_e^2}{\sum(X-\bar{X})^2}}}$$

$$= \frac{17.757}{\sqrt{\frac{2,645,508.80}{1,549,166.68}}}$$

$$= \frac{17.757}{1.307}$$

$$= 13.5861$$

is the observed t corresponding to $\hat{b}$.

The estimated equation, with the attendant t values, can therefore be written as

$$\hat{Y} = 20,357.11 + 17.76X$$
$$(9.4527) \quad (13.5861)$$

where the values in parentheses indicate the t values.

From this compact form, we can test the significance of the estimated parameters, $\hat{a}$ and $\hat{b}$, by comparing the observed t values against their theoretical (critical) values, which can be found in any standard t table. The rule for this comparison is simple: if the observed t is positive and greater than the critical value of t (for a two-tailed test) at a given level of p called the p value, it is considered statistically significant at that level. The opposite is true when it is negative. Since our observed t is greater than the critical value of t at $p=0.05$ (which is fairly standard), with $n-2=12-2=10$ degrees of freedom [i.e., $t(\hat{a})=9.4527>t(a)=2.228$], our estimated parameter $\hat{a}$ is statistically significant. Similarly, since $t(\hat{b})=13.5861>t(b)=2.228$, then for the same level of p and the degrees of freedom, the estimated parameter $\hat{b}$ is also statistically significant. What this means is that there is only a 5 percent chance that these results could have occurred by chance (i.e., due to error). Interestingly, the results would have been significant also at $p=0.01$, given the size of the observed t values.

Confidence Intervals for the Estimated Parameters

To say that our estimates are statistically significant does not mean that they are correct estimates of the true population parameters, a and b. It simply means that our estimates have come from a sample drawn from a population whose parameters are different from zero. Therefore, in order to determine how close the estimated parameters are to the true population parameters, we need to construct a confidence interval for these parameters. In other words, we need to establish, with a certain degree of confidence, two limiting values within which we can expect to find the true value of the population parameter – one high called the upper limit of the interval and one low called the lower limit of the interval, we will call θ.

The procedure for setting up confidence interval is as follows:

$$C[\hat{\theta}-S_{\hat{\theta}}t_{\alpha/2}<\theta<\hat{\theta}+S_{\hat{\theta}}t_{\alpha/2}]=(1-\alpha) \quad [2.31]$$

where C is the confidence statement, $(1-\alpha)$ is the confidence level (i.e., degree of confidence), $\hat{\theta}$ is the estimated value of the parameter θ, $S_{\hat{\theta}}$ is the standard error of the estimate, and $t_{\alpha/2}$ is the critical t value for two tails $(t_{\alpha/2})$ – the probability of obtaining an observed statistic by error for a given degree of freedom.

Using the expression in Equation 2.31, we can construct the confidence intervals for our estimated parameters $\hat{a}$ and $\hat{b}$, say, at the 99 percent level of confidence. That is,

$$C[\hat{a}-S_{\hat{a}}t_{\alpha/2}<a<\hat{a}+S_{\hat{a}}t_{\alpha/2}]=(1-\alpha)$$

$$C[20.357.11-(2{,}153.56)(3.17)<a<20.357.11+(2{,}153.56)(3.17)]=0.99$$

$$C[20.357.11-6{,}826.79<a<20.357.11+6{,}826.79]=0.99$$

$$C[13{,}530.32<a<27{,}183.90]=0.99$$

for $\hat{a}$.

Similarly,

$$C[\hat{b} - S_{\hat{b}} t_{\alpha/2} < b < \hat{b} + S_{\hat{b}} t_{\alpha/2}] = (1 - \alpha)$$

$$C[17.76 - (1.31)(3.17) < b < 17.76 + (1.31)(3.17)] = 0.99$$

$$C[17.76 - 4.15 < b < 17.76 + 4.15] = 0.99$$

$$C[13.61 < b < 21.91] = 0.99$$

for $\hat{b}$.

The results indicate that at the 99 percent level of confidence, the true parameters of the estimated equations, a and b, will lie between \$13,530.32 and \$27,183.90 for a, and between \$13.61 and \$21.91 for b. In other words, we are 99 percent certain that the true parameters of the estimated equations will lie between these two intervals. This is how sure we are about the reliability of our estimates. In general, confidence intervals established at less than the 90 percent level are not considered significant (acceptable).

Goodness of Fit of the Regression Line

Once we have estimated the values of the parameters and set up the equation of the regression line, we need to determine how good the fit is to the sample observations; in other words, how closely the line fits the observed data points (coordinates). In general, the closer the line is to the observed data points, the better the fit. However, to be able to precisely measure this fit we need to use a statistic called the coefficient of determination, or simply R^2. The coefficient of determination measures the amount of variation in the dependent variable that is explained by the variations in the independent variables (assuming there is more than one independent variable in the equation). In other words, it shows how well the observations fit the regression line.

Since the total variation in a regression is equal to the variation explained by the regression line and the error term, that is, total variance equals explained (regression) variance plus error variance, the R^2 can be computed by either of the following two expressions:

$$[1]\ R^2 = \frac{\text{Explained Variance}}{\text{TotalVariance}}$$

$$= \frac{\sum(\hat{Y} - \bar{Y})^2}{\sum(Y - \bar{Y})^2} \qquad [2.32]$$

$$[2]\ R^2 = 1 - \frac{\text{Error Variance}}{\text{Total Variance}}$$

$$= 1 - \frac{\sum(Y - \hat{Y})^2}{\sum(Y - \bar{Y})^2} \qquad [2.33]$$

Now, substituting the respective values from Table 2.3 into the above expressions, we can easily obtain the R^2 for our computer literacy problem, as shown below:

$$[1]\; R^2 = \frac{\sum (\hat{Y} - \bar{Y})^2}{\sum (Y - \bar{Y})^2}$$

$$= \frac{488{,}449{,}950.40}{514{,}916{,}666.73}$$

$$= 0.9486 \approx 0.95$$

$$[2]\; R^2 = 1 - \frac{\sum (Y - \hat{Y})^2}{\sum (Y - \bar{Y})^2}$$

$$= 1 - \frac{26{,}455{,}087.80}{514{,}916{,}666.73}$$

$$= 1 - 0.0514 = 0.9486 \approx 0.95$$

The result indicates that approximately 95 percent of the variation in total cost (Y) is explained by the variation in the number of hours spent (X), which is quite high considering that R^2 ranges between 0 and 1. The remaining 5 percent of the variation remains unaccounted for by the regression line and can be attributed to the error term, *e*. Overall, it appears that our equation was a good fit.

Although a high R^2 reflects the existence of a strong relationship between a dependent and a set of independent variables, it does not say anything about the significance of the coefficient itself. To ensure that our observed R^2 is significant, we use another statistical test called the *F* test. The *F* test tells us, from a statistical point of view, if the estimated regression line is a good fit; in other words, if the model sufficiently describes the variation in the data. The *F* statistic, the basis of this test, is computed by taking the ratio of the variance due to the regression line (i.e., the variance due to the independent variables) to the variance due to the error term, adjusted for appropriate degrees of freedom. That is,

$$F = \frac{\text{Explained Variance} / (k-1)}{\text{Total Variance} / (n-k)}$$

$$= \frac{\sum (\hat{Y} - \bar{Y})^2 / (k-1)}{\sum (Y - \bar{Y})^2 / (n-k)} \qquad [2.34]$$

where *k* is the number of parameters (i.e., the number of variables in a regression equation, including the dependent variable), and *n* is the sample size.

According to Equation 2.34, if the explained variance is the same as the error (unexplained) variance, the ratio will be equal to 1, meaning that the regression line is not a good fit. If it is less than 1, it is much worse and should be considered unacceptable. However, as the explained variance increases in magnitude in relation to the unexplained variance, the F statistic also increases in size. In general, the higher the explained variance in relation to the unexplained variance, the better is the fit. Table 2.5 presents a table, called the ANOVA or F table, to show how the F statistic was computed, based on the information contained in Table 2.3. According to the table, the observed F came out to be 184,434.

To determine if the estimated equation $\hat{Y}=20.357.11+17.757X$ is statistically significant, we need to compare the observed F, as we did for the observed *ts*, against the critical value of F. Since our observed F of 184.634, with one degree of freedom $(k-1)$ for the numerator and 10 degrees $(n-k)$ for the denominator, is greater than the critical value of F of 10.04 at the 0.01 level of p (which can be obtained from any standard F table), we can say with a 99 percent degree of certainty that our estimated equation is statistically significant; that is, there is only 1 percent chance that it could have occurred by chance (i.e., due to error). To further determine if our observed F has been computed correctly, we can add the explained and the error (unexplained) variances together to see if they equal the total variance. If they do, we can say that it has been calculated correctly, which appears to be the case here except for rounding-off errors.

We can now write the estimated model for the computer literacy program in its complete form with the associated statistics, as follows:

$$\hat{Y}=20{,}357.11+17.76X$$
$$(9.4527)\quad(13.5861)$$

$$R^2=0.9486\ F=184{,}634$$

Second-Order Tests

All regression models are based on a set of conditions, or assumptions, called model assumptions. Violations of these assumptions produce results that are often unreliable. Put differently, the estimated results are good as long as the model assumptions are satisfied. To ensure that these assumptions are satisfied, we can conduct a number of tests, commonly known as second-order tests. Important among these assumptions are normality, homoscedasticity, serial independence, and no multicollinearity. All the assumptions, with the exception of no multicollinearity, are based on the distribution of the error term. The following provides a brief discussion of the assumptions.

The assumption of normality. To start with, the normality assumption means that the error term, e, is normally distributed with a zero mean $\mu=0$ and constant variance, σ^2; that is,

$$e \sim N(0, \sigma_e^2) \qquad [2.35]$$

where N indicates the normality of the distribution, and the rest of the terms are the same as before.

Table 2.5 Analysis of Variance (ANOVA) Table

Sources of Variation (A)	*Sum of Squares (B)*	*Degrees of Freedom (C)*	*Mean Squares MS (D) = (B)/(C)*	*F-Value* (MS_{Reg}/MS_{Error})
Due to Regression	$\Sigma(\hat{Y}-\bar{Y})^2 = 488{,}449{,}950.40$	k–1 = 2–1 = 1	488,449,950.40/1 = 488,449,950.40	$= \frac{488{,}449{,}950.40}{2{,}645{,}508.78}$
Due to Error	$\Sigma(Y-\hat{Y})^2 = 26{,}455{,}087.80$	n–k = 12–2 = 10	26,455,087.80/10 = 2,645,508.78	
Due to Data (Total)	$\Sigma(Y-\bar{Y})^2 = 514{,}905{,}038.20$	n–1 = 12–1 = 11	514,905,038.20/11 = 46,809,548.93	= 184.634

The assumption of normality is necessary for two reasons: (1) to conduct the significance tests on the estimated parameters, and (2) to set up the confidence intervals. What this means is that when this assumption is violated, one cannot assess the statistical reliability of the tests of significance or the confidence intervals. For most statistical analysis, however, normality is assured as long as the sample size is large, usually 30 or more (based on a theorem known as the Central Limit Theorem).

The assumption of homoscedasticity. The homoscedasticity assumption is based on the notion that the variance of the error term is the same for all values of the explanatory variables. That is,

$$\mathrm{Var}(e) = E[e_i - E(e)]^2 = E(e_i)^2 = \sigma_e^2 \qquad [2.36]$$

where the term E represents the expected value, e_i is the error term for the ith observation and the rest of the terms are the same as before.

In general, when there is no homoscedasticity (a condition known as heteroscedasticity) it means that the estimated parameters do not have minimum variance, which is a violation of the condition for OLS. In other words, they are inefficient, although they may be unbiased meaning that the difference between the expected value of an estimator and the true parameter is zero. Inefficiency, in this case, means that it is not possible to conduct the tests of significance and construct confidence intervals on the estimated parameters. One way to correct the problem is to transform the original model in such a way that the transformed model will have a constant variance,

The assumption of serial independence. Serial independence means that the error terms of one period are not related to the error terms of the preceding period. In statistical terms, this means that the covariance of the terms is zero. That is,

$$\mathrm{Cov}(e_i e_j) = E[e_i - E(e_i)][e_j - E(e_j)] = 0 \qquad [2.37]$$

where $e_i \neq e_j$.

When the assumption of serial independence is violated, it indicates that there is a serial correlation, called autocorrelation, among the error terms. The presence of autocorrelation suggests that the values of the estimated parameters for any single sample are not correct. In fact, they may be underestimated, but not biased. Underestimation also means that any predictions based on these estimates will be inefficient in that their variances will be much larger compared with predictions based on estimates obtained from alternative methods. However, the problem can be corrected in most instances with the help of a process known as first differencing, according to which the original equation is transformed by subtracting the value of each period for a variable from the value of its preceding period. This allows for removal of any effect that the error terms in the previous period may have on the error terms in the current period.

The assumption of no multicollinearity. Finally, the assumption of no multicollinearity means that the explanatory variables are not perfectly (linearly) correlated (i.e., the correlation between any two explanatory variables is not equal to 1). That is,

$$r_{XiXj} \neq 1 \qquad [2.38]$$

where r stands for correlation, and X_i and X_j are the corresponding explanatory variables. When multicollinearity is present in a relationship, it indicates that the estimates of the parameters are indeterminate. In other words, their standard errors become infinitely large, making the estimated coefficients statistically insignificant. The simplest way to correct the problem is either to increase the size of the sample or to systematically eliminate the redundant variables from the regression model using methods such as step-wise regression, indirect least squares, two-stage least squares, instrumental variables, and principal component analysis, among others. The end result will be a model that is efficient and will include only those variables that will explain most of the variation in the dependent variable.

Although no specific tests were conducted to see if any of the assumptions were violated in the current example, a cursory glance at the results will indicate that normality was not fully assured since our sample size was rather small, especially when we are dealing with time-series data. Also, there could be a minor autocorrelation problem but not severe enough to affect the results, and it does not appear to have any major heteroscedasticity problem that could have seriously affected the efficiency of the model. Since we were dealing with only one explanatory variable, the problem of multicollinearity does not arise. Besides, the significance of the observed t suggests that the standard error was low – an indication commonly associated with the absence of multicollinearity. However, it is worth noting that there are ways in which one can correct the problems when the results do not fully satisfy the model conditions.

Extension of the Simple Linear Model

So far, we have used only one independent variable (i.e., number of hours the students spent after school) to estimate the fixed and the variable cost of the program for the school district. In reality, we do not have one but multiple independent variables such as direct labor (the time spent by the instructors), indirect labor (additional support personnel), direct machine hours (equipment), indirect machine hours (software), and so on that could have some effects on the dependent variable (i.e., the cost of running the program). Theoretically we could have n number of independent variables, instead of one, in which case we will need to expand our simple linear model (with one independent variable) to incorporate multiple independent variables.

The general cost function with multiple independent variables can thus be written as

$$Y = a + b_1X_1 + b_2X_2 + \ldots + b_nX_n + e \qquad [2.39]$$

Equation 2.39 is a typical expression for a regression model with *n* independent variables in linear form. It should be pointed out that when we are dealing with a multiple regression mode it is not possible to use the same computational procedure we used for the simple regression model. It will involve far too many computations that will be difficult to solve by hand. The alternative will be to use matrix algebra or a statistical software that can deal with the multi-dimensionality in a regression problem more efficiently.[3]

Forecasting Future Costs

Once we have estimated the parameters of a cost function for an organization and tested for their statistical significance, we can further extend the discussion by estimating the future costs of operation for the organization. This would, however, require that we have some information on future values of the independent variable(s) of the cost function.

To give an example, let us go back to the computer literacy program again. We begin with the estimated equation:

$$\hat{Y} = 20,357.11 + 17.757X$$

Assume now that we have some information on the number of hours for the month of January, next year. Let us say that it is 1,650. Our objective is to find the corresponding cost for the month of January. Assume that there has not been any change in the cost components of our estimated equation; the total cost for the month of January next year then will be

$$\begin{aligned} \hat{Y}_{January} &= 20,357.11 + 17.757\hat{X} \\ &= 20,357.11 + 17.757(1,650) \\ &= 20,357.11 + 29,299.05 \\ &= 49,656.16 \approx \$49,656.16 \end{aligned}$$

which is about $350 less than it cost in the same month the previous year.

In carrying out the forecast for our example, we assumed that we had some knowledge of the future value of the independent variable, which is the amount of hours spent next January. When the future values of the independent variable in a forecasting model are known a priori, it is called an unconditional forecast. In contrast, when they are not known beforehand, which is frequently the case, it is called conditional forecast. Given a choice, most forecasters would prefer unconditional forecasts to conditional forecasts, but it is not always possible to know with certainty the future values of the independent variable so that one can incorporate this information into the regression model to estimate the future values of the forecast (dependent) variable. As a result, conditional forecasts are more commonly used than unconditional forecasts.

Although they are more frequently used, conditional forecasts have a major weakness in that they have a tendency to add to forecast errors. A forecast error is the difference between actual (observed) and forecast value. The error occurs due to the fact that when one tries to forecast the future values of a forecast variable, given the values of an independent variable, one must first estimate the future values of the independent variable before they can be incorporated into the model. Since the forecast values, which themselves are estimates, depend on the estimated values of the independent variables, the likelihood of a forecast error is higher in conditional forecasts than in unconditional forecasts.

To put the problem differently, error in forecasting is a common occurrence that all forecasters must learn to deal with. Since the problem is inherent in all forecasts, the best one can do is try to minimize the error by using measures that would establish greater confidence in the forecasts. These would include, among others, correct specification of the model, variable transformation, use of lagged as opposed to current data and, more important, constantly updating the data as they become available. Another, perhaps more attractive alternative, would be to use time-series models such as trend line, simple moving average, exponential smoothing, or more advanced methods such as Box-Jenkins (Khan, 1989), which are built on less restrictive assumptions than most regression models. Also the data requirements for many of these models, with the exception of the Box-Jenkins, are not high, meaning they are relatively easy to use, especially in situations where data are not readily available.[4]

Chapter Summary

Cost behavior deals with the way in which costs behave in an organization in response to changes in variables that affect the costs. This chapter has provided a general discussion of cost behavior in government. Several aspects of cost behavior were discussed in the chapter; in particular, cost behavior in the short run and the long run, properties of cost functions, and the distinction between fixed and variable costs. In the short run, costs of factors that go into the production or delivery of a good or service, called input factors such as rent, interest payment on debt, and depreciation on structure are fixed. In the long run, all costs (i.e., all input factors) are variable; as such, costs in the long run are treated more as a planning horizon than as an operating cost for an organization.

An important aspect of cost behavior is cost estimation. The basic idea behind cost estimation is to estimate the relationship between costs and the variables affecting those costs. The chapter has discussed three of the most widely used cost estimation techniques: the graphical method, the high–low method, and the regression method. Of these, the regression method is considered more rigorous and analytically more sophisticated than either the graphical or the high–low method. The is due in part to the fact that (1) it utilizes all the available information in a dataset; (2) the estimated parameters are more precise than those obtained by the other two methods; and (3) the estimated parameters can be tested for statistical significance. The greatest advantage of the regression method is that once the parameters have been estimated and tested for statistical significance, they can be easily used to forecast future costs.

Notes

1 To obtain the estimates of a and b such that the sum of the squared deviations will be the minimum, we take the partial derivative of $\Sigma(\hat{e})^2$ with respect to $\hat{a}$ and $\hat{b}$, and set them equal to 0. That is,

$$\frac{\partial \sum \hat{e}^2}{\partial \hat{a}} = \frac{\partial \sum (Y - \hat{a} - \hat{b}X)^2}{\partial \hat{a}} = 0$$

$$2\sum (Y - \hat{a} - \hat{b}X)(-1) = 0$$

$$-2\sum Y + 2n\hat{a} + 2\hat{b}\sum X = 0 \quad (1)$$

for $\hat{a}$.

Similarly,

$$\frac{\partial \sum \hat{e}^2}{\partial \hat{b}} = \frac{\partial \sum (Y - \hat{a} - \hat{b}X)^2}{\partial \hat{b}} = 0$$

$$2\sum (Y - \hat{a} - \hat{b}X)(-X) = 0$$

$$-2\sum XY + 2\hat{a}\sum X + 2\hat{b}\sum X^2 = 0 \quad (2)$$

for $\hat{b}$.

We can simplify these expressions by dividing them by –2, so that we have

$$\sum Y - n\hat{a} - \hat{b}\sum X = 0 \quad (3)$$

$$\sum XY - \hat{a}\sum X - \hat{b}\sum X^2 = 0 \quad (4)$$

With slight rearrangements, Equations 3 and 4 can be written as

$$\sum Y = n\hat{a} + \hat{b}\sum X \quad (5)$$

$$\sum XY = \hat{a}\sum X + \hat{b}\sum X^2 \quad (6)$$

The new equations (5 and 6) are called the normal equations.

2 To see how estimates are obtained, we can do the following: first, set the deviations of X and Y so that

$$x = X - \bar{X} \quad (7)$$

$$y = Y - \bar{Y} \quad (8)$$

Second, take the first of the two normal equations, Equation 3.20, and divide it by N to obtain

$$\bar{Y} = \hat{a} + \hat{b}\bar{X} \quad (9)$$

Equation (9) indicates that the means of both X and Y lie on the regression line.

Third, we take Equation 2.14 and subtract Equation (3) from it, so that

$$(Y-\bar{Y}) = \hat{a} - \hat{a} + \hat{b}X - \hat{b}\bar{X} + \hat{e}$$

$$= \hat{b}(X-\bar{X}) + \hat{e} \tag{10}$$

$$y = \hat{b}x + \hat{e} \tag{11}$$

From the above, we can obtain the function of the variable b (to be minimized) by squaring and summing over N, so that

$$\sum (y-\hat{b}x)^2 = \sum \hat{e}^2 \tag{12}$$

Finally, we take the derivative of this function with respect to b, and set it equal to 0 to obtain the estimate of b. That is,

$$\frac{d\sum (y-\hat{b}x)^2}{d\hat{b}} = 2\sum x(y-\hat{b}x) = 0$$

$$\sum xy - \hat{b}\sum x^2 = 0$$

$$\sum xy = \hat{b}\sum x^2$$

$$\therefore \hat{b} = \frac{\sum xy}{\sum x^2} \tag{13}$$

Now, substituting the above into Equation (3) would produce the estimate of a. That is,

$$\hat{a} = \bar{Y} - \hat{b}\bar{X} \tag{14}$$

which means that once we have estimated the value of b (the slope), we can obtain the value of the intercept, a, without difficulty.

3 Before one can obtain the estimates of the parameters of the general linear model in Equation 2.39 using matrix algebra, it is necessary to set up the model in a format that is suitable for matrix formulation. Note that our model has n independent variables and one dependent variable, which means $n+1$ variables altogether and m observations. Since there are m observations in the model, we can present it in terms of m equations, one for each observation. That is,

$$Y_1 = a + b_1X_{11} + b_2X_{21} + b_3X_{31} + \ldots + b_nX_{n1} + e_1 \tag{15}$$

$$Y_2 = a + b_1X_{12} + b_2X_{22} + b_3X_{32} + \ldots + b_nX_{n2} + e_2 \tag{16}$$

$$Y_3 = a + b_1X_{13} + b_2X_{23} + b_3X_{33} + \ldots + b_nX_{n3} + e_3 \tag{17}$$

....

....

....

$$Y_m = a + b_1X_{1m} + b_2X_{2m} + b_3X_{3m} + \ldots + b_nX_{nm} + e_m \tag{18}$$

We can put these equations in matrix form so that we have

$$\mathbf{Y} = \mathbf{Xb} + \mathbf{e} \tag{19}$$

where $\mathbf{Y}$ is an $m \times 1$ column vector of observations (for Y, the dependent variable), $\mathbf{X}$ is an $m \times n$ matrix of observations (where Xs are the independent variables), $\mathbf{b}$ is an $n \times 1$ column vector of parameters, and $\mathbf{e}$ is an $m \times 1$ column vector of error terms. It should be worth noting that there is a column of 1s in the $\mathbf{X}$ matrix that stands for the constant term (the intercept).

To obtain the estimates of b, we need to minimize the sum of the squares of the residuals, $\Sigma(\hat{e})^2$, similar to Equation 2.17. That is,

$$\text{Minimize} \sum \hat{e}^2 = \hat{e}'\hat{e} \tag{20}$$

Note that $\hat{e}'$ is the transpose (where rows become columns and columns become rows) of the vector of residuals (i.e., the estimated error terms, $\hat{e}$), where $\hat{e}$ can be written as

$$\hat{e} = Y - \hat{Y} \tag{21}$$

For convenience, we can simplify Equation (21) by rewriting and explaining and expanding it as

$$\hat{e} = Y - \hat{Y}$$

$$\hat{Y} = X\hat{b} + \hat{e} - \hat{e}$$

$$\therefore \hat{e} = Y - X\hat{b} \tag{22}$$

Next, we take the vector of the estimated error terms $\hat{e}$ in Equation (22) and substitute it into the right-hand side of Equation (20) for minimization such that

$$\text{Minimize } \hat{e}'\hat{e} = (Y - X\hat{b})'(Y - Xb)$$

$$= Y'Y - X'\hat{b}'Y - X\hat{b}Y' + X'\hat{b}X\hat{b}$$

$$= Y''Y - 2X'\hat{b}'Y + X'\hat{b}'X\hat{b} \tag{23}$$

Equation (23) is the function that we finally need to minimize. Since $\hat{b}'X'Y'$ is a scalar (constant), it is equal to its transpose, $\hat{b}XY'$. As before, to minimize the sum of the squared residuals we take the partial derivative of $\hat{e}'e$ with respect to $\hat{b}$ and set it equal to 0. That is,

$$\frac{\partial(\hat{e}'e)}{\partial\hat{b}} = -2X'Y + 2X'X\hat{b} = 0$$

Finally, we divide both sides of the derivation above by 2, and rearrange the terms by changing the sides to obtain the estimates of our regression parameters. That is,

$$X'Y = X'X\hat{b}$$

$$\hat{b} = (X'X)^{-1}(X'Y) \tag{24}$$

where $\mathbf{(X'X)}^{-1}$ is called the inverse (Gilchrist, 1976) of $\mathbf{(X'X)}$ and $\hat{b}$ is the vector of estimated parameters (i.e., the estimated coefficients for the regression model in Equation 3.39).

4 For lengthy discussion of time-series forecasting, see Montgomery and Johnson (1976), Box and Jenkins (1976), Bails and Peppers (1982), Pindyck and Rubinfeld (1991), Makridakis, Wheelwright, and Hyndman (1998), and Green (2003).

References

Bails, D. G., & Peppers. L. C. (1982). *Business forecasting: Forecasting techniques and application.* Englewood-Cliffs, NJ: Prentice-Hall.

Box, G. E. P., & Jenkins, G. M. (1976). *Time series analysis: Forecasting and control.* San Francisco, CA: Holden-Day.

Gilchrist, W. (1976). *Statistical forecasting.* New York: Wiley & Sons.

Green, W. H. (2003). *Econometric analysis*. New York: Prentice Hall.

Khan, A. (1989). Forecasting a local government budget with time-series analysis. *State and Local Government Review, 21*, 123–129.

Makridakis, S. G., Wheelwright, S. C., & Hyndman, R. J. (1998). *Forecasting: Methods and applications*. New York: Wiley & Sons.

Montgomery, D. C., & Johnson, L. A. (1976). *Forecasting time-series analysis.* New York: McGraw-Hill.

Pindyck, R. S., & Rubinfeld, D. L. (1991). *Econometric models and economic forecasts.* New York: McGraw-Hill.

3 Cost Analysis

An important financial consideration for firms and businesses in the private sector is to be able to determine the level of operation at which they will have maximum profit on their investments, or earn just enough income to break even. A government does not operate to maximize profit, but it is just as important for a government to know in advance the level of operation at which a project or activity will be economically and financially viable. However, in order to determine the level at which an operation will be viable one must be able to measure the cost and return associated with the operation in precise monetary terms. Unfortunately, for a majority of public goods and services, with the exception of those considered as enterprise goods, it is difficult to measure the costs and returns in precise monetary terms because of the nature of the goods that separate them from private goods. Nevertheless, where it is possible to measure the costs and returns with some degree of precision, efforts should be made to do so to ensure the viability of a government activity.

This chapter presents several simple yet useful tools of cost analysis that have received considerable attention in cost studies. They are: break-even analysis, differential cost analysis, payback period, buy or lease, benefit–cost analysis, and life-cycle costing. Of these, break-even analysis has been extensively used in business and benefit–cost analysis in government. In fact, all five techniques have been used to varying degrees in business, as well as in government, to determine the level of operation at which a given project or activity will be economically and financially viable.

Break-Even Analysis

Developed originally by Rautenstrausch in the late 1930s, break-even analysis is considered as one of the most popular techniques used in cost analysis (Rautenstrausch, 1939). Its purpose is to integrate the cost, revenue, and output of an activity in order to determine the effect it will have on alternative courses of action. The basic objective is to find a level of operation, for a given cost, at which there will be neither a gain nor loss; that is, it will break even. The fundamental rule that underlies this objective is that at the lowest level of an activity, cost exceeds revenue, producing a loss, but as the level of activity increases, revenue will increase at a faster rate than cost, resulting in a situation where the two would become equal, producing a break-even point. If the level of activity continues to increase, revenue will eventually exceed cost, producing a net gain until a point comes when it will begin to diminish again. This fundamental law of operation that serves as the basis for break-even analysis is universal, but for the law to be of any practical significance one must be able to demonstrate how this relationship works in reality. This section presents examples of linear as well as nonlinear break-even analysis to provide support to this basic tenet and, in particular, to illustrate the relationship between cost, revenue, and output.

Linear Break-Even Analysis

We begin our discussion with a simple assumption that there is a linear relationship between cost, revenue, and output level of an activity. The linearity assumption sets up a mathematical relationship between cost, in particular fixed and variable costs, revenue, and output, that can be expressed in terms of a straight line, provided that the following conditions hold: (1) the variable cost is constant and hence is linearly dependent on output; (2) the fixed cost is independent of output; (3) there are no other financial costs; (4) output can be increased without significantly affecting the cost structure; and (5) all output units are charged or sold at the same price.

These conditions imply a constancy of rate scale that does not vary for different levels of output. What this means is that the price a government charges for a unit of good or service it provides, or the cost it incurs in providing the unit of the good, does not change regardless of the quantity provided. Since the price does not change, the total revenue from an operation can be represented by a linear curve emanating from the point of origin (as in a two-dimensional plane). Similarly, since the cost per unit does not change, the total cost of operation can also be represented by a linear curve emanating from the point of origin. Figure 3.1 shows this relationship between total revenue, total cost, and break-even quantity.

According to the figure, since both revenue and cost curves are straight lines, there is only one break-even point, which occurs at Q^*. It is the point at which total revenue (TR) equals total cost (TC). The shaded areas above and below the total cost line represent net gain and loss, respectively.

Break-Even Quantity, Price, and Revenue

The mathematical relationships underlying the cost and revenue functions described above provide a basis for developing a set of simple algebraic expressions that can be used to deal with various break-even problems such as break-even quantity, break-even price, break-even revenue, and so forth. We begin with the simplest one, the break-even quantity, as shown below:

$$TR = pQ \quad [3.1]$$

$$TC = FC + VC(Q) \quad [3.2]$$

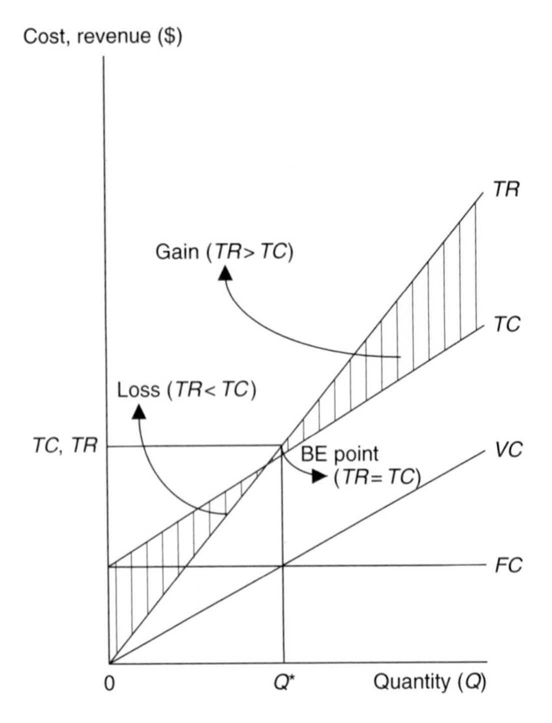

Figure 3.1 A Linear Break-Even Chart

where p is the price per unit of output, Q is the total quantity, FC is the fixed cost, and VC is the per unit variable cost associated with that output.

From these two simple equations, one can obtain the break-even quantity by setting $TR = TC$, since at the break-even point total revenue must always be equal to total cost. We can formally express this relationship as

$$TR = TC$$

$$TR - TC = 0$$

$$pQ - [FC + VC(Q)] = 0$$

$$pQ - [VC(Q)] = FC$$

$$Q(p - VC) = FC$$

$$\therefore Q^* = FC/(p - VC) \tag{3.3}$$

where Q^* is the break-even quantity and $(p - VC)$ is the contribution margin (discussed later).

With slight algebraic manipulation, Equation 3.3 yields the break-even price, corresponding to Q^*. That is,

$$Q^* = FC/(p - VC)$$
$$Q^*(p - VC) = FC$$
$$pQ^* - VC(Q^*) = FC$$
$$pQ^* = FC + VC(Q^*)$$
$$\therefore p^* = (FC/Q^*) + VC \quad [3.4]$$

where p^* is the break-even price.

Since it is derived from Q^*, at the break-even price, p^*, total revenue must be the same as total cost. We can use a simple example to illustrate how these relationships work. Suppose that the number of citations given by the police department of a large metropolitan government for minor traffic infractions such as not stopping at the stop sign is 100,000 per year. Suppose also that the fixed cost associated with this activity is \$2 million, with a variable cost of \$25 per citation. The department charges \$50 across the board for each citation, regardless of the nature of infraction. Assume that the department is working at 100 percent capacity. Our objective is to do two things: find the break-even quantity for the department and determine the net revenue from operation at the current rate of 100,000 citations per year.

Let us start with the break-even quantity for the department. Since we already know the price one will pay for an infraction, we can determine the break-even quantity by directly applying Equation 3.3 to the problem. That is,

$$Q^* = FC/(p - VC)$$
$$= \$2{,}000{,}000/(\$50 - \$25)$$
$$= \$2{,}000{,}000/\$25$$
$$= 80{,}000$$

which is 80,000 citations per year.

This is the number of citations the department must produce to break even. At this rate, there should not be any gain or loss for the department, meaning that its total revenue must be equal to total cost. To see if this is true for the current example, we can substitute the value of Q^* into Equations 3.1 and 3.2, then set them equal to each other, so that

$$TR = TC$$
$$pQ^* = FC + VC(Q^*)$$
$$(\$50)(80{,}000) = \$2{,}000{,}000 + (\$25)(80{,}000)$$
$$(\$50)(80{,}000) = \$2{,}000{,}000 + \$2{,}000{,}000$$
$$\$4{,}000{,}000 = \$4{,}000{,}000$$
$$\therefore TR = TC = \$4{,}000{,}000$$

The result produces a sum of \$4 million for both *TR* and *TC*, which means that at the break-even point of 80,000 citations per year, total revenue is equal to total cost. In other words, at the break-even quantity, Q^*, the net revenue is 0. That is,

$$\pi = TR - TC = \$4,000,000 - \$4,000,000 = 0 \quad [3.5]$$

where π is the net revenue (profit).

On the other hand, the net revenue at the current level of output is positive, not zero, since the department is producing 20,000 more citations than the break-even quantity of 80,000 [100,000–80,000 = 20,000]. The difference between the two output quantities, therefore, will produce a net revenue of \$500,000 for the department, as can be seen from the following calculations:

$$\begin{aligned}\pi &= TR - TC \\ &= pQ - [FC + VC(Q)] \\ &= [(\$50)(100,000)] - [\$2,000,000 + (\$25)(100,000)] \\ &= \$5,000,000 - \$4,500,000 \\ &= \$500,000\end{aligned}$$

Note that we could also have arrived at the same result by subtracting the net revenue at the break-even level of citations from the net revenue at the current level of citations. That is,

$$\begin{aligned}\pi &= [pQ - (FC + VC(Q))] - [pQ^* - (FC + VC(Q^*))] \\ &= [(\$50)(100,000) - (\$2,000,000 + (\$25)(100,000))] \\ &\quad - [(\$50)(80,000) - (\$2,000,000 + (\$25)(80,000))] \\ &= [\$5,000,000 - \$4,500,000] - [\$4,000,000 - \$4,000,000] \\ &= \$500,000 - \$0 = \$500,000\end{aligned}$$

Since they produce the same result, it does not matter which expression is used.

Changes in Capacity

Earlier in the example we made an assumption that the department is working at 100 percent capacity. In reality, very few organizations work at their maximum capacity level. For instance, if we assume that the department is working at 90 percent capacity instead of 100 percent, the net revenue from operation will be \$250,000 instead of \$500,000. That is,

$$\begin{aligned}\pi &= [(0.90)(\$50)(100,000)] - [\$2,000,000 + (0.90)(\$25)(100,000)] \\ &= \$4,500,000 - [\$2,000,000 + \$2,250,000] \\ &= \$4,500,000 - \$4,250,000 \\ &= \$250,000\end{aligned}$$

If, on the other hand, we assume that its capacity increased to 105 percent from the current 100 percent, the net revenue for the department will increase from $500,000 to $625,000, provided that no changes took place in its operating costs or the fines it levies for an infraction. That is,

$$\begin{aligned}\pi &= [(1.05)(\$50)(100{,}000)] - [\$2{,}000{,}000 + (1.05)(\$25)(100{,}000)] \\ &= \$5{,}250{,}000 - [\$2{,}000{,}000 + \$2{,}625{,}000] \\ &= \$5{,}250{,}000 - \$4{,}625{,}000 \\ &= \$625{,}000\end{aligned}$$

However, at the break-even level of output, where the net revenue is zero, the capacity level for the department is 80 percent $[(80{,}000/100{,}000) = 0.80)]$. In other words, to break even the department only needs to work at 80 percent of its capacity, as shown below:

$$\begin{aligned}\pi &= [(0.80)(\$50)(100{,}000)] - [\$2{,}000{,}000 + (0.80)(\$25)(100{,}000)] \\ &= \$4{,}000{,}000 - [\$2{,}000{,}000 + \$2{,}000{,}000] \\ &= \$4{,}000{,}000 - \$4{,}000{,}000 \\ &= 0\end{aligned}$$

The result indicates that any drop in capacity from the current level of 80 percent will produce a loss (i.e., net negative return) for the department.

Changes in Revenue, Output, and Price

Financial necessities often require that a government must find ways to increase its revenue to meet a specified goal or target. Theoretically a government can increase its revenue in one of three ways: (1) by raising taxes, (2) by increasing the price of goods and services it provides, and (3) by increasing its output. Since most governments are reluctant to increase revenue by increasing price or raising taxes, the alternative is to increase output for those goods and services where the potential exists for an increase. For instance, a government can install more parking meters, issue more permits, give more citations, produce more electricity, collect refuse more frequently, add more hospital beds, and so on to increase the output quantity, as long as it is possible to establish that such efforts will be economically viable (i.e., they will increase the net revenue of the government).

Operationally, then, how does one determine the level of output that will generate a desired level of net revenue? To determine the level of output that will produce a desired level of net revenue, π', the convention is to treat the target level as an increment of fixed costs (i.e., a constant), since, like any fixed cost, it will not change in the short run. We can easily extend the expression in Equation 3.3 to obtain this output quantity, as shown below:

$$Q' = (FC + \pi')/(p - VC) \qquad [3.6]$$

where Q' is the projected break-even quantity and the rest of the terms are the same as before.

To illustrate the point above, let us return to the traffic citation example for a moment. Suppose that the department now sets as its goal net revenue of $1 million ($\pi'$) instead of the $500,000 it obtained earlier from 100,000 citations. Assuming no changes in price or costs and all other conditions remaining the same, the projected break-even quantity, Q', will thus be 120,000. That is,

$$\begin{aligned} Q' &= (FC + \pi')/(p - VC) \\ &= (\$2{,}000{,}000 + \$1{,}000{,}000)/(\$50 - \$25) \\ &= \$3{,}000{,}000/\$25 \\ &= 120{,}000 \end{aligned}$$

This is the total number of citations the department must give to generate the target net revenue of $1 million. That is,

$$\begin{aligned} \pi' &= TR - TC \\ &= pQ - [FC + VC(Q)] \\ &= (\$50)(120{,}000) - [\$2{,}000{,}000 + (\$25)(120{,}000)] \\ &= \$6{,}000{,}000 - [\$2{,}000{,}000 + \$3{,}000{,}000] \\ &= \$6{,}000{,}000 - \$5{,}000{,}000 \\ &= \$1{,}000{,}000 \end{aligned}$$

A question that becomes relevant here is what if the department does not have the physical capacity to raise its output from 100,000 citations per year to 120,000 but, at the same time, it needs to increase its net revenue to $1 million. This would obviously mean that the department must increase the fine it presently levies for each citation in order to bring the revenue to the desired level, while keeping the level of output (i.e., the number of citations) constant at 100,000. To achieve this, we can easily reorganize the expression in Equation 3.6 for projected output, so that

$$\begin{aligned} &Q = (FC + \pi')/(p' - VC) \\ &(p' - VC)Q = (FC + \pi') \\ &\therefore p'Q - VC(Q) = (FC + \pi') \\ &p'Q = (FC + \pi') + VC(Q) \\ &p' = [(FC + \pi')/Q] + VC \qquad [3.7] \end{aligned}$$

where p' is the new or projected price, and the rest of the terms are the same as before.

Applying the expression in Equation 3.7 to the citation problem, we will obtain a new price for the department, as shown below:

$$\begin{aligned} p' &= [(FC + \pi')/Q] + VC \\ &= [(\$2{,}000{,}000 + \$1{,}000{,}000)/100{,}000] + \$25 \\ &= (\$3{,}000{,}000/100{,}000) + \$25 \\ &= \$30 + \$25 \\ &= \$55 \end{aligned}$$

which is \$55. This is the price the department must charge for each citation to produce a net revenue of \$1 million. To see if the result is correct (i.e., if it would produce the target revenue), we can substitute the value of p' into the expression in Equation 3.5 for net revenue. That is,

$$\begin{aligned} \pi &= TR - TC \\ &= p'Q - [FC + (VC)(Q)] \\ &= [(\$55)(100{,}000)] - [\$2{,}000{,}000 + (\$25)(100{,}000)] \\ &= \$5{,}500{,}000 - \$4{,}500{,}000 \\ &= \$1{,}000{,}000 \end{aligned}$$

The result yields a net revenue of \$1 million, as expected.

Nonlinear Break-Even Analysis

Although it is relatively easy to deal with a linear break-even analysis, the real relationship between cost, revenue, and output may not be linear; it may be nonlinear. The notion that per unit price or cost will always be a constant, regardless of the quantity of output consumed, is not a realistic assumption since, in reality, costs related to labor, maintenance, and utility change disproportionately, rather than in fixed, constant amounts. To give an example, recall the citation problem again. Assume now that we have a total revenue and a total cost function for citations given by a single officer, rather than the entire department. We can express the two functions in terms of two simple equations, one linear and one nonlinear (quadratic), as given below:

$$TR = 50Q \qquad [3.8]$$

$$TC = 0.08Q^2 - 10Q + 750 \qquad [3.9]$$

where the price of an infraction is \$50, as before, with a fixed cost of \$750 and a variable cost that varies with different levels of output of the quadratic cost function. As before, our objective is to obtain a break-even quantity for the officer, given the information we have on the price and the cost of operation for the department.

To obtain the break-even quantity for the officer, we set up a net revenue function by rearranging the terms of Equations 3.8 and 3.9, so that

$$\begin{aligned}\pi &= TR - TC\\ &= 50Q - (0.08Q^2 - 10Q + 750)\\ &= 50Q - 0.08Q^2 + 10Q - 750\\ &= -0.08Q^2 + 60Q - 750\end{aligned}$$

The function resembles a typical quadratic function of the form

$$aQ^2 + bQ + c = O \quad [3.10]$$

where a and b are the parameters (constants) associated with Q. What this means is that Equation 3.10 has two solutions, called roots, one high and one low, that are determined by a standard expression used for solving quadratic equations, given by the expression

$$Q = \frac{-b \pm \sqrt{(b^2 - 4ac)}}{2a} \quad [3.11]$$

When the expression in Equation 3.11 is applied to the rearranged net revenue function, it produces two break-even quantities – one low and one high, as shown below:

$$\begin{aligned}Q &= [-b \pm \surd(b^2 - 4ac)]/2a\\ &= [-60 \pm \surd(60)^2 - (4)(-0.08)(-750]/(2)(-0.08)\\ &= [-60 \pm \surd(3{,}600 - 240)]/-0.16\\ &= [-60 \pm \surd 3{,}360]/-0.16\\ &= [-60 \pm 57.9655]/-0.16\end{aligned}$$

$$\begin{aligned}\therefore\ Q_1^* &= -2.0345/-0.16 = 12.7156 \approx 13.0\\ Q_2^* &= -117.9655/-0.16 = 737.2844 \approx 737.0\end{aligned}$$

That is, 13 and 737 citations, respectively, since you cannot have fractional citations.

Although both solutions produce positive break-even quantities, it is obvious that the department will not go with the lower quantity of 13 citations because it makes no sense to issue only 13 citations for the year for a single officer; it will also not generate enough revenue for the department. Therefore, the second solution, $Q_2=737$ will be the preferred break-even quantity.

To see if the break-even points obtained for the problem are correct, we can substitute these quantities into the expression for $\pi=TR-TC$; if the result produces a net revenue of zero it should be considered correct. We can start with $Q_2^*=737.2844$. That is,

$$\pi=TR-TC$$

$$=50Q-[0.08Q^2-10Q+750]$$

$$=(50)(737.2844)-[(0.08)(737.2844)^2-(10)(737.2844)+750]$$

$$=\$36.864.22-[\$43{,}487.0629-\$7{,}372.844+\$750]$$

$$=\$36{,}864.22-\$36{,}864.22=0$$

which comes out to be 0, as expected. We could do the same for $Q_1^*=12.7156$, but it would not be necessary since we do not think the department will go with 13 citations.

Other Uses of Break-Even Analysis

The break-even analysis we have discussed up to this point deals with a single activity, but it can also be used for comparing multiple activities with varying levels of fixed and variable costs. Consider an example of a city that is planning to replace the city incinerator for solid waste disposal. Projects such as incinerators are highly capital intensive in that they require large initial costs (capital) which often run into millions of dollars. Fortunately for most governments, the high initial cost is frequently offset by a low operating cost that is directly related to output; that is, at low levels of output they are costly, but as the level of output increases the cost goes down, resulting mostly from economies of scale (i.e., decreasing average cost of production).

To illustrate the case in point discussed above, assume that the city in question is considering offers from two private firms (X and Y), both wanting to supply the city with a new incinerator. Assume further that we have two cost functions for the incinerator based on the information provided by the firms. For convenience, we present the cost functions in terms of two linear equations, one for X and the other for Y, as shown below:

$$TC_X=20{,}500{,}000+25Q \quad [3.12]$$

$$TC_Y=25{,}000{,}000+10Q \quad [3.13]$$

According to the equations, project Y appears to be more capital intensive but has a lower operating cost: \$10, compared to \$25 for project X. On the other hand, both projects have some advantages and disadvantages; as such, we need to have some guidelines or criteria by which we can make a comparative assessment of the projects. The criterion most frequently used in this case is the fixed cost efficiency (defined as the relationship between output and the fixed cost of a project or activity). To determine this efficiency, the rule of thumb is to find an equilibrium quantity, Q^*, and· use that as a basis for comparison among projects.

We can obtain the equilibrium quantity for our two projects by setting the two equations equal to each other, as before, then solving the expression for Q^*, as shown below:

$$TC_X = TC_Y$$

$$20{,}500{,}000 + 25Q = 25{,}000{,}000 + 10Q$$

$$25Q - 10Q = 25{,}000{,}000 - 20{,}500{,}000$$

$$15Q = 4{,}500{,}000$$

$$\therefore\ Q^* = 300{,}000 \text{ tons/year}$$

The result produces an equilibrium quantity of 300,000 tons per year.

What the result means is that at an output level of less than 300,000 tons per year, project X will be more efficient since it has a lower fixed cost. The explanation for this is quite simple: with low fixed costs, a high variable cost project is preferred at low levels of output because the gain from low fixed costs initially offsets the high variable costs. As output level increases, fixed costs tend to decrease in relative importance for both project scales, eventually reversing the benefits of the smaller scale. In other words, the gain in low variable cost for the higher fixed cost project will more than offset the difference in fixed costs at Q greater than 300,000. Put differently, at any given level of output, the lower the fixed cost, the smaller the fraction of total revenue necessary to recover these costs.

Operating Leverage

We can extend the discussion by introducing into the above example an important break-even concept called the operating leverage. Operating leverage is the measure of the extent to which fixed production facilities affect the operation of a project. In other words, it measures the sensitivity of an operation resulting from the use of fixed, rather than variable inputs. The degree of operating leverage for break-even analysis, for any amount of output, can be measured by a ratio, given by the return on sunk (irrecoverable) costs of an operation and its net revenue.

We can formally express this ratio (i.e., the operating leverage) as

$$OL = [TR - VC(Q)] / \pi \qquad [3.14a]$$

$$= [pQ - VC(Q)] / \pi$$

$$= [pQ - VC(Q)] / (TR - TC)$$

$$= [pQ - VC(Q)] / [pQ - (FC + VC(Q))]$$

$$= [pQ - VC(Q)] / [pQ - FC - VC(Q)]$$

$$= [pQ - VC(Q)] / [(pQ - VC(Q)) - FC]$$

$$= Q(p - VC) / [Q(p - VC) - FC] \quad [3.14b]$$

where OL is the operating leverage, $Q(p-VC)$ is the sunk cost, and $Q(p-VC)-FC$ is the net return (π).

Note that the expression $(p-VC)$, mentioned earlier, is the contribution margin, obtained by taking the difference between price and variable cost. Contribution margin measures the contribution this difference makes toward recovering the fixed cost of an operation, as well as any net gain from it. For instance, if a government charges $10 (price) for a fishing permit and it costs the government $3 in variable costs, then each permit issued recovers its variable cost plus $7. The $7 is the contribution margin [$p-VC=\$10-\$3=\$7$], since it contributes to the recovery of the fixed cost plus any net gain from operation. Thus, if the fixed cost of issuing a permit is $5, the net gain ($\pi$) for the government for the permit will be $2; that is, $P-(FC+VC)=\$10-(\$5+\$3)=\$2$.

Let us now return to the incinerator example. Assume that it costs the government $150 for each ton of waste disposed. Thus, at an output level of 300,000 tons per year, project X will have an operating leverage of 2.21 with a contribution margin of $125, as can be seen from the following calculations:

$$OLx = [300{,}000(\$150 - \$25)] / [300{,}000(\$150 - \$25) - \$20{,}500{,}000]$$

$$= \$37{,}500{,}000 / (\$37{,}500{,}000 - \$20{,}500{,}000)$$

$$= \$37{,}500{,}000 / \$17{,}000{,}000$$

$$= 2.21$$

On the other hand, at the same level of output, project Y will have an operating leverage of 2.47 and a contribution margin of $140, as shown below:

$$OLy = [300{,}000(\$150 - \$10)] / [300{,}000(\$150 - \$10) - \$25{,}000{,}000]$$

$$= \$42{,}000{,}000 / (\$42{,}000{,}000 - \$25{,}000{,}000)$$

$$= \$42{,}000{,}000 / \$17{,}000{,}000$$

$$= 2.47$$

We can interpret these results the same way as one would interpret the elasticity of revenue resulting from a change in output. As noted previously, the elasticity means the degree to which a variable changes in response to a change in another variable. The higher the degree of responsiveness, the greater is the elasticity. Thus, in the current example, a 1 percent increase in output over the equilibrium quantity of 300,000 tons per year will result in 2.21 percent increase in revenue for project *X*. At the same level of output for project *Y*, which has a slightly higher operating leverage, a 1 percent increase in output will result in a 2.47 percent increase in revenue. In general, operating leverage is greater for capital-intensive operations than it is for labor-intensive operations, meaning that capital-intensive operations will have a higher break-even point than labor-intensive operations. This is clearly evident in the break-even quantity produced by the two projects, as shown below:

$$Q^*X = FC/(P-VC) = \$20{,}500{,}000/(\$150-\$25) = 164{,}000 \text{ tons/year}$$
$$Q^*Y = FC/(P-VC) = \$25{,}000{,}000/(\$150-\$10) = 178{,}571 \text{ tons/year}$$

As the results indicate, project *Y*, which is the more capital intensive of the two, has a higher break-even point than project *X*. In other words, how much cost advantage a capital-intensive operation will enjoy depends on the level of output; that is, at higher levels of output, capital-intensive operations have an advantage over labor-intensive operations, while at lower levels of output labor-intensive operations have the advantage over capital-intensive operations.

Break-Even Analysis with Multiple Services

An interesting aspect of operating leverage, in particular the contribution margin, is that it can be used to determine break-even points for not one, but multiple services at the same time. Let us look at a simple example to illustrate this. Suppose that the outpatient clinic of a local county hospital provides three different types of services: *X*, *Y*, and *Z*. Service *X* serves 1,000 patients per year, Service *Y* serves 3,000 patients, and Service *Z* serves 5,000 patients. The price (fee) the clinic charges for the three services per visit is \$100 for *X*, \$75 for *Y*, and \$50 for *Z*. The variable costs for the services, respectively, are \$60, \$40, and \$25, with a fixed cost of \$180,000 (for all three services combined). Our goal is to determine (1) the contribution margin for each service, (2) the respective break-even quantity, and (3) the break-revenue for each program. Table 3.1 shows the results.

According to the table, the contribution margin (*CM*) for the three services, with corresponding break-even quantity (Q^*) and revenue (R^*), are as follows:

Service *X*: $CM_X = \$40$ (40 percent); $Q_X^* = FC/CM = \$180{,}000/\$40 = 4{,}500$;
$R_X^* = \$100 \times 4{,}500 = \$450{,}000$

Service *Y*: $CM_Y = \$35$ (47 percent); $Q_Y^* = FC/CM = \$180{,}000/\$35 \approx 5{,}143$;
$R_Y^* = \$75 \times 5{,}143 = \$385{,}725$

Service *Z*: $CM_Z = \$25$ (50 percent); $Q_Z^* = FC/CM = \$180{,}000/\$25 = 7{,}200$;
$R_Z^* = \$50 \times 7{,}200 = \$360{,}000$

Table 3.1 Break-Even Analysis for Multiple Services

Service X	*Units (Q)*	*Price-VC-CM ($)*	*Total Revenue ($)*
Revenue	1,000	100	100,000
Variable Cost	1,000	60	60,000
CM	1,000	40	40,000
CM(%)	$40/$100=0.40 (40%)		
Service Y	*Units*	*Price-VC-CM ($)*	*Total Revenue ($)*
Revenue	3,000	75	225,000
Variable Cost	3,000	40	120,000
CM	3,000	35	105,000
CM(%)	$35/$75=0.47 (47%)		
Service Z	*Units*	*Price-VC-CM ($)*	*Total revenue ($)*
Revenue	5,000	50	250,000
Variable Cost	5,000	25	125,000
CM	5,000	25	125,000
CM(%)	$25/$50=0.50 (50%)		
Service: X + Y + Z	*Units*	*Price ($)*	*Total Revenue ($)*
Revenue	9,000[1]	63.89[3]	575,000[2]
Variable Cost	9,000	33.89[5]	305,000[4]
CM	9,000	30.00	270,000
CM(%)	$30/$63.89=0.47 (47%)		

Notes
1 $Q_{X+Y+Z}=1{,}000_X + 3{,}000_Y + 5{,}000_Z=9{,}000$
2 $TR_{X+Y+Z}=\$100{,}000 + \$225{,}000 + \$250{,}000=\$575{,}000$
3 $P_{X+Y+Z}=TR/Q=\$575{,}000/9{,}000=\$63.89$
4 $VC_{X+Y+Z}=\$60{,}000 + \$120{,}000 + \$125{,}000=\$305.000$
5 $VC_{(Average)}=\$305{,}000/9{,}000=\$33.89.$

Given the information above, we can easily obtain the contribution margin, the break-even quantity, and the break-even revenue for the clinic as a whole, taking all three services together, as shown below:

$$\text{Service } X+Y+Z\text{: } CM=\$30;\ Q^*=FC/CM=\$180{,}000/\$30=6{,}000;$$
$$R_{(X+Y+Z)}^* = \$63.89\times 6{,}000=\$383{,}340$$

We can now use this information to construct different scenarios for the clinic. Let us look at three such scenarios; for instance, what will be the break-even quantity and revenue, if (1) the fixed cost of operation were to increase by 10 percent, (2) the variable cost were to increase by 5 percent across the board, and (3) the prices were to increase by 10 percent across the board, with a safety margin (*SM*) of 5 percent?

The results of the scenarios are shown below:

1 Change in $FC=\$180{,}000+(\$180{,}000\times 0.10)=\$180{,}000+\$18{,}000=\$198{,}000$
$Q_{(X+Y+Z)}^*=\$198{,}000/\$30=6{,}600;\ R_{(X+Y+Z)}^*=\$63.89\times 6{,}600=\$421{,}674$

2 Change in Variable Cost, $VC_{(X+Y+Z)}$=\$305,000+(\$305,000 × 0.05)=\$320,250; VC_{Ave}=\$320,250/9,000=\$35.58; $Price_{(X+Y+Z)}$=\$63.89 (from Table 3.1); $CM_{(X+Y+Z)}$=(p-VC)=\$63.89-\$35.58=\$28.31; $Q_{(X+Y+Z)}$*=FC/(p-VC)=FC/CM=\$180,000/\$28.31=6,358.1773≈6,358; $R_{(X+Y+Z)}$*=pQ*=\$63.89×6,358.1773=\$406,223.95;
Proof: At break-even, TR=TC; that is,
pQ*=FC+VC(Q*)
or (\$63.89×6,358.1773)=\$180,000+(\$35.58×6,358.1773)
or \$406,223.95=\$406,223.95

3a Change in Price, P=\$63.89+(63.89×0.10)=\$63.89+\$6.39=\$70.28; $VC_{(X+Y+Z)}$=\$33.89 (from Table 3.1); $CM_{(X+Y+Z)}$=(p-VC)=\$70.28–\$33.89=\$36.39; $Q_{(X+Y+Z)}$*=FC/(p-VC)=FC/CM=\$180,000/\$36.39=4,946.4139≈4,946; $R_{(X+Y+Z)}$*=pQ*=\$70.28×4,946.4139=\$347,633.97
Proof: At break-even, TR=TC; that is,
pQ*=FC+VC(Q*)
or (\$70.28×4,946.4139)=\$180,000+(\$33.89×4,946.4139)
or \$347,633.97=\$347,633.97

3b With 5 Percent Safety Margin, SM: $R_{(X+Y+Z)*}$=\$347,633.97+(\$347,633.97×0.05)
=\$347,633.97+\$17,381.70=\$365,015.67
Alternatively, $R_{(X+Y+Z)}$*=[4,946.4139+(4,946.4139×0.05)]×\$70.28
=(4,946.4139+247.3207)×\$70.28
=\$365,015.67

The results presented above show how break-even analysis can be used under different conditions. We could easily extend the analysis to include additional services the clinic provides and develop a comprehensive picture of financial viability of the organization. This is where the appeal of the method lies, in that it is simple in approach yet produces useful information about the financial viability of an organization under different conditions.

Differential-Cost Analysis

Like break-even analysis, differential cost analysis deals with cost, revenue, and output of an activity. Its purpose is to use cost and other related information to determine the course of action that will produce the best possible return from that decision for an organization. For instance, when a government wants to initiate a new program or eliminate an existing one, or when it wants to reduce a service charge or buy a piece of equipment, it must compare these decisions against the existing conditions to determine if the net return is higher in each instance to justify that course of action. Without this comparison, it will be difficult for a government to determine the relative worth of its decisions. In general, a decision is considered worthwhile if it produces a return that is higher than the return produced under the existing conditions. This section discusses a situation in which differential cost analysis makes most sense: to determine the effects of a change in cost on net return from an operation.

Earlier in the chapter, we defined net revenue as the difference between total revenue and total cost, but in differential cost analysis it is defined as the difference between two differentials: differential revenue and differential cost. Differential revenue is the difference between a current revenue and the revenue one will earn from an alternative. Similarly, differential cost can be defined as the difference between current cost and the cost under an alternative. Thus, in calculating net revenue, the emphasis is placed on the differential rather than on the simple difference between cost and revenue.

The calculation of net return from these differentials is simple, provided that one has all the information one needs on differential costs and revenues. Equation 3.15 presents a simple expression for calculating this net return:

$$\pi_D = [\Delta R \times T] - [C_0 + (\Delta C \times T)] \quad [3.15]$$

where π_D is the net return from differentials, ΔR is the change in revenue, C_0 is the one-time cost such as construction cost, ΔC is the change in cost, excluding the one-time cost, and T is the time.

We can use a simple example to illustrate this. Suppose that a large metropolitan government owns and operates a parking garage with a capacity for parking 150 cars at any given time. The government charges \$3.75 per day for parking, which runs from 7:00 in the morning to 7:00 in the evening. The garage remains closed after that time. The number of cars parked on average is 200 each day, including weekends. Suppose now that the demand for parking space has been steadily increasing in recent months. To deal with the problem, the city has three options: (1) outsource the garage, (2) privatize it (i.e., sell it to a private provider, say, for \$1 million), and (3) upgrade it. Let us assume that, for now, the government is not considering options (1) and (2), since the garage provides a consistent source of income for the city, which leaves it with the third option: upgrade it. Upgrading will involve increasing the capacity by another 150 parking spaces; thus, raising the number of cars that can be parked on average to 400 per day.

Assume that it will cost the city \$750,000 to upgrade the garage, which it will borrow as an intermediate-term loan from a local bank payable in three years called the cost of funds. The city is expected to retire the principal on or before the end of this period, along with the interest it will accrue for the period of the loan. Assume further that the city plans to increase the parking fee by 75 cents to recover some of the costs of construction, thereby raising the current price from \$3.75 to \$4.50 per day. Table 3.2 shows a three-year differential return, including the time it will take to even out the costs of upgrading.

According to the table, the city revenue from the upgrade will increase by \$383,250 per year, resulting in part from the price increase and in part from the revenue generated from additional parking spaces. That is,

$$Ru = RAP - REC = [(400)(365)(\$3.75) + (400)(365)(\$0.75)] - [(200)(365)] (\$3.75) = [(\$547{,}500 + \$109{,}500) - \$273{,}750] = \$383{,}250$$

Table 3.2 Differential-Cost Analysis

	Existing Condition ($)	*Alternative Proposal ($)*	*Difference ($)*
I. Revenue from Parking Space	273,750[1]	547,500[2]	273,750
Price Increase	0	109,500[3]	109,500
Total Revenue/Year	273,750	657,000	383,250
II. Cost Due to Construction	0	750,000	750,000
Interest Payment	0	56,250	56,250
Repair & Maintenance	2,500	5,000	2,500
Payroll Increase	18,000	36,000	18,000
Miscellaneous	1,000	2,000	1,000
Total Cost/Year	21,500	99,250*	77,750*
III. Three-Year Differential			
Total Differential Revenue ($383,250 × 3 = $1,149,750)			1,149,750
Total Differential Cost [$750,000 + (77,750 × 3)] = $983,250			(983,250)
Net Return from Differential			166,500

Notes
*Excludes the construction cost; 1 (200)(365)($3.75); 2 (400)(365)($3.75); 3 (400)(365)($0.75).

where *Ru* is the revenue from the upgrade, R_{AP} is the revenue from an alternative proposal, and R_{EC} is the revenue from the existing condition.

The operating cost will also increase by $77,750 per year, much of which will be due to increase in payroll, costs of additional repair and maintenance, and interest payments on debt. The table also shows a three-year net return from differentials, which stands at $166,500 (after all the adjustments have been made); that is, $\pi = [\Delta R \times T] - [C_0 + (\Delta C \times T)] = (\$383{,}250 \times 3) - [\$750{,}000 + (\$77{,}750 \times 3)] = \$1{,}149{,}750 - \$983{,}250 = \$166{,}500$. Does this mean that the city will be better off by not selling the garage? The answer is "yes," since it will take the city about three years to break even and another two years or so to recover the $1 million it will have earned if it had decided to sell it, while retaining ownership of the garage (assuming no changes in the cost of operation).

Payback Period

An important conclusion that emerges from the preceding example is that it will take the city about three years to recover the cost of the project. In other words, it is the amount of time the city will need for the earnings (cash inflows) from the garage to equal the cost of its expansion (cash outflows). In cost analysis, this is known as the payback period. Payback period means the time during which cash inflows from a project operation will equal its cash outflows. In other words, it is the time when the net cash flow (the difference between cash inflows and cash outflows) from an operation will be zero.

The definition of payback presented above has an interesting implication in cost analysis in that it focuses primarily on recovery time (i.e., the time it takes to realize the full cost of a project), but not what happens after the costs have been recovered. Since it is mostly concerned with recovery time, payback is often used as a benchmark for project acceptance. The benchmark, in this case, means the maximum number of years it will take an organization to recover the cost of an investment. Thus, if it takes five years to recover the cost of an investment project, it is the time that should be used as the benchmark for that project. However, the time it takes to recover the cost of an investment varies, depending on the size of the project and the flow of funds. For instance, a small project with low positive net flow can take a much longer time to realize the full cost of its investment than a large project with high positive net flow.

There are several advantages of using payback in cost analysis. One, it is simple to use since it requires only three sets of information: inflows, outflows, and the initial cost of investment. Two, it can help an organization determine how fast it can recover the cost of a project to avoid the problem of holding up cash for a long period of time. This is particularly important for organizations that are strapped for cash. Three, and most important, it is flexible, meaning that it can be used as a supplementary tool or in combination with other decision tools such as discounted cash flows.

Payback period is easy to calculate if the return from an investment is uniform, in which case it can be obtained by simply dividing the cost of initial investment by the sum of cash inflows; that is, payback period = initial investment / sum of cash inflows. Suppose that a government wants to invest \$25,000 in a project that will produce an annual return of \$5,000 per year; it will take exactly five years to recover the cost of the investment [$\$25{,}000/\$5{,}000_{\text{Year}}=5$ y ears]. However, cash inflows are seldom uniform for most investments. When the inflows are not uniform, the calculation of the recovery time gets somewhat more involved, as can be seen in the following expression:

$$PBP = MRP_{CNNF} + \frac{|CNNF|_{MRP}}{INF_{MRP+1}} \quad [3.16]$$

where *PBP* is the payback period, MRP_{NCNF} is the most recent period with negative cumulative net flow, $|CNF|_{MRP}$ is the absolute value of cumulative net flow corresponding to MRP with negative cumulative net flow, and INF_{MRP+1} is the inflow in the period immediately following the most recent period with negative cumulative net flow.

Let us look at a simple example to illustrate this. Suppose that a state government wants to upgrade its network system (hardware, as well as software) that is fast becoming obsolete. The new system will be used to produce information for use by individual agencies and departments of the government. The system will cost the government \$1.2 million, which it expects to recover from the revenue it will receive from the various user departments. Table 3.3 shows the expected cash flows (inflows and outflows) from the upgrading of the system.

Table 3.3 Expected Outflows, Inflows, and Net Flows

Year	*Expected Outflows ($)*	*Expected Inflows ($)*	*Cumulative Net Flow ($)*	*Cumulative Net Flow (Adjusted) ($)*
0	(1,200,000)	–	(1,200,000)	(1,200,000)
1	–	125,000	(1,075,000)	(1,195,000)
2	–	170,000	(905,000)	(1,144,500)
3	–	200,000	(705,000)	(1,058,950)
4	–	215,000	(490,000)	(949,845)
5	–	250,000	(240,000)	(794,830)
6	–	250,000	10,000	(624,313)
7	–	350,000	260,000	(336,744)
8	–	375,000	635,000	4,582

As the table shows, it will take the government about six years to recover the invested funds, which is shown under the column Cumulative Net Flow. However, to be precise about the number of years it will take the government to recover the cost we can directly apply Equation 3.16 to the cash flow data. That is:

$$PBP = MRP_{NCNF} + \frac{|CNF|_{MRP}}{INF_{MRP+1}}$$

$$= 5 + (|\$240{,}000| / \$250{,}000)$$

$$= 5 + 0.96$$

$$= 5.96 \text{ years}$$

In other words, it will take 5.96 years to recover the cost of the project.

In the example presented above, we assumed that the government has the necessary funds to undertake the investment project. In the event it does not, the government can borrow the funds with an interest, known as the cost of funds. Let us say that the government borrows the funds from a local bank at a 10 percent rate of interest. Therefore, to recover the full cost of investment, the government needs to include both the cost of funds and the amount borrowed, adjusted for flow of funds. This is shown in Table 3.3 under Adjusted Cumulative Net Flow.

As the table shows, the net flow for the current year will be $1.2 million, since there is no cash inflow during this period and the interest payment does not start until the first year. The cost of funds corresponding to net flow in the first year will be $120,000, obtained by multiplying the net flow by the interest rate; that is, $1{,}200{,}000 \times 0.10 = \$120{,}000$. Thus, to obtain the cumulative net flow for the first year we will need to add this cost of funds to the net flow of $1.2 million and subtract the total from the inflow for the first year, which will produce a net flow of $1,195,000. That is,

$$CNF_1 = (CNF_0 + CF_0) - EIF_1$$
$$= [\$1{,}200{,}000 + (\$1{,}200{,}000 \times 0.10)] - \$125{,}000$$
$$= (\$1{,}200{,}000 + \$120{,}000) - \$125{,}000$$
$$= \$1{,}320{,}000 - \$125{,}000$$
$$= \$1{,}195{,}000$$

where CNF_1 is the cumulative net flow at time 1, CNF_0 is the cumulative net flow at time 0, CF_0 is the cost of funds at time 0, and EIF_1 is the expected inflow at time 1.

We do the same for the second year. That is,

$$CNF_2 = (CNF_1 + CF_1) - EIF_0$$
$$= [\$1{,}195{,}000 + (\$1{,}195{,}000 \times 0.10)] - \$170{,}000$$
$$= (\$1{,}195{,}000 + \$119{,}500) - \$170{,}000$$
$$= \$1{,}314{,}500 - \$170{,}000$$
$$= \$1{,}144{,}500$$

and continue to repeat the process until the costs are fully recovered.

As shown in Table 3.3, the recovery of the invested funds takes place in the eighth year. In other words, it will take the government about eight years to recover the full cost of investment. We could have also used Equation 3.16 (with a minor change in the denominator) to obtain the recovery time, as shown below:

$$PBP = MRP_{NCNF} + \frac{|CNF|_{MRP}}{CNF_{MRP+1}} \quad [3.17]$$

$$= 7 + (|\$336{,}744| / \$635{,}000)$$
$$= 7 + 0.5303$$
$$= 7.5303 \text{ years}$$

which will be 7.5303 years. Note that *CNF* is the cumulative net flow.

Looking at the cash inflows for the investment problem in our example (Table 3.3), it seems obvious that it will continue to produce positive returns beyond the payback period but, as we noted earlier, payback is not concerned with what happens once the cost has been recovered. This is a major weakness of payback since it fails to take into account the returns that most projects continue to generate beyond the period of cost recovery. Furthermore, it takes a fragmented approach to project investment. What this means is that it does not consider the overall liquidity or cash flow position of an organization, but rather focuses on the cash flow resulting from a single project.

Buy or Lease

Buying or leasing is a special case of differential cost analysis in which costs are compared not between an existing condition and an alternative, but between two alternatives – buy or lease. Like any business organization, when a government decides to acquire an asset it has the option to buy or lease. Which option it should use depends on a number of factors such as the cost of the asset, the length of time for which it will be used, the salvage value it will have at the end of its useful life, and so on. Although as a financial option leasing has been around for a long time, more governments are using it today than ever before because of its viability as an alternative to buying.

Leasing can be defined as a process whereby the owner of an asset, called the lessor, enters into an agreement (lease) with the user of the asset, called the lessee, to allow the latter to use the asset for a specified time at an agreed cost. Theoretically the lease cost must be less than the purchase price to justify the option to lease, but there are situations where buying may be a better alternative to leasing even if it costs a government more initially, provided that in the long run it will save the government more money.

There are several explanations as to why leasing is considered a better alternative to buying:

1. It provides a viable alternative for acquiring assets for governments that have limited resources. Without the opportunity to lease, many of these governments will not be able to obtain the facilities and equipment they need.
2. It can conserve or free-up existing resources of a government that can be used to meet other financial needs.
3. It offers a fast and flexible alternative to raising taxes or borrowing funds to pay for the facilities and equipment a government needs.
4. It provides 100 percent financing in most cases. In addition to that, costs such as delivery fees and installation charges are frequently included in lease payments, thereby reducing the cost burden for the government.
5. It avoids the risk of obsolescence. In an age of rapid technological changes, ownership does not have a distinct advantage every time a new and improved product is introduced in the market.
6. It does not affect a government's ability to borrow and does not reduce its credit rating.

Although the advantages are obvious, a government considering leasing as an option must do some financial analysis to see if it has a definite cost advantage over buying. In reality, the actual analysis does not have to be all that complicated and can be accomplished in three simple steps: (1) determine the costs of buying, as well as leasing; (2) calculate the difference between the two options; and (3) select the one that produces a lower cost to the government. We can formally express this relationship between buying and leasing in terms of a simple mathematical expression, as shown in Equation 3.18:

$$C_W = \left[\left(P_0 + \sum_{t=0}^{T} O_t\right) - SV_T\right] - \sum_{t=0}^{T}(L_t - O_t) \quad [3.18]$$

where C_W is the net cost of ownership, P_0 is the purchase price at time 0, L_t is the cost of leasing at time *t*, O_t is the operating cost at time *t* for the duration of the time the asset will be in use, SV_T is the salvage value at time *T* (for $t=0, 1, 2, \ldots, T$).

Equation 3.18 has two parts: the first part represents the cost of ownership and the second part the cost of leasing. While the equation seems simple enough to be useful for everyday purposes, it has one major weakness in that it does not take into consideration the time value of money, which is important if the item in question has a useful life that extends over several years.

Time Value of Money

Since leasing involves a commitment of resources that extends over several years into the future, it is necessary to introduce the notion of the time value of money. The time value of money simply means the value a given sum of money will have over time. In general, the value of money decreases as time progresses, indicating that money loses its value over time. Several factors contribute to this apparent loss of value of money over time, such as inflation (it will cost us more to buy the same bundle of goods and services tomorrow than it costs us today), our unwillingness to defer current consumption in favor of consumption in the future (we are better off consuming the resources today than postponing them for tomorrow because they will cost more), uncertainty (we do not know what awaits us in the future), and so forth.

Since money loses its value over time, it is necessary to make some adjustments in the cost of ownership to get an accurate picture of its real worth. The simplest way to accomplish this is to discount the cost of ownership (C_W) by an appropriate discount rate (i.e., the rate by which we discount a stream of costs and benefits or returns). The result would produce a discounted value of ownership called the present value (PV) of ownership.[1] By present value, we mean the value a future stream of costs and returns will produce in today's terms – but why in today's terms? Because when we make a decision to acquire an asset it involves a resource commitment at the time when the decision is made to acquire it, although the actual returns from the decision will take place over a period of time.

To give an example, suppose that we invest an amount, say, *X*, at a 5 percent rate of interest (discount rate) that will produce exactly \$100 per year from now, the PV of \$100 today will be:

$$\begin{aligned} PV_X &= \frac{X_1}{(1+i)^1} \\ &= \$100/(1+0.05)^1 \\ &= \$100/1.05 \\ &= \$95.24 \end{aligned} \quad [3.19a]$$

In other words, this is how much $100 next year will be worth today, which is the same as saying that if we invest $95.24 today at 5 percent rate of discount, it will produce $100 next year (i.e., the value a current investment will produce at some future point in time called the future value).

Similarly, if we invest an amount, say, *X*, today that will produce exactly $100 two years from now, the PV of $100 today will be $90.70. That is,

$$PV_X = \frac{X_2}{(1+i)^2} \quad [3.19b]$$

$$= \$100/(1+0.05)^2$$

$$= \$100/1.1025$$

$$= \$90.70$$

and so forth.

To put the discussion above in present value terms, any amount invested today that will produce a return at some future point in time will be worth much less when expressed in today's terms. We can now formally extend the expression in Equation 3.18 in present value terms to determine the net cost of ownership as

$$NPV_W = [P_0 + \sum_{t=0}^{T} \frac{O_t}{(1+i)^t} + \frac{SV_T}{(1+i)^T}] - [\sum_{t=0}^{T} \frac{(L_t - O_t)}{(1+i)^t}] \quad [3.20]$$

where NPV_W is the net present value of ownership (defined as the difference between the present value of purchase, i.e., ownership, and the present value of lease), *i* is the rate of discount, and the rest of the terms are the same as before.

Note that the expression $O_t/(1+i)^t$ in Equation 3.20 can also be written as (0_t) $[1/(1+i)^t]$, where $1/(1+i)$ is called the discount factor (*DF*). A discount factor is the value by which the costs and returns of an asset are multiplied to obtain a discounted stream of costs and returns. In general, if the net present value of ownership is positive, leasing should be preferred because it will cost more to own than lease. On the other hand, if it is negative, purchasing should be preferred because it will cost more to lease than to own.

We can use a simple example to illustrate this. Suppose that the public works department of a local government needs a heavy-duty truck for carrying out its normal repair and maintenance work. If the department has to buy the truck, it will cost $65,000. If it has to lease, the rental cost will be $18,000 per year, payable at the end of each period for three years (assuming the department will lease the truck for three years). If purchased, there is a salvage value of $23,500 at the end of the asset's useful life. There is also an operating cost of, say, $5,000, regardless of whether the vehicle is purchased or leased. Assume a discount rate of 9 percent. Table 3.4 shows the present value of ownership for the problem.

Table 3.4 Present Value of Ownership and Lease

Year	*Purchase Price ($)*	*Operating Cost ($)*	*Salvage Value ($)*	*Discount Factor (DF)*	*PV of Purchase ($)*
0	65,000	–	–	–	65,000.00
1	–	5,000	–	0.917431*	4,587.40
2	–	5,000	–	0.841680	4,208.40
3	–	5,000	–	0.772183	3,860.92
4	–	–	23,500	0.708425	16,647.99
Year	*Lease Price ($)*	*Operating Cost ($)*	*Total Cost ($)*	*Discount Factor*	*PV of Lease ($)*
0	–	–	–	–	–
1	18,000	5,000	23,000	0.917431	21,100.91
2	18,000	5,000	23,000	0.841680	19,358.64
3	18,000	5,000	23,000	0.772183	17,760.21
4	–	–	–	–	–

Note
* $1/(1+i)^1 = 1/(1+0.09)^1 = 1/(1.09)^1 = 0.917431$

Based on the information presented in the table, the net present value of ownership will be $2,788.97, obtained by taking the difference between the sum of the present value of purchase and the sum of the present value of lease. That is,

$$
\begin{aligned}
NPV_W &= \Sigma PV \text{ of purchase} - \Sigma PV \text{ of lease} \\
&= [(\$65{,}000 + \$4{,}587.40 + \$4{,}208.40 + \$3{,}860.92) - \$16{,}647.00] \\
&\quad - [\$21{,}100.91 + \$19{,}358.64 + \$17{,}760.21] \\
&= \$61{,}008.73 - \$58{,}219.76 \\
&= \$2{,}788.97
\end{aligned}
$$

where NPV_W is the net present value of ownership.

Now to determine whether the department should buy or lease the vehicle, we can apply the decision rule we established earlier; that is, if the net present value of ownership is positive, it should lease; if not, it should purchase. Since the net present value of ownership is positive – in other words, since the cost of ownership is higher than the cost of lease – the department should go with the option to lease.

We have taken a rather simplistic approach here to determine the cost of ownership. Although somewhat easy to understand, there is a weakness in this approach in that it tends to ignore intangible factors such as convenience, location, and personal relationship between a lessee and a lessor that can affect the final cost. Another problem with this approach is that it is too broad and fails to include a host of other requirements that may also have an important determining influence on the final cost. For instance, a government may be required to pledge collateral against an existing asset before acquiring the asset, or there may be restrictions on the use of the asset once acquired, or there may be a special clause requiring the government to buy safety insurance to protect the asset against potential loss, theft, or damage. In each of these instances, there is a monetary implication that must be incorporated in the final calculation before making a decision on whether to buy or lease.

Benefit–Cost Analysis

One of the earliest decision tools used in cost analysis with a long history that goes back at least to the seventeenth century is benefit–cost analysis (BCA) and it still remains among the most popular decision tools in government (Thompson, 1982). BCA is particularly suitable where resources are scarce and the demand for goods and services exceeds the available resources. Constrained by resource limitations, the rational choice facing a decision maker is to undertake those activities that will produce the greatest amount of return for the resources utilized. This fundamental rule of efficiency that guides the allocation decision of a household or firm also guides a government to allocate its resources among competing needs and interests. Without this measure of efficiency, there will be very little basis for determining how best a government can utilize its resources or what it can do to improve its allocation decisions in the future.

As with any decision tool involving (project) investments, BCA follows a number of distinct steps. These steps typically are (1) define the goal and objective of analysis, (2) identify the project or projects to be analyzed, (3) determine the costs and benefits associated with each project, (4) find the present value of costs and benefits for each project, (5) evaluate the projects using one or more decision criteria, and (6) select the project(s).

Decision Criteria in BCA

Three sets of criteria are generally used in BCA to determine whether a project or activity is worth undertaking: benefit–cost ratio (B:C), net present value (NPV), and internal rate of return (IRR). A benefit–cost ratio is simply the ratio of the sum of benefits to the sum of costs, and when the time value of money is incorporated it becomes the ratio of the sum of present value of benefits (PV_B) to the sum of present value of costs (PV_C). The ratio indicates the return the funds invested in a project or projects will produce for an organization. In general, the higher the ratio the greater the return and the more attractive the investment. NPV is the difference between a discounted stream of benefits and costs (i.e., the difference between the sum of present value of benefits and the sum of present value of costs). A positive NPV indicates the benefits are greater than the costs, thereby making a project worth undertaking, while a negative NPV indicates the opposite. Like benefit–cost ratio, the higher the NPV, the greater the attractiveness of an investment. Finally, the IRR is the rate at which the sum of the discounted stream of benefits equals the sum of the discounted stream of costs. In other words, it is the rate at which the NPV will be zero. In general, the higher the rate of return, the greater the attractiveness of a project, as the rate is frequently associated with the growth rate of a project.

The following provides a brief summary of the three decision criteria, and the corresponding decision rules:

(1) Decision criterion: B:C

B:C $=\Sigma B/\Sigma C$ (without the time value of money)
B:C $=\Sigma PV_B/\Sigma PV_C$ (with the time value of money)

Decision rules:

1. For a single project: select the project if its B:C > 1.
2. For multiple projects: select the project with the highest positive *B:C* (assuming projects are mutually exclusive); if not, select those with B:C > 1.

(2) Decision criterion: NB, NPV

NB (Net Benefit) $=\Sigma B-\Sigma C$ (without the time value of money)
NPV $=\Sigma PV_B-\Sigma PV_C$ (with the time value of money)

Decision rules:

1. For a single project: select the project if its NPV > 0.
2. For multiple projects: select the project with the highest positive NPV (assuming the projects are mutually exclusive); if not, select those with positive NPV.

(3) Decision criterion: IRR

IRR: $\Sigma PV_B=\Sigma PV_C$
IRR: NPV $=\Sigma PV_B-\Sigma PV_C=0$

Decision rules:
For both single and multiple projects, the rule of thumb is to use IRR if it is higher than the required rate of return (known commonly as cost of capital).

Several things are worth noting about the decision criteria used in BCA. For instance, when B:C is used the rule of thumb is to select a project or projects as long as the ratio is greater than 1, B:C > 1, meaning that a dollar invested would produce more than a dollar worth of return. Thus, B:C = 1.67 would mean that for each dollar invested it will produce a return of \$1.67, indicating a profit or net positive return of \$0.67. In the case of government, the ratio can be set at greater than or equal to 1; that is, B:C ≥ 1, which makes sense because, unlike private firms and businesses, a government does not operate to make profit. However, for goods and services that are provided in a business-like manner, most governments would prefer a net positive return.

Like B:C, the NPV tells us whether an investment will create value for an organization, but its greatest advantage lies in the fact that it explicitly takes into consideration the time value of money, indicating that money loses its value over time (i.e., a dollar in the future is worth less than a dollar today). As such, it produces a more realistic picture of whether or not a project is worth undertaking. Finally, the IRR is considered good for a single project but not for multiple projects, because when multiple projects are involved it is difficult to determine which IRR would produce the correct result since we can have more than one rate for a project, depending on the nature of cash flows (i.e., costs and benefits). In general, an IRR would produce a correct result when there is an initial positive cash flow, followed by a series of negative flows; it tends to produce mixed results with inconsistent IRRs when the cash flows are mixed, such as some positives and some negatives, and in no precise order.[2]

To illustrate how the rules are used, in particular the NPV since it is most frequently used in government, let us look at an example in which a government is considering two mutually exclusive projects, *A* and *B*, each with a life span of five years. Let us say that the initial cost of construction for both projects is $1 million. Both projects have a one-time cost, but produce benefits that occur at different amounts over time. We will use NPV, assuming a discount rate of 6 percent. Which of the two projects is worth undertaking for the government?

Since our example involves a one-time cost and does not include any salvage value at the end of the useful lives of either of the two projects, we can write the NPV as

$$NPV = \sum_{t=0}^{T=5} \frac{B_t}{(1+i)^t} - C_0 \qquad [3.21]$$

where C_0 is the one-time cost and the rest of the terms are the same as before.

It is worth noting that a one-time cost means that there are no subsequent costs a project would incur, except for normal repair and maintenance which, in principle, should be treated as operating costs for the government. However, if costs occur over the entire length of a project's useful life, rather than at one time, then they should be discounted the same way as one would discount a stream of benefits.[3] Table 3.5 shows the benefits and costs for both projects, along with their corresponding NPV and B:C.

According to the table, Project B produces a higher B:C and also has a higher NPV; therefore, it should be selected. Assume now that Project A costs $900,000, while B costs $1 million and the benefit streams remain the same for both, as before. The resulting NPV and B:C for Project A will thus be: NPV_A= $1,566,308 − $900,000 = $666,308; $B:C_A$ (without the time value of money) = $1,850,000 / $900,000 = 2.0556; and $B:C_A$ (with the time value of money) = $1,566,308 / $900,000 = 1.7403. This creates an interesting problem for the decision makers since Project A has a lower NPV than B, as before, but produces a higher B:C, with and without the time value of money. The final decision will depend on what the decision makers are trying to achieve for the projects. If the objective is to have maximum NPV, then they should select Project B since it has a higher NPV. If, on the other hand, the objective is to get the most return

Table 3.5 Benefit–Cost Analysis with Time Value of Money

Year	*Cost ($)*	*Benefit ($)*	*Discount Factor*	*PV of Benefit ($)*
Project A				
0	1,000,000	–	–	–
1	–	700,000	0.9434	660,380
2	–	450,000	0.8900	400,500
3	–	325,000	0.8396	272,870
4	–	275,000	0.7921	217,828
5	–	100,000	0.7473	74,730
Total	1,000,000	1,850,000	–	1,566,308
Project B				
0	1,000,000	–	–	–
1	–	125,000	0.9434	117,925
2	–	200,000	0.8900	178,000
3	–	475,000	0.8396	398,810
4	–	750,000	0.7921	594,075
5	–	525,000	0.7473	392,333
Total	1,000,000	1,925,000	–	1,681,143

Notes
$\text{NPV}_A = \$1{,}566.308 - \$1{,}000{,}000 = \$566{,}308$.
$\text{NPV}_B = \$1{,}681{,}143 - \$1{,}000{,}000 = \$681{,}143$.
$\text{B:C}_A = \$1{,}850{,}000 / \$1{,}000{,}000 = 1.8500$ (without time value of money).
$\text{B:C}_A = \$1{,}566{,}308 / \$1{,}000{,}000 = 1.5663$ (with time value of money).
$\text{B:C}_B = \$1{,}925{,}000 / \$1{,}000{,}000 = 1.9250$ (without time value of money).
$\text{B:C}_B = \$1{,}681{,}143 / \$1{,}000{,}000 = 1.6811$ (with time value of money).

for each dollar of investment, Project A should be selected since it has the higher B:C of the two.

Note that no attempt was made here to calculate the IRR for the two projects. However, one way to obtain the rate would be to use trial and error until a rate is found at which NPV = 0. It can also be obtained using a spreadsheet such as Microsoft Excel or any standard financial calculator.

A Note on Accounting Rate of Return

On occasion, cost and managerial accountants use a method called accounting rate of return (ARR). It is much simpler than the methods discussed above – in particular the IRR – in that it measures the return on an investment not in terms of cash flows, but in terms of income an investment will produce for an organization. Its simplicity also comes from the fact that it does not use discounted streams of costs and returns; instead, it takes into consideration depreciation to provide some recognition to the role time plays in an asset's useful life.

The following provides the expression for ARR:

$$\text{ARR} = AVE_{INC} / Inv_0 \quad [3.22]$$

where AVE_{INC} is the average income and Inv_0 is the initial investment.

To give an example, suppose that a government is planning to invest in a project that will require an initial outlay of \$100,000 with no salvage value. The project has a useful life of five years and will produce cash flows of: $\$20{,}000_1$, $\$30{,}000_2$, $\$40{,}000_3$, $\$30{,}000_4$, and $\$30{,}000_5$, respectively, with a total cash flow of \$150,000. The average cash flow for the project will be \$30,000 [(\$20,000+\$30,000+\$40,000+\$30,000+\$30,000)/5=\$30,000]. The average depreciation for the project, based on straight-line depreciation, will be \$20,000 [\$100,000/5]. The difference between the two will produce the average (net) income for the government from the investment at \$10,000 [\$30,000−\$20,000]. We can now use this information to obtain the ARR by dividing the average (net) income by the outlay, which will be 10 percent; that is, \$10,000/\$100,000=0.10.

In the example presented above, we assumed that there is no salvage value, but if there is a salvage value at the end of the useful life of the project we can easily incorporate it in the expression in Equation 3.22, so that

$$\text{ARR} = AVE_{INC}/(INV_0 + SV) \qquad [3.23]$$

where SV is the salvage value and the rest of the terms are the same as before. The new denominator is generally known as average investment.

Some Common Concerns with BCA

In spite of its long history and use, there are several problems inherent in BCA. The first and foremost is the requirement that the costs and benefits are expressed in precise dollar terms to make easy comparisons of the alternatives. The problem with this is that it is not always possible to find the dollar value of costs and benefits. However, the problem can be considerably overcome with surrogate or proxy variables and, where possible, with imputed or shadow prices for costs and benefits (Khan & Farias, 1996). When dollar values of costs and benefits are not available, a good alternative would be to use cost-effectiveness analysis, which does not require one to express the costs and benefits, especially the benefits in precise dollar terms; they can be expressed in their original units (Cellini & Kee, 2010).

Another frequent problem in BCA is that the projects being compared may not have identical lifespans, unlike the example we discussed above, which can make comparison difficult. One way to overcome the problem would be to take a number of short-lived projects that are similar, and compare them against several larger projects such that they would produce an approximate common time (Aronson & Schwartz, 1981).[4]

While it is generally assumed that the costs and benefits associated with a project are known with certainty, it is quite possible that there will be risks and uncertainties associated with them, in which case they must be expressed in terms of probabilities, and their present value computations must also reflect this (Staehr, 2006). Perhaps the real challenge in BCA is determining the discount rate because what rate one would use can have a significant effect on the selection of a project (Mikesell, 1977). This is where IRR has an advantage in that the rate is determined "objectively" rather than by the decision makers, which may explain why it is preferred by private firms and businesses (although in recent years there has been a shift toward NPV). In general, low discount rates tend to make a project more attractive than high discount rates, because at lower rates the consumers vis-à-vis the public will be more willing to give up their current consumption in favor of consumption in the future, and vice versa.

Life-Cycle Costing

Earlier in our discussion on buy or lease, we compared the cost of ownership against the cost of lease to determine the best course of action for a government. Life-cycle costing (LCC) is an extension of cost of ownership in that it takes into consideration the total cost of ownership of an asset, from the initial cost of acquisition to all other costs the asset will incur during its entire lifespan until it is replaced, sold, or no longer in use. In other words, it takes into consideration all possible costs, including those that are hidden and not readily apparent at the time of acquisition. Formally, it can be defined as the total cost of ownership of an asset, including the costs of acquisition, operation, and maintenance (Lee, 1996). As a concept, LCC is nothing new; it was introduced by the US Logistics Management Institute in 1965 as part of defense-related expenditure analysis (US LMI, 1965). This was followed by a number of publications by the Defense Department on the subject in the 1970s and 1980s. Today, it is widely used by all levels of government, as well as by various firms and businesses in the country.

Stages of the Life Cycle

All physical assets – in particular machines, tools, buildings, and equipment – go through several stages during their lifespans, which begin with their acquisition, construction, or installation, and end when they are no longer in use. This is called the physical life of an asset. The stages typically begin with planning and development, followed by acquisition and break-in, growth, maturity, and decline (Figure 3.2). They are most productive during their maturity when they are operating at their fullest capacity. There is a direct relationship between the physical life of assets and how they are maintained: assets that are not maintained properly will deteriorate at a faster rate than those that are maintained properly and, as such, will cost an organization more over the longer term.

Assets also have an economic life that is directly related to ownership cost and the cost of operation. In general, the cost of ownership tends to decrease as the operating cost begins to increase with time. Economic life represents the time period when the two become equal. Put differently, when the operating cost of an asset exceeds the cost of ownership, it is costing an organization more to operate the asset than to own it, and vice versa. Ideally, an asset should be replaced before it reaches its economic life (Douglas, 1978).

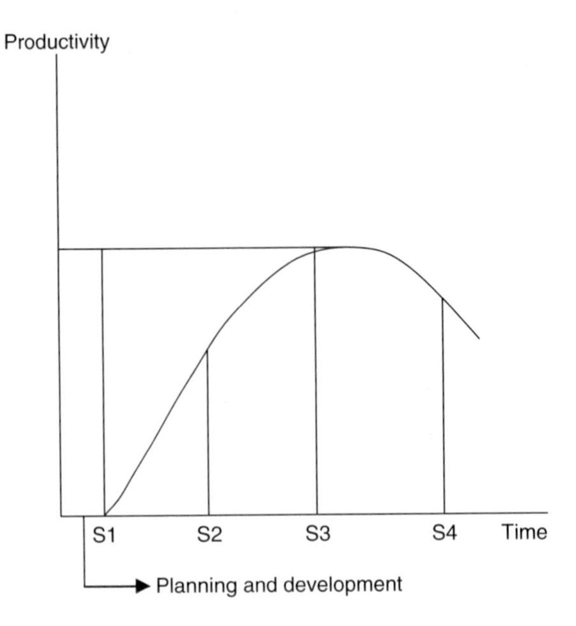

Figure 3.2 Physical Life of an Asset

From a cost perspective, the life cycle has a direct cost effect on an organization. During the early years of an asset's life, the net return from operation is usually positive (assuming the operation has costs, as well as benefits that can be measured in dollar or related terms) which continues to grow until maturity, at which point it becomes maximum and then begins to decline with age. This relationship between net return and the physical life of an asset is commonly known as profit life (Gransberg, Popescu, & Ryan, 2006).

Life-Cycle Costing Process

LCC requires that before an asset is acquired or installed, one performs a cost analysis over the lifespan of the asset. Four sets of costs are generally involved in this process: (1) cost of initial investment, (2) operating costs, (3) maintenance costs, and (4) replacement costs. There are also preliminary costs that an organization will incur before an investment decision is made, which must be included in cost calculations. This would include costs of planning, information gathering, and training personnel who will be operating the asset, and so forth. If the asset has a residual or salvage value at the end of its useful life, it must also be included in the analysis. Like most economic decisions, the process begins with a clear statement of objectives, identifying alternatives (assets), gathering information, determining the LCC of each alternative, and estimating the payback period.

We can formally express the LCC of an asset, as follows:

$$\text{LCC} = C_0 + \left[\sum_{t=1}^{T}\frac{OC_t}{(1+i)^t} + \sum_{t=1}^{T}\frac{MC_t}{(1+i)^t} + \sum_{t=1}^{T}\frac{RC_t}{(1+i)^t}\right] - \frac{SV_T}{(1+i)^T} \qquad [3.24]$$

where LCC is the life-cycle cost, C_0 is the initial cost, *OC* is the operating cost at time *t*, *MC* is the maintenance cost at time *t*, *RC* is the replacement cost at time *t*, SV_T is the salvage value at time *T* (for $t = 1, 2, 3, \ldots, T$), and *i* is the interest rate.

The initial investment costs include the full value of an asset at the time of its acquisition, construction (if it is physical plant or a building), and installation. The level of detail of the costs will depend on the nature of the asset. Obviously, the cost of constructing a building or a physical plant will be different from the cost of purchasing a utility tractor, a garbage truck, a petrol vehicle, or a large piece of equipment. Some assets may require getting help from a professional or consultant, in which case the cost of consulting should be included in the initial costs of investment. Operating costs, as distinct from maintenance costs, typically include the costs of utility and personnel such as custodial services. They are also called variable costs because they depend on the volume or quantity used. Ideally, only the costs directly associated with the operation should be included; indirect costs should be included if it is possible to segregate them by cost items associated with the asset. Since these costs take place over the lifespan of the asset, it is important to discount them to obtain their present value.

Maintenance costs, which also include repair costs, are those needed for the upkeep of an asset. As noted earlier, well-maintained assets keep the operating costs low and lengthen their useful lives. Examples of maintenance costs are repairing a broken window or door, replacing a tire, repairing the leak on the roof, and so forth. Some maintenance costs are recurring and some infrequent, and some difficult to predict. Regular inspection of assets can minimize the likelihood of incurrence of repairs, especially those that are difficult to predict. Since maintenance must be done on a regular basis, it is often treated as an annual cost. It can also be treated as a fixed cost if an organization contracts out the service to a firm. Like the operating costs, all maintenance costs should be discounted to their present value before they are included in the overall cost calculation.

Replacement costs include the costs of replacing an asset or a part of it in the future; they are anticipated expenditures that are required to maintain the operation of an asset. Replacement items usually have a shorter life than the life of the facility where they are installed, such as a water heater or a boiler-burner for a heating system. It is expected that these items will fail at some future point in time and need to be replaced to keep the facility operational. As before, all replacement costs must be discounted to their present value before adding them to the LCC.

Finally, the residual or salvage value is the value an asset has at the end of its useful life. Another term that is used interchangeably with residual value is scrap value. Regardless of the term used, it represents the likely value of an asset at the end of its useful life. The value depends on the nature of the asset and the level of detail involved. From an accounting point of view, it depends on how an asset is depreciated.

To give an example, suppose that a government is planning to purchase a dump truck and is seeking bids from two vendors: *A* and *B*. Table 3.6 presents the LCC of the two vehicles. As the table shows, the bid price for vehicle 1 (vendor A) is $10,000 less than the price for vehicle 2 (vendor B). Both vehicles have identical lifespans of 10 years and both are expected to operate 1,800 hours per year for each of the 10 years. However, the vehicles vary in terms of their costs of operation and maintenance, as well as their resale value. It is these differences in operation and maintenance costs and the resale value that frequently contribute to the differences in LCCs when comparing multiple bids. The LCC of the two vehicles, obtained by adding all the costs minus the resale value, is higher for vehicle 1 by $30,000, even though the bid price for the vehicle was less than vehicle 2. Since the LCC is much higher for vehicle 1, the government should go with vehicle 2.

While life-cycle costing serves as a useful tool for a government when making a decision on whether to acquire an asset or replace an existing one, it has some limitations. First and foremost is the data requirement; the government must have detailed information on all the costs an asset will produce during its lifespan. Second, although no attempt has been made in the current example, the costs must be discounted to their present value. Third, the costs of operation, maintenance over the lifespan of an asset, and its resale value are essentially estimates, meaning that failure to correctly estimate these costs can have a significant impact on the decision. Finally, since governments are often required by law to use bids when purchasing or acquiring an asset, the bid process must be carefully observed to ensure full compliance with the legal procedures as to vendor selection, opening bids, and awarding contracts.

Chapter Summary

All organizations need to analyze their cost behavior on a regular basis to ensure that there are no cost overruns and that they are making cost decisions that are

Table 3.6 Life-Cycle Costs

Parameters	*Vendor A (Vehicle 1)*	*Vendor B (Vehicle 2)*
Bid cost	$90,000	$100,000
Annual usage	1,800 Hours	1,800 Hours
Useful life	10 Years	10 Years
Operating costs		
Labor cost	$450,000	$450,000
Fuel cost	$65,000	$55,000
Maintenance cost		
Repair and maintenance	$30,000	$15,000
(Less) resale value	$10,000	$25,000
Life-cycle cost*	$625,000	$595,000

Note
*Undiscounted.

financially viable. While there are different ways to analyze the cost behavior of an organization, this chapter has presented several measures that are commonly used in cost analysis. They are break-even analysis, differential cost analysis, buy or lease, benefit–cost analysis, and life-cycle costing. Of these, break-even analysis is among the oldest and also one of the most widely used techniques in cost analysis. Its primary purpose is to bring together the cost, revenue, and output of an activity to determine the best course of action for a government.

Like break-even analysis, differential cost analysis is used to determine an appropriate course of action, based on cost, revenue, and output level of an activity, but the two are not quite the same. The advantage of using differential cost analysis, as opposed to break-even analysis, is that it gives the decision makers an opportunity to evaluate the viability of an alternative decision against the existing condition. Payback period, on the other hand, measures the length of time it takes to recover the cost of an investment. In general, the lower the recovery time, the better off an organization is, especially if it cannot tie up resources on an investment for a long time. While buying or leasing have their advantages and disadvantages, much depends on the need of the organization, the purpose for which an asset will be used, and the savings it will produce for the organization in the long term. The long-term effect of the decision also means that it must take into account the time value of money.

In addition to the above, the chapter has also presented two other methods frequently used in cost analysis: BCA and LCC. While both are important as analytical tools for decision making, BCA has been more extensively used in government than any of the other tools discussed here. Somewhat more complex analytically, BCA is not difficult to use provided that one is able to measure the costs and benefits in monetary terms. The problem can be avoided by using surrogates or proxy variables, but one must be careful how the variables are selected. However, the real challenge in BCA is the selection of the discount rate since an incorrect selection of the rate can result in incorrect choices. Finally, LCC provides useful information about the costs an asset will produce for an organization over its lifespan. Since it deals with assets with a lifespan of several years, it requires the use of the time value of money, similar to buy or lease and BCA.

Notes

1 As opposed to *PV*, which measures the current value of a future stream of costs and returns, the future value (FV) measures the value a current investment will produce at some future point in time. The calculation of FV is given by the expression:

$$\mathrm{FV} = \mathrm{PV}(i+i)^t \quad (1)$$

Thus, if we invest \$95.24 today at a 5 percent rate of interest, it will produce exactly \$100 one year from now; that is, $\mathrm{FV} = \$95.24(1+0.05)^1 = \$100$. Similarly, if we invest \$90.70 today at a 5 percent rate of interest, it will produce \$100 two years from now; that is, $\$90.70(1+0.05)^2 = \$90.70(1.1025) \approx \$100$, and so forth.

2 When projects have uneven cash flows the convention is to use the modified internal rate of return (MIRR); it is given by the expression:

$$\text{MIRR} = \sqrt[n]{\frac{\sum_{t=1}^{T} FVCINFL}{IINVEST}} - 1 \quad (2)$$

where n is the number of periods of investment, *FVCINFL* is the future value of cash inflows, and *IINVEST* is the initial investment, assuming one-time investment (outlay). If investments would take place over a period of time, Equation (2) will become

$$\text{MIRR} = \sqrt[n]{\frac{\sum_{t=1}^{T} FVCINFL}{PVCOFL}} - 1 \quad (3)$$

where *PVCOFL* is the present value of cash outflows, including initial investment, and the rest of the terms are the same as before.

3 If costs are incurred over time (i.e., for multiple periods), Equation 3.21 will become

$$\text{NPV} = \sum_{t=0}^{T} \frac{B_t}{(1+i)^t} - \sum_{t=0}^{T} \frac{C_t}{(1+i)^t} \quad (3)$$

and if there is a salvage value (*SV*) at the end of the useful life of a project, then Equation 3 will become

$$\text{NPV} = \left[\sum_{t=0}^{T} \frac{B_t}{(1+i)^t} - \frac{SV_T}{(1+i)^T} \right] - \sum_{t=0}^{T} \frac{C_t}{(1+i)^t} \quad (4)$$

4 To give an example, suppose that we have two comparable projects, X with a life of three years and Y with a life of four years. Together, they produce a common life of 12 years; that is, if project X is repeated four times and project Y is repeated three times, both will produce equal useful lives. Next, we find the NPV for each project and select the one with the higher NPV. The term commonly used to describe this approach is replacement chain method, also known as the common-life approach.

References

Aronson, R. J., & Schwartz, E. (1981). Capital budgeting. In R. J. Aronson & E. Schwartz (Eds.), *Management policies in local government finance* (pp. 433–457). Washington, DC: ICMA.

Cellini, S. R., & Kee, J. E. (2010). Cost-effectiveness and cost–benefit analysis. In J. S. Wholey, H. P. Hatry, & K. E. Newcomer (Eds.), *Handbook of practical program evaluation* (3rd ed.; pp. 493–530). New York: Jossey-Bass.

Douglas, J. (1978). Equipment costs by current methods. *Journal of Construction Division, 104*(C02), 191–225.

Gransberg, D., Popescu, R., & Ryan, R. (2006). *Construction equipment management for engineers, estimators, and owners*. Boca Raton, FL: Taylor & Francis.

Khan, A., & Farias, C. (1996). Benefit–cost analysis. In J. M. Shafritz (Ed.) *The international encyclopedia of public policy and administration* (pp. 173–180). New York: Henry-Holt.

Lee, R. C. (1996). Life-cycle costing. In J. Rabin, W. Bartley Hildreth, & G. Miller (Eds.), *Budgeting: Formulation and execution* (pp. 420–423). Athens, GA: Carl Vinson Institute of Government, University of Georgia.

Mikesell, R. F. (1977). *The rate of discount for evaluating public projects.* Washington, DC: American Enterprise Institute for Studies in Public Policy.

Rautenstrausch, W. (1939). *Economics of business enterprise.* New York: Wiley & Sons.

Staehr, K. (2006). *Risk and uncertainty in cost–benefit analysis*. Copenhagen: Institute for Miljovurdering.

Thompson, M. S. (1982). *Benefit–cost analysis for program evaluation.* Beverley Hills, CA: Sage.

US Logistics Management Institute (LMI). (1965). *Life-cycle costing in equipment procurement*. Washington, DC: US Logistics Management Institute.

4 Cost Accounting

There is not a single organization anywhere that is not concerned about costs and how to account for them. As organizations become complex, requiring more diverse and intelligent information to survive, the need for cost accounting becomes even more important. The primary objective of cost accounting is to provide cost and related information that is essential for an organization to determine how to run its internal operations; in particular, how to assign costs, how to allocate and control costs, and what measures to take to improve cost performance. In government where efficiency and accountability have always been a concern, cost accounting can fill an important gap by producing information that is critical to the success of the organization.

This chapter presents several topics that, according to conventional wisdom, constitute the essence of a cost accounting system. Although as a system cost accounting is broad and covers a wide range of methods and tools, this chapter focuses on those that provide the nuts and bolts of cost accounting for an organization. The chapter begins with a brief background discussion of cost accounting before discussing the specific methods in some detail.

Background Discussion

Cost accounting is the process of accounting for costs an organization incurs in carrying out its normal, everyday operations. The process is complex, involving a number of distinct, relatively detailed, and well-defined activities. These activities include gathering, recording, analyzing, summarizing, evaluating, and interpreting cost and related information, and communicating the results to those who would directly benefit from them. In government, this would mostly include the chief administrator, elected officials, various agency heads, internal auditors, plus any other individual who is directly involved in the routine operations of the government. Although individuals or organizations outside of government may benefit from the information the process generates, it is primarily used for internal consumption.

Cost accounting is often discussed in conjunction with two other fields of accounting – financial and managerial. Although the three are discussed in the same vein, they are not quite the same, especially financially accounting. What distinguishes them is the basis of use (i.e., who uses the information generated by the accounting process and the purposes it serves). For instance, financial accounting produces information that is primarily used for external consumption to meet the needs of outside individuals and organizations, such as bond rating institutions, investors in government securities, concerned citizens, and other external evaluators. Much of the information financial accounting produces is presented in the form of financial statements that contain detailed accounts of the financial position of a government. To ensure that these statements are presented fairly, consistently, and accurately, they are prepared according to a set of prescribed guidelines, known as generally accepted accounting principles (GAAP).

GAAP consists of a set of rules, conventions, and procedures that define the accepted accounting principles in operational terms. In government, the responsibility for establishing these rules falls on the Governmental Accounting Standards Board (GASB), an autonomous organization headed by a chairman, a full-time director, and several part-time members (Ingram, 1991). The board reports directly to the central accounting organization, called the Financial Accounting Foundation (FAF). The FAF, in turn, selects its members and raises funds to support its activities and those of its private sector counterpart – the Financial Accounting Standards Board (FASB).

In contrast, managerial accounting produces information that is mostly used by agencies, internal auditors, elected officials, and others who are directly involved in the day-to-day operation of a government. Since the information produced by managerial accounting is largely used for internal consumption, it is not governed by the GAAP, and, as such, is considered more flexible than financial accounting. It is also more eclectic, drawing heavily from other disciplines such as planning, economics, mathematics, operations research, finance, and management. In that regard, there is a similarity between cost and managerial accounting – both rely heavily on other disciplines for substantive growth, both provide information that is used for internal consumption, and in both cases their practices are not governed by GAAP. Because of these apparent similarities, the two fields are often treated as one, with one major difference: cost accounting is more concerned with routine operations of an organization, whereas managerial accounting is more finance-oriented.

Cost Accounting System

All costs boil down to three things: materials, labor, and overhead. A responsible manager must know how to keep track of these costs because they accumulate. In the private sector, managers keep track of their costs by assigning them to an individual consumer, a cost unit, or to a unit of a good produced. When costs are assigned to an individual consumer, a cost unit, or to a unit of a good produced, it is called unit costing. Unit costing is a useful method when applied to goods that are divisible and there is no free-rider problem such that one is able to trace the costs to an individual costing unit. Most private goods, as well as public goods that are provided in a business-like manner, would fall into this category.

Interestingly, for a majority of public goods, with the exception of those that are provided in a business-like manner, it is difficult to determine the exact quantity an individual consumes of these goods. As a result, public organizations often maintain cost information based on a specific job, service, or activity rather than on a per capita or per unit basis, although the latter may be implicit in most cost calculations. For instance, a federal agency may want to know the cost it would incur in running a job-training program, or a state agency may want to know the cost of upgrading the interstate highway. Similarly, a local school district may want to know the cost it would incur in running a computer literacy program for disadvantaged children, or the cost a city recreation department would incur in organizing a summer youth festival.

In reality, one should not have much difficulty in assigning costs to a specific activity, especially if the costs are direct. When costs are indirect it becomes difficult to assign them to specific activities, in which case one must use some rational basis for cost allocation. Proper assignment of costs is important for an organization to improve performance, ensure control, and increase efficiency in cost management. This process of recording costs as they accumulate, determining the costs incurred for direct labor, direct materials, and overheads, known commonly as cost measurement, and finally assigning costs to a specific job, service, or activity, and allocating them using established methods or some rational basis to ensure control and improve performance, constitutes the essence of what one would call a cost accounting system.[1] The four topics that are to be discussed next – job costing, process costing, cost allocation, and activity-based costing (as an extension of cost allocation) – along with cost control (discussed in the next chapter) are integral to this system.

Job Costing

Job costing is an accounting system that traces costs to a specific job, service, or activity. Tracing costs to a specific job makes it possible for an organization to determine the costs it will accumulate separately for materials, labor, and overheads. When materials and labor used in a job can be directly traced to it, they are called direct materials and direct labor. Indirect materials and labor that cannot be directly traced to a job are generally treated as overheads. Job costing begins when an operating department receives instructions from the responsible authority to begin work on a job. During this stage, materials are issued, labor is expended, and overhead is incurred. Three things must take place to complete the process: a number of source documents must be prepared, the information must be entered into a journal, and the overhead rates must be determined.

Prepare the Source Documents

Source documents are materials used to accumulate the costs for an individual job. The most frequently used source document is the job cost sheet, prepared primarily to accumulate and summarize cost information on direct materials, direct labor, and overheads. A typical job cost sheet includes the following information: job identification number, job description, beginning and completion dates, and costs of materials, labor, and overheads. The actual entries are made from materials requisition forms, labor cost sheets, and from other evidence of costs. A materials requisition form provides a detailed account of the type and quantity of materials that are to be drawn from a central location such as a central warehouse, and the job to which the materials are to be charged. The labor cost sheet specifies the number of hours the labor worked, the rate by which the labor was paid, and the total cost of the labor used. Both the requisition form and the labor cost sheet serve as a means for controlling the flow of materials and labor into the job completion process; in particular, for making entries into the job cost sheet.

Table 4.1 presents a simple example of a job cost sheet for street repair for a local government. According to the table, it cost the government $220,000 in total repair, including $45,000 in direct materials, $125,000 in direct labor, and $50,000 in overheads. The table also provides information on the exact nature of the job performed, the beginning and completion dates, the job requisition number, and the department responsible for the job.

As noted earlier, job costing does not begin until the operating department receives notification that a service order has been issued for the job. Once the direct materials have been issued and the labor costs identified, the operating department makes entries into the job cost sheet, charging the job noted on the sheet with the costs of materials and labor used. When the job is completed, the total costs of materials and labor plus the overheads are summarized and presented on the sheet.

Table 4.1 A Job Cost Sheet for Street Repair*

Summary Information	
Job Summary	
Job number	037-PW
Date started	August 15, 2017
Date completed	August 31, 2017
Job description	Repairing asphalt/concrete, and leveling
Job location	19th Street and Flint
Relevant Information	
Requisition number	012-PW
Responsible department	Public works
Cost Summary (Actual Cost)	
Direct materials	$45,000
Direct labor	125,000
Overhead ($35,000/labor plus $15,000/ materials)	50,000
Total cost	$220,000

Note
*Year ended: December 31, 2017.

Journalize the Information

As the information on materials, labor, and overhead costs becomes available, they are entered into a journal called the book of initial entry. The purpose of journalizing the entries is to show the flow of costs as they accumulate. Several basic transactions are recorded in a journal, such as the purchase of materials, issuance of direct and indirect materials, accrual of direct and indirect labor, payment of labor, incurrence of overheads other than materials and labor, and so on.

To enter the transactions in a journal, the convention is to use a double-entry accounting system called T-accounts. A T-account has two sides – a left side called debit, and a right side called credit. In general, when assets are acquired they are posted on the debit side and liabilities on the credit side. Since costs in general are considered a liability, any increase in costs is recorded on the credit side. By the same token, any decrease in costs is recorded on the debit side since it represents a decrease in liability. Convention dictates that when liabilities are incurred by an organization, which should also apply to government, they should be recognized as such and vouchers should be issued before paying the liabilities in cash. Therefore, the journal entries must include a recognition of liability when vouchers are issued and a payment of liability when a cash payment is made (Lynn & Norville, 1984).

The following shows a typical T-account:

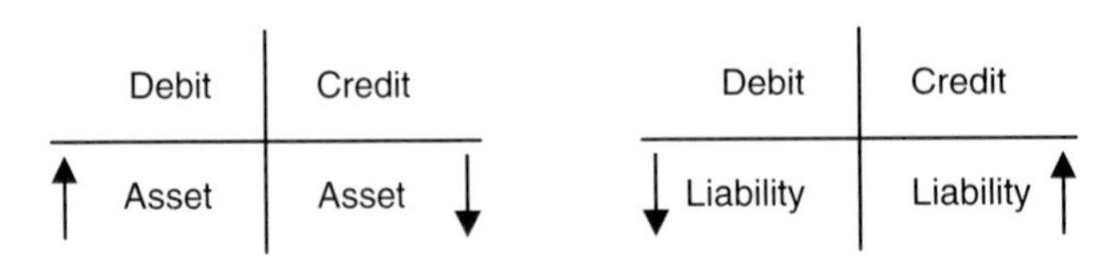

The greatest advantage of a double-entry accounting system is that the two sides of the account must be equal, meaning that the entries must be balanced. Thus, if a government purchases a vehicle, it will reflect an increase in its asset position; as such, it should be debited. At the same time, it will also increase the liability of the government if it will have to pay for the vehicle using credit, or decrease its asset position if it will have to pay using cash, which is an asset and, therefore, it should be credited. These transactions, as they take place, are recorded in a journal called the book of initial entry. If the vehicle is paid in full during the accounting period, the two sides should even out; on the other hand, if the payment is not made in full, the difference will appear as "balance" on the liability side of the ledger. The latter is known as the book of final entry.

To give an example of how this applies to job costing, let us go back to the street repair problem. Suppose that the central warehouse of the government purchases materials for the job in the amount of $90,000, which include $65,000 in direct and $25,000 in indirect materials, and pays for them in cash. It is not unusual for a warehouse to have these materials in its inventory, in which case the journal entry would recognize them as a simple inventory account with no outstanding obligation or payment of liability to make.

Assume for the first scenario that the government does not have the materials in its inventory and that the warehouse had to purchase them. When materials are acquired, they add to the asset position of the government and, as such, will be debited. In government, as in the private sector, all transactions (except for petty cash transactions) require that vouchers be issued before making any payments. Since vouchers are liabilities, any increase in vouchers will increase the liability and will be credited. Likewise, when vouchers are paid they will decrease the liability position for vouchers and will be debited. Finally, when cash payments are made it will decrease the asset position of the government since cash is an asset and, as such, will be credited.

The journal entries for the transactions can thus be shown as

	Debit	Credit
1. Purchase of materials (direct)	65,000	
Purchase of materials (indirect)	25,000	
Vouchers payable		90,000
To record purchase of materials		

	Debit	Credit
2. Vouchers payable for materials (direct)	65,000	
Vouchers payable for materials (indirect)	25,000	
Cash		90,000
To record payment of materials		

Convention also dictates that when direct materials are issued for a job, they are recorded in an account called Work in Process (WIP), which is essentially an inventory account indicating goods and services in the production or delivery process that have not been completed yet. In other words, WIP accounts include materials and partially completed products that are at various stages of production; they do not generally include raw materials at the beginning of a production cycle, or inventory at the end of the production cycle. Indirect materials, on the other hand, are recorded in the overhead account because all costs associated with a job, except for direct materials and labor, are accumulated in this account. The journal entries for the WIP account for materials, based on the amount issued and using the same rules for recording transactions, can be presented as:

	Debit	Credit
WIP materials (direct)	45,000	
Overhead materials (indirect)	15,000	
Inventory of materials (direct)		45,000
Inventory of materials (indirect)		15,000
To record issuance of materials		

Similarly, the direct labor costs accumulated for a job are recorded in the WIP account as debit and the indirect labor costs are debited to the overhead account. As before, to pay for the labor, vouchers must be issued (i.e., recognition of liability) and paid out in cash (i.e., payment of liability). The entries for the labor cost transactions can thus be shown as

	Debit	Credit
1. WIP labor (direct)	125,000	
Overhead labor (indirect)	35,000	
Vouchers payable		160,000
To record the costs of labor		

	Debit	Credit
2. Vouchers payable for labor (direct)	125,000	
Vouchers payable for overhead labor (indirect)	35,000	
Cash		160,000
To record payment of labor		

The process can be continued until every single transaction related to the job is exhausted. Once the journal entries are completed, they are posted onto a ledger (not shown here). A ledger, as noted earlier, is a group of accounts in which the data from the journal entries are recorded and summarized by accounts. In cost accounting, they are commonly known as cost ledgers, which include a principal ledger and a group of subsidiary ledgers. The principal ledger includes general information on various cost accounts, while the subsidiary ledgers include detailed information on these accounts which allow an organization to more effectively control its financial information. For instance, in the current example, the job cost sheet will serve as the subsidiary ledger containing the details for the WIP general ledger account. It is worth noting that double-entry accounting is not required for cost reporting, but has become a standard practice over the years because it helps prepare cost journals and ledgers.

Determine the Overhead Rates

As noted previously, overhead costs are indirect costs that cannot be attributed to a specific job. Some overheads, especially those related to utilities, are not always known until the end of a given period. Thus, rather than hold the completed jobs in an inventory until all costs can be traced to it, it may be necessary to develop a system of allocating overhead costs to a job on a predetermined basis. The simplest way to do so is to divide the estimated overhead cost for, say, an entire year by an appropriate base such as the amount of direct materials, direct labor hours, and so forth.

The following provides a simple expression for calculating the predetermined overhead rate for direct labor hours:

$$OR = \frac{TOC}{\sum_{i=1}^{n} DLH_i} \qquad [4.1]$$

where OR is the overhead rate, TOC is the total overhead cost for the organization as a whole, DLH_i is the direct labor hour attributable to job i (for $i = 1, 2, 3, \ldots, n$).

It is worth noting that when the actual data on overhead costs and indirect labor are not available one can use estimates of these costs, similar to the expression in Equation 4.1, where the numerator in Equation 4.1 will represent the estimated total overhead costs and the denominator the estimated total direct labor hours. To give an example, suppose that our government in the street repair problem has estimated that its total overhead cost and direct labor hours for the year will be $15,850,000 and $634,000, respectively. Therefore, its predetermined overhead rate per direct labor hour for the year, based on this, will be $25 per hour, as shown below:

$$OR = \frac{TOC}{\sum_{i=1}^{n} DLH_i} = \frac{\$15,850,000}{634,000} = \$25$$

Let us further suppose that the street repair job requires a total of 400 hours of indirect labor and $25,000 in fringe benefits, mostly health and accident insurance. The total overhead cost for the repair, including the labor cost of $25 per hour, will therefore be $35,000 [$(400 \times \$25) + \$25,000 = \$10,000 + \$25,000 = \$35,000$]. Since the selection of a predetermined rate requires that estimates be used for both the numerator (total overhead cost) and the denominator (direct labor hours), practitioners often use a rule of thumb to determine this rate by taking the ratio of an agency budget to the total budget of a government (Kelly, 1984). For instance, if an agency has an annual budget of $2,382,500 that is approximately 5 percent of the total budget of the government, the overhead rate will also be 5 percent and the overhead allocation for the agency will be $119,125 [$\$2,382,500 \times 0.05 = \$119,125$]. The same procedure will also be used to allocate the overhead costs for every agency in the government.

While the method suggested above is simple, there is a problem with it in that if the agency budget constitutes an unusually large (or small) fraction of the total budget it may over- or underestimate the overhead allocation. Also, when money is frequently transferred between funds, a common practice in government, it may misrepresent the overhead allocation by the amount of the transfer. The alternative would be to remove the effect of interfund transfers before the method can be used. The general position taken here, in particular for the predetermined rate, is that overheads should be allocated on the basis of the factors or activities that cause these costs. In other words, the more an activity causes an overhead, the higher the cost that should be charged to it. The trick is to select a base such as labor or machine hours that is commonly used by an organization so that its overhead application will be equitable among jobs. Since the focus is on the base, another term used for it is activity-based costing, discussed later in the chapter.

Process Costing

Process costing is a system of accounting in which costs are assigned equally to all units of a good produced by an organization. Since costs are equally assigned to all units of the good produced, it is assumed that the units produced are homogeneous. As a result, process costing is useful in situations in which there is a continuous flow of production such as gas, water, and electricity, where product variation is almost nonexistent. However, it can also be applied to situations where the units of production are discrete rather than continuous, such as routine citations for traffic infractions, trash collection, vaccination for the common cold, and so on, as long as one is able to measure them in distinct units and their essential characteristics remain the same throughout the production or delivery process.

Since it is expected that the units produced in process costing must be homogeneous or, in the least, their essential characteristics must remain the same, one accumulates costs by department or agency rather than by job and assigns these costs equally to all units that pass through the department. Thus, in process costing, instead of using a job cost sheet one uses a set of reports called schedules. Three types of schedule are commonly used for this purpose: a schedule of equivalent production; a schedule of cost analysis; and a schedule of cost summary.

Schedule of Equivalent Production

When costs are assigned equally to all units of a good or service as they pass through a department, convention dictates that only those units that have been completed are taken into consideration. In process costing this would include completed as well as partially completed units. Together, these units (completed and partially completed) are called equivalent units. The rationale for including partially completed units in cost assignment is that even though they have not been fully completed, the department or departments involved in the process did incur some costs (materials, labor, and overhead) for the portion of the job that has been completed. Therefore, to accurately measure the amount of goods produced during a given period one must include both completed as well as partially completed units.

Two methods are commonly used for calculating the equivalent units: a weighted average method and a first-in, first-out (FIFO) method. Both methods are simple in concept and can be used together or interchangeably for the same problem. The following sections provide a brief account of the two methods.

Weighted Average Method. According to this method, the equivalent units are calculated as the number of units completed during a given period, plus the units that are incomplete multiplied by the percentage of their completion. In other words, equivalent units include both completed and partially completed units. To give an example, consider the federal Drug Enforcement Agency that deals with the Drug Interdiction Program, which is a continuous process of interrupting drug trafficking into the country. The process includes a number of steps that usually starts with queuing, followed by detecting, sorting, and monitoring, to intercepting, and finally to apprehending and prosecuting. The steps do not necessarily have to take place in sequence; but for the sake of argument we will assume they take place in sequence and the measures used in each step are fairly consistent.

Let us suppose that the government has been investigating 125 drug interdiction cases this past year, of which 45 have been resolved completely (i.e., prosecuted) by the government and 80 remain unresolved by the end of the year. Since 45 cases were fully completed, we can safely assume that these cases received 100 percent of materials, labor, and overheads needed to complete them. Of the cases that remain unresolved, assume that 75 percent were completed as to materials and 60 percent as to direct labor and overheads, called conversion, when the year ended. To obtain the equivalent units for the remaining cases, we use a standard expression, given by

$$EU=(UP)(PC) \quad [4.2]$$

where EU is the total equivalent units, UP is the units remaining in process, and PC is the percentage of completion.

The equivalent units for the department, therefore, will be 60 for materials and 48 for conversion, obtained by applying the expression in Equation 4.2; that is, $UP \times PC = 80 \times 0.75 = 60$ and $UP \times PC = 80 \times 0.60 = 48$, respectively. However, in order to obtain the total equivalent units, it is necessary to revise Equation 4.2 so that the new equation will include the units that have already been completed, as well as those remaining in process, as shown below:

$$EU=UC+(UP \times PC) \quad [4.3]$$

where EU is the equivalent units, UC is the units completed, and the rest of the terms are the same as before.

Applying the revised expression in Equation 4.3 to the problem, we can obtain the respective equivalent units. That is,

$$\begin{aligned} EU &= UC+(UP \times PC) \\ &= 45+(80 \times 0.75) \\ &= 45+60 \\ &= 105 \text{ units} \end{aligned}$$

for materials, and

$$EU = UC + (UP \times PC)$$
$$= 45 + (80 \times 0.60)$$
$$= 45 + 48$$
$$= 93 \text{ units}$$

for conversion.

Table 4.2 shows the total equivalent units for materials and conversion, together with information on the units to be accounted for by the department. Note that we made an implicit assumption in the example that all 125 cases of interdictions occurred during the accounting period, which means that there were no carryovers from the previous year. As a result, no units appear under the columns for materials and conversion for the beginning inventory, but if we change the assumption that there were carryovers from the previous year, the result would be different. This is where the second method, FIFO, comes in.

First-in, First-out Method. Unlike the weighted average method, in which the units are completed simultaneously, the calculation of equivalent units under the FIFO method takes into consideration the beginning inventory (i.e., the carryovers from previous periods). This is particularly important in situations such as homicide investigations, drug interdictions, treatments of patients for certain

Table 4.2 Schedule of Equivalent Production*

Stages of Completion	*Units to be Completed*	*Equivalent Units*	
		Materials	*Conversion*
I Without Beginning Inventory			
Beginning inventory	–	–	–
Units started and completed	45	45	45
Ending inventory	80	–	–
Work in progress	–		–
Materials at 75%	–	60	–
Conversion at 60%			48
Total units	125	105	93
II With Beginning Inventory			
Beginning inventory	15	15	15
Units started and completed	30	30	30
Ending inventory		–	–
Work in progress	80		–
Materials at 60%	–	48	
Conversion at 25%	–	–	20
Total units	125	93	65

Note
*Year ended: December 31, 2017.

diseases, and so forth where the results (outputs) of a process do not always get completed within a given period and where there are frequent carryovers from previous periods. In other words, there is a continuous flow of output in the production (delivery) process such that the units remaining in process at the end of one period become the beginning inventory in the next period. Thus, in calculating the equivalent units for an organization one needs to take into consideration the units in the beginning inventory. In order to do so, we need to reformulate the expression in Equation 4.3 for equivalent units by taking the beginning inventory into consideration, as shown below:

$$EU = BI + UC + (UP \times PC) \qquad [4.4]$$

where *BI* represents the beginning inventory and the rest of the terms are the same as before.

Table 4.2 also shows the equivalent units for the department based on the FIFO method (lower half). According to the table, of the 45 units that were fully completed, let us assume that 15 were in the beginning inventory. That means all 15 units, like the remaining 30 in this group, received full materials, labor, and overhead support needed for their completion. As a result, these 15 units also appear under the materials and conversion columns for the beginning inventory.

As for the units in ending inventory, assume that 60 percent were completed with regard to materials and 25 percent with regard to conversion, thereby producing a total of 93 and 65 equivalent units for the respective categories, as shown below:

$$\begin{aligned} EU &= BI + UC + (UP \times PC) \\ &= 15 + 30 + (80 \times 0.6) \\ &= 15 + 30 + 48 \\ &= 93 \text{ units} \end{aligned}$$

for materials, and

$$\begin{aligned} EU &= BI + UC + (UP \times PC) \\ &= 15 + 30 + (80 \times 0.25) \\ &= 15 + 30 + 20 \\ &= 65 \text{ units} \end{aligned}$$

for conversion.

Note that these results also appear under the materials and the conversion columns in Table 4.2. It should be pointed out that there is a fundamental difference in application when these methods are used in government – in particular FIFO – as opposed to when they are used in the private sector. For instance, when FIFO is used in the private sector, the units in beginning inventory are usually completed first, especially if the products are perishable or tend to become obsolete quickly, contributing to loss in value the longer they are held in inventory and thus need to be transferred out before beginning to work on the current inventory; hence the term first-in first-out. However, in government, the units in beginning inventory do not necessarily have to be completed before dealing with the units that began in the current period. They can be addressed simultaneously or in any sequence but, from an accounting point of view, they must be recognized in calculating the total equivalent units for a given period.

Last-in First-out Method (LIFO). In process costing it may be necessary sometimes to complete the activities that enter the process first before completing those already in process, in which case the method that would be appropriate is LIFO. In the private sector, LIFO is primarily used in situations where the cost of acquiring inventories increases quickly either due to inflation or some other factors that make the income statements appear higher than they should be.

To give an example, suppose that a firm acquires an inventory at the beginning of the year consisting of 500 units of a good, say *X*, at a cost of $10 per unit, for a total cost of $5,000. It sells the good for $13 per unit, thereby producing an income of $6,500. Let us say that the cost of the good increases to $12 per unit, which will increase the cost of the inventory at the current price to $6,000 [$12 × 500 = $6,000], while the firm's income statement will show the income at $6,500. This means that it will have to pay more in taxes on its income, while its net income decreases by $500 [$6,500 – $6,000 = $500]. From an income perspective, the firm will be better off selling the last batches as fast as possible to minimize the effect on income, since it had to purchase them at a higher price.[2] Theoretically, if price remains the same over time and the inventories are not perishable or do not become obsolete quickly, it makes very little difference which method is used.

In the case of government, on the other hand, there are cases that by their very nature need to be addressed first before those already in process, either because of urgency or severity of the cases. Criminal investigations involving homicides, treatment of patients in an emergency room, and public services with mandates are good examples of where LIFO could be easily applied.

Schedule of Cost Analysis

Once we know the number of equivalent units in an inventory, we can compute the cost of total equivalent units, as well as the cost per equivalent unit for a department. The procedure for calculating these costs is quite simple: the former is obtained by simply adding all the costs in an inventory, whereas the latter is obtained by dividing the total cost by the number of equivalent units. The procedure applies equally to both methods – weighted average and FIFO.

Table 4.3 presents a schedule of cost analysis for the department with and without the beginning inventory. As the table shows, the total cost for the department without the beginning inventory was $6.89 million, which was obtained by adding

Table 4.3 Schedule of Cost Analysis*

Cost ($)			
Cost Description	*Materials*	*Conversion*	*Total*
I Without Beginning Inventory			
Costs for:			
Beginning inventory	–	–	–
Current period	1,564,250	5,325,750	6,890,000
Equivalent units	105	93	–
Cost/equivalent unit	14,897.62	57,266.13	72,163.75
II With Beginning Inventory			
Costs for:			
Beginning inventory	340,054.35	1,401,513.19	1,741,567.54
Current period	1,224,195.66	3,924,236.82	5,148,432.48
Total	1,564,250.01	5,325,750.01	6,890,000.02
Equivalent units	93	65	–
Cost/equivalent unit	16,819.89	81,934.62	98,754.51

Note
*Year ended: December 31, 2017.

the costs of materials ($1,564,250) and conversion ($5,325,750). In general, the total cost of materials and conversion are obtained by multiplying the number of equivalent units by the corresponding cost per (equivalent) unit. Since we did not have the data for cost per equivalent unit, we divided the total cost for each category (assumed given here, for convenience) by their corresponding equivalent units to obtain the cost per equivalent unit of $14,897.62 for materials and $57,266.13 for conversion. However, as one would expect, when the costs for beginning inventory are included in the schedule, it changes the picture somewhat. Assume that the total cost to be accounted for remains the same at $6.89 million, the cost per equivalent unit (with the beginning inventory included) will now increase to $16,819.89 for materials and $81,934.62 for conversion, as one would expect.

To see if our calculation is correct, we multiply the total cost for each category by the cost per equivalent unit and add them together, which produces a total cost of $6.89 million, except for the rounding-off error, indicating the calculation to be correct, as shown below:

$$\begin{aligned} TC &= (CPEU_{Mat} \times EU) + (CPEU_{Con} \times EU) \\ &= (\$16{,}819.89 \times 93) + (\$81{,}934.62 \times 65) \\ &= \$1{,}564{,}249.77 + \$5{,}325{,}750.30 \\ &= \$6{,}890{,}000.07 \approx \$6{,}890{,}000.00 \end{aligned} \quad [4.5]$$

where TC is the total cost, $CPEU$ is the cost per equivalent unit, and EU is the equivalent unit.

Schedule of Cost Summary

The final phase in process costing involves the construction of a schedule of cost summary, showing the costs for both beginning and ending inventories. Information contained in this schedule comes from the schedules of equivalent production and the unit of cost analysis. Table 4.4 presents this schedule for materials, as well as conversion costs.

As shown in Table 4.4, the cost of completed units without the beginning inventory is $3,247,368.75, which is obtained by multiplying the cost per unit by the number of units fully completed; that is, $72,163.75 × 45 = $3,247,368.75. Similarly, the cost of units in process is obtained by multiplying the cost per equivalent unit by the number of equivalent units for materials and conversion. The result produces $893,857.20 in materials costs and $2,748,724.20 in conversion costs, with a total cost of $3,642,631.40. Together with the costs of equivalent units, they produce a total of $6.89 million, which is the total cost of operation for the department (after adjusting for rounding-off errors).

Similarly, the costs of equivalent, with the beginning inventory included, also turns out to be $6.89 million. A breakdown of these costs will include $1,481,317.60 for completed units in the beginning inventory, $2,962,635.60 for completed units that began in the current period, and $2,446,047.12 for partially completed units in the ending inventory, including $807,354.72 in materials costs and $1,638,692.40 in conversion costs.

It is important to note that as long as there are units remaining in a WIP account, they will automatically become a part of the beginning inventory for the next period. Consequently, one must transfer the costs of completed units out

Table 4.4 Schedule of Cost Summary*

Cost Description	*Costs of Units Completed ($)*	*Costs of Units in Process ($)*
I Without Beginning Inventory		
Beginning inventory	–	–
Units started and completed [45 × $72,163.75]	3,247,368.75	–
Ending inventory (WIP)		
Materials: 60 × $14,897.62	–	893,857.20
Conversion: 48 × 57,266.13	–	2,748,774.20
Total	3,247,368.75	3,642,631.40
II With Beginning Inventory		
Beginning inventory [15 × $98,754.52]	1,481,317.60	
Units started and completed [30 × $98,754.52]	2,962,635.60	
Ending inventory (WIP)		
Materials: 48 × $16,819.89	–	807,354.72
Conversion:20 × $81,934.62	–	1,638,692.40
Total	4,443,953.20	2,446,047.12

Note
*Year ended: December 31, 2017.

of WIP inventory to show the balance at the end of the period. Take the cost for the department with beginning inventory as a case in point: the total cost of operation for the year was $6.89 million, of which the costs of completed units, including those that appear in the beginning inventory, were $4,443,953.20; that is, 1,481,317.60+2,962,635.60=$4,443,953.20. The difference between $6.89 million and $4,443,953.20, which is approximately $2,446,047.12, then becomes the costs associated with the beginning inventory for next year. This should be added to next year's materials and conversion costs for all units (completed, as well as partially completed) to produce the total cost to be accounted for that period, and the process will continue.

In dealing with the problem in the current example, we made an assumption that we have a single production unit (the Drug Enforcement Agency) that is responsible for drug interdictions, among others. The drawback of this assumption is that since we were dealing with a single production unit, we needed only one WIP inventory. In reality, there is usually more than one production unit where the output from one department would pass to the next, and eventually to the final inventory account. When more than one unit is involved, the process can be complicated as it would require that the accounting system maintain as many WIP inventory accounts as there are units in the process. However, this will require more work, since a separate set of schedules will have to be prepared for each unit and the corresponding WIP will have to be reconciled in the final account.

In spite of this apparent complexity, process costing is simple and easy to calculate and, more important, it allows an organization to allocate costs to processes so as to have an accurate account of costs. Also, since cost data are easily available for each stage of the process, it makes it possible for an organization to better manage its costs of operation.

Cost Allocation

Operationally, all agencies and departments within an organization serve two distinct purposes: those that provide services to other departments, called service or provider departments; and those that receive services from the service departments, called user departments. In providing services, like any provider, the service departments incur costs that must be allocated to the user departments and, where appropriate, to the recipients of the service. For instance, a hospital may want to allocate the costs of patient admissions to patient wards, or to outpatient services, and eventually to the patients. Similarly, an IT department of a government may want to allocate the costs of information services it provides to various user departments. Cost allocation deals with the process of allocating costs from a service department to one or more user departments for which it is possible to measure costs separately. When costs are measured separately, it enables one to keep track of the origin and destination of costs which, in turn, makes it possible for an organization to identify the points of inefficiency in service provision.

All cost allocations are based on a number of key considerations that serve as guidelines for allocating costs to various user departments. Important among these considerations are:

1 Costs should be allocated according to the degree of association or benefit received, which means that only those user departments that benefit from a given service should be included in cost allocation.
2 The inefficiencies in resource use of the service department should not be passed on to the user and other departments since the inefficiencies do not produce any benefit for these departments.
3 Most important, the costs charged by a service department to a user department must be independent of the activities of all other departments. In other words, the benefit received by a user department for which a cost is charged must be unique to that department.

Cost Allocation Process

As noted earlier, cost allocation is the process of allocating costs to a cost object (which could be anything – a department, an agency, a good, a service, or an activity) for which costs are measured separately. Measuring costs separately makes it possible for an organization to allocate costs properly and, more important, to meet legal and reporting requirements, especially where contractual obligations are concerned – a common practice in government.

Several steps are typically involved in a cost allocation process (Oliver, 2000). By and large, they include the following: (1) determine the cost object; (2) define the allocation objective; (3) identify the costs, in particular indirect costs as the direct costs are easier to determine and allocate; (4) select the allocation base, which serves as the basis for allocating costs or a group of costs, called a cost pool, to a cost object; (5) calculate the allocation rate, which can be easily obtained, for instance, by dividing the total cost by the total quantity; and (6) assign the costs, using any of the available methods.

Of the steps mentioned above, the cost allocation base deserves special attention because how the base is selected can have a significant impact on the allocation process and, more important, on the overall cost management of an organization. Shim and Siegel (1991) suggest several criteria for selecting the base, which can also apply to government. They include, among others, the cause-and-effect relationship, benefits received, fairness, and ability to pay. The cause-and-effect relationship establishes a logical connection between a cost object and the incurrence of costs. It provides the rationale for allocating costs in that there has to be a direct relationship between resource consumption and the costs charged to the cost object. For instance, if a user department consumes k amount of resources, the cost manager can do a demand analysis to determine what contributes to the level of consumption and its effect on the cost, and how to control costs if demand changes.

The benefits received criterion is self-explanatory; it simply says that cost should be allocated on the basis of the quantity of service received or resource consumed by a user department. To give an example, suppose that a government introduces an organization-wide job-training program that will benefit all the departments. The government can allocate the costs of the program to the participating departments based on the number of employees that received training for each department. Thus, if a department sends more of its employees for training it will be charged more, and vice versa.

The fairness criterion suggests that the basis of allocation should be fair. Although fairness is difficult to define in precise terms, we can use a common sense explanation to indicate what it means. Returning to the training program example, allocating costs based on head counts (i.e., number of employees who participated in the program from each department) would be a fair way to allocate costs so that no one department bears a disproportionate share of the costs.

Finally, the ability to bear criterion indicates that the users' departments must have the ability to bear the costs. Again, going back to the training program example, if a user department does not have the ability to bear the cost yet needs the program, the government must find a way to assume the cost or distribute it between the participating departments if, in the judgment of the decision makers, the program will benefit all the participating departments; that is, the government as a whole will benefit from the program. In other words, in the aggregate, government will be better off from the program, as will the individual departments.

Methods of Cost Allocation

Cost allocation is relatively simple if cost flows in one direction from a single service department to one or more user departments, but when several service departments are involved in a process (i.e., when cost flows in multiple directions), allocation can become complex. This section presents three simple but frequently used methods that can deal with different levels of complexity in cost allocation: direct method, step method, and reciprocal method.

DIRECT METHOD

In a direct method, costs are allocated directly from a service department to one or more user departments. In general, when *n*-user departments receive service from a service department, the costs are allocated in direct proportion to the amount of service they receive. The proportionality factor, which serves as a basis for cost allocation, is called the cost allocation base. The cost allocation base for a service department reflects the amount of service it provides to other departments. In most instances, the base is selected in such a way that it is relatively easy to measure. Examples of cost allocation base will be the number of labor hours, number of machine hours, number of employees, units of goods, and so on.

To give an example, consider a case in which the IT department of a government which, for convenience we will call *X*, provides service to four user departments within the government: D_1, D_2, D_3, and D_4. Let us say that it cost the department $1.5 million last year to provide the service. The total amount of service provided by the department was 100,000 units (cost allocation base) at a cost of $15 per unit. This included the cost of both capital (machine and equipment) and labor used by the department. The distribution of service for the four departments, let us say, was 20,000 units for D_1, 30,000 units for D_2, 40,000 units for D_3, and 10,000 units for D_4.

We can use a simple expression to determine the cost allocation for our hypothetical departments, based on the proportion of service each department receives from *X*, as shown below:

$$TC_X = \sum_{j=1}^{m} c_j Q_j$$

$$= c_1 Q_1 + c_2 Q_2 + c_3 Q_3 + \ldots + c_m Q_m \qquad [4.6]$$

where TC_X is the total cost of service for department *X*, c_j is the cost per unit of the *j*th department, and Q_j is the quantity of service received by the *j*th department (for $j = 1, 2, 3, \ldots, m$).

Note that we treated *c* as a constant on the assumption that each department must pay at the same rate, regardless of the quantity of service they received. Since *c* is a constant, we can rewrite Equation 4.6 by factoring it out to obtain the cost share for each department, as shown below:

$$TC_X = c \sum_{j=1}^{4} Q_j$$

$$= cQ_1 + cQ_2 + cQ_3 + cQ_4$$

$$= (\$15)(20{,}000_1) + (\$15)(30{,}000_2) + (\$15)(40{,}000_3) + (\$15)(10{,}000_4)$$

$$= \$300{,}000_1 + \$450{,}000_2 + \$600{,}000_3 + \$150{,}000_4$$

$$= \$1{,}500{,}000 \qquad [4.7]$$

where the subscripts 1, 2, 3, and 4 represent the departments D_1, D_2, D_3, and D_4, respectively.

As the results show, of the $1.5 million in total cost the service department incurred, $300,000 are allocated to D_1 for its share of the service, $450,000 to D_2, $600,000 to D_3, and $150,000 to D_4 for their respective share of the service. In other words, these are the cost shares of the user departments, based on the amount of service each received from *X*.

In the example presented above, we assumed that the constant term *c*, which is cost per unit of service, included both fixed and variable costs. While the fixed cost remains constant regardless of the quantity, the variable cost, by definition, must vary in relation to the quantity of service received by the user departments. As such, we need to distribute the variable cost among the four departments in proportion to their share of the service, so that it does not fall disproportionately on any one department.

The following expression shows the distribution of costs for various user departments with both fixed and variable costs included:

$$TC_X = FC + \sum_{j=1}^{m} VC_j f_j \qquad [4.8]$$

where FC is the fixed cost, VC is the variable cost, and f is the fraction of service received by each user department.

Since the fixed cost (FC) remains the same regardless of the quantity, each user department must pay exactly the same amount. The variable cost (VC) then must be allocated between the departments in proportion to their share of the service from X. Equation 4.8 can thus be rewritten to reflect this proportionality factor:

$$TC_X = FC + VC \sum_{j=1}^{m} f_j \qquad [4.9]$$

Let us suppose that of the \$1.5 million in total cost of service, \$300,000 or 20 percent of the total cost or 20 percent of the cost per unit, which is \$3 out of \$15, is the fixed cost (i.e., $(\$15 \times 100{,}000)(0.2) = \$1{,}500{,}000 \times 0.2 = \$300{,}000$) to be shared equally between the four user departments (i.e., $\$300{,}000/4 = \$75{,}000$). The remaining \$1.2 million will then be the variable cost, divided between the departments in proportion to their share of the service.

The following provides a breakdown of cost share for the four user departments:

$$\begin{aligned}
TC_X &= FC + VC \sum_{j=1}^{m} f_j \\
&= (FC + VC \times f_1) + (FC + VC \times f_2) + (FC + VC \times f_3) + (FC + VC \times f_4) \\
&= [\$75{,}000_1 + (\$1{,}200{,}000 \times 0.2_1)] + [\$75{,}000_2 + (\$1{,}200{,}000 \times 0.3_2)] \\
&\quad + [\$75{,}000_3 + (\$1{,}200{,}000 \times 0.4_3)] + [\$75{,}000_4 + (\$1{,}200{,}000 \times 0.1_1)] \\
&= (\$75{,}000_1 + \$240{,}000_1) + (\$75{,}000_2 + \$360{,}000_2) \\
&\quad + (\$75{,}000 + \$480{,}000_3) + (\$75{,}000_4 + \$120{,}000_4) \\
&= \$315{,}000_1 + \$435{,}000_2 + \$555{,}000_3 + \$195{,}000_4 \\
&= \$1{,}500{,}000
\end{aligned}$$

where \$75,000 is the share of the fixed cost for each department. Interestingly, we could have obtained the same result by directly multiplying the variable cost by the quantity, as shown below:

$$TC_X = FC + VC\sum_{j=1}^{m} Q_j \qquad [4.10]$$

$$= (FC + VC \times Q_1) + (FC + VC \times Q_2) + (FC + VC \times Q_3) + (FC + VC \times Q_4)$$

$$= [\$75{,}000 + (\$12 \times 20{,}000_1)] + [\$75{,}000 + (\$12 \times 30{,}000_2)]$$

$$= [\$75{,}000_1 + (\$12 \times 20{,}000)] + [\$75{,}000_2 + (\$12 \times 30{,}000x)]$$

$$+ [\$75{,}000_3 + (\$12 \times 40{,}000 \times 0.4_3)] + [\$75{,}000_4 + (\$12 \times 10{,}000)]$$

$$= (\$75{,}000_1 + \$240{,}000_1) + (\$75{,}000_2 + \$360{,}000_2)$$

$$+ (\$75{,}000 + \$480{,}000) + (\$75{,}000_4 + \$120{,}000_4)$$

$$= \$315{,}000_1 + \$435{,}000_2 + \$555{,}000_3 + \$195{,}000_4$$

$$= \$1{,}500{,}000$$

Table 4.5 shows the distribution of cost share for the four departments. As the table shows, when the variable cost is taken into consideration, the cost share of each user department comes out to be different from those observed when no distinction was made between fixed and variable costs. For instance, the cost share of D_1 and D_4 increased, respectively, by \$15,000 and \$45,000, while it decreased by \$15,000 for D_2 and by \$45,000 for D_3. Had the proportionality factor in variable cost not been taken into consideration, D_2 and D_3 would have paid \$60,000 more than their share of the cost [($450,000_2 - $435,000_2) + ($600,000_3 - $555,000_3 = $15,000_2 + $45,000_3 = $60,000), and D_1 and D_4 would have paid \$60,000 less than their share of the cost [($300,000_1 - $315,000_1) + ($150,000_4 - $195,000_4) = (-$15,000_1) + (-$45,000_4) = -$60,000].

Let us further assume that it is unfair to ask the user departments to pay in equal amounts for the fixed cost if they received different amounts of service, in which case we need to divide the fixed cost by respective share of the service for each department. Equation 4.11 shows the revised allocation for the four departments:

$$TC_X = FC\sum_{j=1}^{m} f_j + VC\sum_{j=1}^{m} f_j \qquad [4.11]$$

$$= [(FC \times f_1) + (VC \times f_1)] + [(FC \times f_2) + (VC \times f_2)] + [(FC \times f_3) + (VC \times f_3)]$$

$$+ [(FC \times f_4) + (VC \times f_4)]$$

$$= [(\$300{,}000 \times 0.2_1) + (\$1{,}200{,}000 \times 0.2_1)] + [(\$300{,}000_2 \times 0.3_2)$$

$$+ (\$1{,}200{,}000 \times 0.3_2)] + [(\$300{,}000_3 \times 0.4_3) + (\$1{,}200{,}000 \times 0.4_3)]$$

$$+ [(\$300{,}000_4 \times 0.1_4) + (\$1{,}200{,}000 \times 0.1_4)]$$

$$= (\$60{,}000_1 + \$240{,}000_1) + (\$90{,}000_2 + \$360{,}000_2) + (\$120{,}000_3$$

$$+ \$480{,}000_3) + (\$30{,}000_4 + \$120{,}000_4)$$

$$= \$300{,}000_1 + \$450{,}000_2 + \$600{,}000_3 + \$150{,}000_4$$

$$= \$1{,}500{,}000$$

Table 4.5 Distribution of Cost Share for the User Departments*

User Departments	*Fixed Cost ($)*	*Variable Cost ($)*	*Total Cost ($)*	*Revised Fixed Cost ($)*	*Total Cost ($)*
D_1	75,000	240,000	315,000	60,000	300,000
D_2	75,000	360,000	435,000	90,000	450,000
D_3	75,000	480,000	555,000	120,000	600,000
D_4	75,000	120,000	195,000	30,000	150,000
Total cost ($)	300,000	1,200,000	1,500,000	300,000	1,500,000

Note
*Year ended: December 31, 2017.

As the results show, Departments 1 and 4 will now pay a little less, while Departments 2 and 4 will pay a little more than when the proportionality factor was not included in the fixed cost. In other words, without the proportionality factor, per capita share would have been higher for Departments 1 and 4, given their share of the service.

STEP-DOWN METHOD

While it is simple to use, the direct method has a major weakness in that it does not take into consideration the reciprocal relationship between departments. By reciprocal, we mean a manner of service provision in which a department not only receives, but also provides services to other departments. This is where the step method, also known as the sequential method, has an advantage over the direct method in that it takes this relationship explicitly into consideration.

As one would expect, since it deals with a reciprocal relationship, the allocation process will be somewhat more involved in the step method. According to this method, a department is selected first and a pro rata (in proportion or proportionate) share of its cost is allocated to the rest of the departments. Next, a second department is selected and its costs, including those allocated from the first department, are then allocated to all the remaining departments, except for the one allocated first. The process is continued until all service costs have been fully allocated. What this means is that the allocations are made in such a way that the departments whose costs have been previously assigned do not absorb any costs of the departments being allocated. This makes it possible to avoid any problem such as double counting that may result from reciprocal allocation. However, the use of the method requires that some order be established in allocating departments. Two approaches have been traditionally used for this purpose: (1) Select a department that serves the largest percentage (fraction) of departments; (2) select the departments at random. The latter is recommended only in situations where the percentage or fraction of departments served is evenly distributed.

To illustrate how the method is used, we can look at the service allocation problem again. Assume that three of the four departments (D_1, D_2, and D_3) are service departments and the fourth department, D'_4, is a user department (we used a prime to distinguish it from the service departments), meaning that it does not provide any service to the other departments. In other words, the three service departments provide service to each other, as well as to the fourth department, while the fourth department receives services provided by the three departments but does not provide any service in return.

Table 4.6 shows the services provided by the three departments to each other, as well as to the user department. For instance, department D_1 provides 20,000 units of service to D_2, 30,000 units to D_3, and 15,000 units to D'_4. Similarly, Department D_2 provides 25,000 units of service to D_1, 45,000 units to D_3, and 30 units to D'_4, and so on. The zeroes in the principal diagonal indicate that the service by a department to itself is ignored. The table also shows the costs prior to allocation that are to be allocated among the departments.

We begin with the order of the departments first. Let us say that we use the first approach. Thus, based on the largest amount of services provided by a service department, we select the order D_1, D_2, and D_3 such that (20,000+30,000)/65,000 =0.77 for D_1>(25,000+45,000)/100,000=0.70 for D_2>(20,000/45,000)=0.44 for D_3. This means that D_1 provides more service than D_2, which provides more service than D_3. Accordingly, all three departments (D_2, D_3, and D'_4) utilizing D_1's service shared its costs of $450,000. The distributions of these costs to the three departments, respectively, are $138,460 for D_2, $207,690 for D_3, and $103,850 for $D'_{4,}$ obtained in the following way: $450,000×(20/65)=$138,460$_2$, $450,000×(30/65)=$207,690$_3$, and $450,000×(15/65)=$103,850$_4$.

Similarly, the departments utilizing D_2's service (D_3 and D'_4) shared its costs of $688,460, which included the original cost of $550,000, plus the $138,460 that came from D_1 [$550,000+$138,460=$688,460]. Note that D_1 was excluded from D_2's cost allocation to avoid the problem of reciprocity, mentioned earlier. Finally, both D_1 and D_2 were excluded in D_3's allocation of $1,370,770, which included the original cost of $750,000, plus the costs that came from D_1 and D_2 [($207,690+$413,080)+$750,000=$1,370,770]. To see if our allocation was correct, we can add the service costs assigned to D'_4 by all three departments. If the sum of these costs equals the sum of the costs prior to allocation, the allocation should be considered correct, which appears to be the case here.

RECIPROCAL METHOD

As opposed to the step method, the reciprocal method uses a more direct approach to cost allocation. Somewhat analytically more involved than the previous two methods, the reciprocal method considers all the service departments together by presenting the problem in terms of a system of simultaneous equations. The purpose of these equations is to find solutions for the cost of individual service departments. Once the solutions to the equations have been found, they can be used for allocating each department independently to all user departments.

To illustrate the use of the method, let us go back to the problem again. We start with the three service departments, D_1, D_2, and D_3, as before. To be consistent

Table 4.6 Step-Down and Reciprocal Methods*

User Departments	*Service Departments (Units of Service Provided, in 1,000)*			*Cost Prior to Allocation ($1,000)*
	D_1	D_2	D_3	
D_1	0	25	20	450
D_2	20	0	0	550
D_3	30	45	0	750
D'_4	15	30	25	700
Total	65	100	45	2,450

Step-Down	*Cost Allocation by Department ($000)*			
	D_1	D_2	D_3	D'_4
Cost prior to allocation				
D_1(20/65, 30/65,15/65)	450	550	750	700
D_2(45/75,30/75)	(450.00)	138.46	207.69	103.85
D_3(25/25)	0	(688.46)	413.08	275.38
Total	0	0	(1,370.77)	1,370.77
	0	0	0	2,450.00

Reciprocal	*Cost Allocation by Department ($000)*			
	D_1	D_2	D_3	D'_4
Cost prior to allocation	450	550	750	700
D_1(0.31, 0.46, 0.23)	(1,557.511)	482.289	716.455	358.228
D_2(0.25, 0.45, 0.30)	258.207	(1,032.829)	464.773	309.849
D_3(0.44, 0.00, 0.56)	849.927	0	(1,931.653)	1,081.726
Total	0	0	0	2,449.803

Note
*Year ended: December 31, 2017.

with our earlier discussion, we begin with D_1, although it does not make any difference which department we start with since the order is not important under the reciprocal method. Our objective is to set up an equation for D_1 that will include its original costs of $450,000 plus a share of the costs in D_2, and D_3. According to Table 4.6, D_1 should absorb 25 percent of D_2's cost [25,000/100,000=0.25], and 44 percent of D_3's [20,000/45,000=0.44]. Therefore, the equation for D_1 will be

$$D_1 = 450{,}000 + 0.250D_2 + 0.440D_3 \quad [4.12]$$

Similarly, the equations for D_2 and D_3 will be

$$D_2 = 550{,}000 + 0.310D_1 \quad [4.13]$$

$$D_3 = 750{,}000 + 0.460D_1 + 0.450D_2 \quad [4.14]$$

We can now solve these equations simultaneously to find the costs for each service department by substituting Equations 4.12 and 4.13 into Equation 4.14, first for D_1:

$$\begin{aligned} D_1 &= 450{,}000 + 0.250\,(550{,}000 + 0.310D_1) + 0.440\,(750{,}000 + 0.460D_1 + 0.450D_2) \\ &= 450{,}000 + 137{,}500 + 0.0775D_1 + 330{,}000 + 0.202D_1 + 0.198D_2 \\ &= 917{,}500 + 0.280D_1 + 0.198D_2 \end{aligned}$$

Substituting Equation 4.13 into the right-hand side of the solved equation above for D_2, we get

$$\begin{aligned} D_1 &= 917{,}500 + 0.280D_1 + 0.198\,(550{,}000 + 0.310D_1) \\ &= 917{,}500 + 0.280D_1 + 108{,}900 + 0.061D_1 \\ &= 1{,}026{,}400 + 0.341D_1 \end{aligned}$$

With slight algebraic manipulation, D_1 can be reduced to

$$\begin{aligned} D_1 - 0.341D_1 &= 1{,}026{,}400 \\ 0.659D_1 &= 1{,}026{,}400 \\ \therefore D_1 &= \$1{,}557{,}511.38 \approx \$1{,}577{,}511 \end{aligned}$$

Now, substituting the above result into Equation 4.13 we obtain the solution for D_2:

$$\begin{aligned} D_2 &= 550{,}000 + 0.31(1{,}557{,}511.38) \\ &= 550{,}000 + 482{,}828.53 \\ &= \$1{,}032{,}828.53 \approx \$1{,}032{,}829 \end{aligned}$$

Similarly, substituting the values of D_1 and D_2 into Equation 4.11 we get the following value of D_3:

$$D_3 = 750{,}000 + 0.46(1{,}558{,}220.70) + 0.45(1{,}033{,}048.42)$$
$$= 750{,}000 + 716{,}781.52 + 464{,}871.78$$
$$= \$1{,}931.228.07 \approx \$1{,}931{,}228$$

The results obtained for D_1, D_2, and D_3 thus represent the cost amount for each of the service departments. Put simply, these are the amounts that are to be allocated to respective departments in proportion to the services they received from the service departments.

Table 4.6 shows the reciprocal allocations (bottom section). As the table shows, of the $1,558,221 obtained for D_1, $483,048 are allocated to D_2, $716,782 allocated to D_3, and $358,390 to D'_4. The process is repeated for D_2 and D_3 to show similar allocations of service costs. Since the sum of the costs allocated to D'_4 equals the total cost of service prior to allocation, except for rounding-off errors, we can assume the results to be correct.

Reciprocal Allocation by Matrix Inversion. The procedure we used to solve the system of equations above is called the substitution method, which works fine as long as the number of equations to be solved remains small. When the number increases in size (i.e., when there are many more service departments with complex reciprocal relationships), it becomes difficult to use the method because of the amount of substitutions that will be required. The alternative is to use matrix algebra; in particular, matrix inversion.

Matrix inversion is a process in which a square matrix (one in which the number of rows is equal to the number of columns), when inverted and subsequently multiplied by the original matrix, produces an identity matrix (one in which the principal diagonals are all 1s and the off-diagonal elements are all 0s). To see how the method applies to reciprocal allocation, we begin with the equations for the three service departments and present them in terms of a system of simultaneous equations, as shown below:

$$D_1 = 450{,}000 + 0.25D_2 + 0.44D_3 \qquad [4.15]$$

$$D_2 = 550{,}000 + 0.31D_1 + 0.00D_3 \qquad [4.16]$$

$$D_3 = 750{,}000 + 0.46D_1 + 0.45D_2 \qquad [4.17]$$

With slight reorganization, the equations can be written as:

$$1.00D_1 - 0.25D_2 - 0.44D_3 = 450{,}000 \qquad [4.18]$$

$$-0.31D_1 + 1.00D_2 + 0.00D_3 = 550{,}000 \qquad [4.19]$$

$$-0.46D_1 - 0.45D_2 + 1.00D_3 = 750{,}000 \qquad [4.20]$$

Using the standard matrix notations, the equations can be further written as:

$$AD = C \qquad [4.21]$$

where A is the matrix of coefficients, D is a column vector of service departments, and C is a column vector of costs corresponding to each of these departments.

Expanding the terms of the expression in Equation 4.21, we can write it in full as:

$$\begin{bmatrix} 1.00 & -0.25 & -0.44 \\ -0.31 & 1.00 & 0.00 \\ -0.46 & -0.45 & 1.00 \end{bmatrix} \begin{bmatrix} D_1 \\ D_2 \\ D_3 \end{bmatrix} = \begin{bmatrix} 450,000 \\ 550,000 \\ 750.000 \end{bmatrix}$$

To obtain the inverse of matrix A we do the following: first, find a matrix of cofactors (i.e., submatrices with plus and minus signs), A^c, which are obtained by taking the determinants (i.e., single values or scalars found only in square matrices) of the submatrices by removing a row and a column corresponding to an element of the matrix, each time a determinant is calculated for one of the submatrices. Thus, for a 3×3 matrix, there will be nine determinants corresponding to nine submatrices.

The following shows the cofactor matrix for the allocation problem, A^C:

$$A^C = \begin{bmatrix} 1.0000 & 0.3100 & 0.5995 \\ 0.4480 & 0.7976 & 0.5650 \\ 0.4400 & 0.1364 & 0.9225 \end{bmatrix}$$

Second, we take the transpose of this matrix by changing the rows into columns, called an adjoint matrix, $(A^c)^T$. That is,

$$(A^C)^T = \begin{bmatrix} 1.0000 & 0.4400 & 0.4400 \\ 0.3100 & 0.7976 & 0.1364 \\ 0.5995 & 0.5650 & 0.9225 \end{bmatrix}$$

Third, and finally, we divide the adjoint matrix by the determinant of the original matrix, $|A|$, to obtain the inverse matrix, which turns out to be 0.6587.[3] Therefore, dividing the adjoint matrix, $(A^c)^T$, by 0.6587 (the determinant of A) will produce the inverse matrix for the problem, as shown below:

$$(A^C)^T / |A| = \begin{bmatrix} 1.5181 & 0.6803 & 0.6680 \\ 0.4706 & 1.2109 & 0.2071 \\ 0.9101 & 0.8578 & 1.4005 \end{bmatrix} = A^{-1}$$

Next, to obtain the value of the column vector C (i.e., the cost of each service department), we multiply the inverse matrix by C. That is:

$$A^{-1}C = \begin{bmatrix} 1.5181 & 0.6803 & 0.6680 \\ 0.4706 & 1.2109 & 0.2071 \\ 0.9101 & 0.8578 & 1.4005 \end{bmatrix} \begin{bmatrix} 450,000 \\ 550,000 \\ 750,000 \end{bmatrix} = \begin{bmatrix} 1,558,310 \\ 1,033,090 \\ 1,931,710 \end{bmatrix} = \begin{bmatrix} D_1 \\ D_2 \\ D_3 \end{bmatrix}$$

The result shows the costs associated with each service department, which, respectively, are \$1,558.310 for D_1, \$1,033,090 for D_2, and \$1,931,710 for D_3. As expected, the results came out to be almost the same as those obtained by the substitution method (Table 4.6), except for rounding-off errors.

Activity-Based Costing

Activity-based costing (ABC) was introduced in the early 1980s as an alternative to traditional costing methods with substantial indirect costs. In traditional costing, costs usually go into a cost pool and from there they are allocated to specific cost objects, which can fail to produce an accurate estimate of costs. ABC corrects the problem by using activities as the building blocks of costing, which makes it easier to relate costs to activities and then allocate them to corresponding cost objects, rather than directly from a cost pool (resources) to cost objects. Activities are essentially processes that consume substantial resources to produce a good or deliver a service. The primary objective of an activity is to convert inputs (i.e., resources such as labor, capital, materials, and supplies) into outputs (i.e., goods and services).

The rationale for using activities as the basis for costing is quite simple: if one can think of an organization as an entity consisting of a number of activities, then assigning costs directly from activities to cost objects makes better sense than allocating them indirectly from a cost pool. Operationally, it is consistent with the basic structure of a government, where the agencies and departments are organized by functions, programs, and activities. Perhaps the greatest advantage of the method is that since the costs are allocated directly from activities to cost objects, it is possible to turn some of the indirect costs into direct costs (Michel, 2004). Figure 4.1 shows a simple ABC structure with an activity pool, cost drivers, and cost objects.

Conceptually we can think of the activities in ABC as an activity pool, similar to a cost pool under the traditional costing methods, except that they are associated with individual activities that support a good or service. Like the cost pool, in an activity pool the costs flow from an activity to a cost object that is directly associated with it. However, the direct costs remain the same under both traditional costing and ABC; it is the indirect cost that is allocated directly from the activities to their respective cost objects that makes ABC an attractive alternative to conventional costing methods.

ABC Process

As with any system, implementing ABC may require some changes in the management process. For instance, it may be necessary for the management to

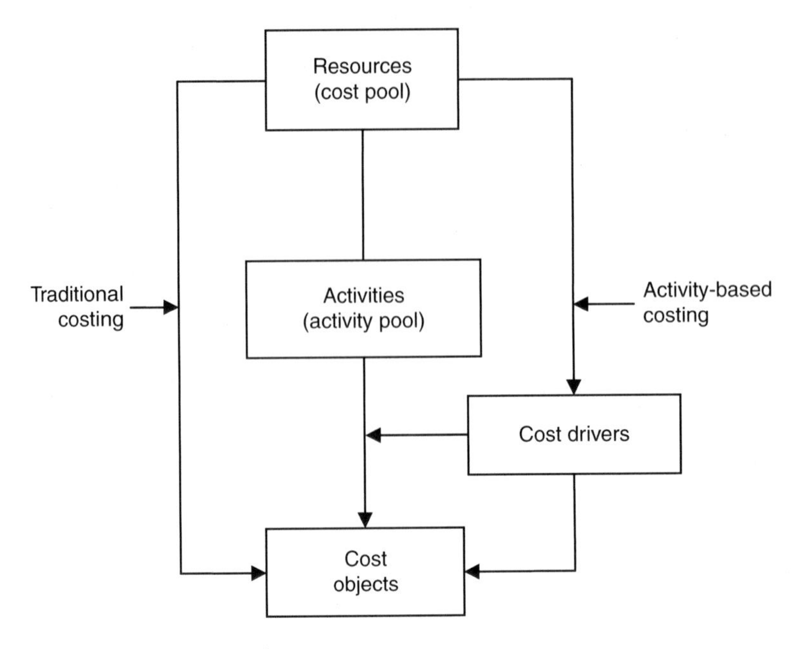

Figure 4.1 A Typical ABC Process

conduct a thorough analysis of the strengths and weaknesses of the decision and, if possible, do a cost–benefit analysis to ensure that the long-term benefits of implementing ABC will outweigh its costs. Once a decision has been made to implement, the process must begin with a number of clearly defined steps that are easy to follow. In some cases, it may be useful to conduct a pilot study, similar to pre-testing in statistical sampling, involving a segment of the organization in which the method will be applied before implementing it for the entire organization. This will allow the organization to work out any obstacles that may impede a successful implementation of the system.

The following steps are typically involved in the development of ABC for an organization:

1 make a list of all goods and services by functional areas where the system will be used;
2 identify specific activities for each good and service;
3 classify each activity using some established criteria;
4 determine the total cost, direct as well as indirect, for each activity;
5 identify the appropriate cost driver for each activity;
6 determine the cost driver level;
7 calculate the cost driver rate; and
8 apply the cost of an activity to the good or service, based on the level of activity usage.

Several things are worth noting in the process described above – in particular, how the activities are identified, how they are classified, how the cost drivers and their levels are determined and, finally, how the cost driver rates are calculated. Activities are generally identified by a process known as tracing that determines how much of each activity is consumed by the final user. Tracing costs to the final user serves two important purposes: one, it allows us to understand the cost structure and, two, it helps us determine if an alternative activity is superior (i.e., would be better) to the current activity (Brimson & Antos, 1994).

Although there are no hard-and-fast rules for classifying activities, the convention is to use a cost hierarchy that relates costs of activities to the goods and services an organization provides. Four levels of costs are generally considered in a cost hierarchy: unit level, batch level, service level, and facility level. The unit level costs, the lowest level in the hierarchy, are the costs of activities performed on each unit of a good or service. The batch level costs, next in the hierarchy, are the costs of activities performed on a group of units rather than on an individual unit of a good or service. The third level in the hierarchy, the service level costs, takes into consideration the costs of activities performed for a service rather than individual units or groups of units of the service. Finally, the facility level costs, the highest level in the hierarchy, include costs of activities that support an entire functional area rather than goods and services. In general, facility level costs are used in situations where the activities cannot be easily traced to a specific cost object. On the other hand, where goods are divisible and it is possible to trace the costs of activities to a specific cost object, the unit level makes the most sense.

Activities can also be classified based on whether they are recurring or nonrecurring, or if they are primary or secondary. In general, primary activities are those that contribute directly to the mission of an organization, while the secondary activities are those that support the primary activities. In using the secondary activities, one has to be careful to make sure that redundant activities are not included in the process. One way to determine that is to ascertain the degree of their influence in primary activities vis-à-vis the mission of the organization.

The Role of Cost Drivers

Three terms are critical to understanding ABC and how it is used in practice: cost drivers, cost driver level, and cost driver rate. Cost drivers serve as the link between the activity cost pool, in particular between an activity and its associated cost object that drives the cost. In other words, the cost drivers cause costs to increase or decrease. For instance, if the purchasing department of a government buys items that are of poor quality but inexpensive in that they will save the government money in the short run, but will increase its operating costs in the long run, it may not serve the mission of the government if the goal is to ensure efficiency in service provision.

To ensure that the drivers can be operationalized, they are frequently expressed in numerical terms such as machine hours, direct labor hours, number of purchase orders, square feet of space, and so forth. Cost driver levels, on the other hand, are aggregates; that is, the total quantity for each cost driver such as total number of machine hours, total number of purchase orders, total number of employees, and so forth. Finally, the cost driver rate is the cost per unit of an activity for a cost driver, obtained usually by dividing the total cost of an activity by the appropriate cost driver level.

To illustrate how the method is used, let us go back to the street repair problem we introduced at the beginning of the chapter under job costing. Table 4.7 shows both direct and indirect costs of the repair. According to the table, the direct costs include both direct labor and direct materials, while the indirect costs are organized by activities which, for the current problem, would include insurance coverage, support personnel, utility expenses, vehicle insurance, and vehicle maintenance. The cost driver levels would be insurance coverage by months, support personnel by hours, utility expenses by months, vehicle insurance and vehicle maintenance by miles. Finally, the cost driver rates were based, for convenience, on the rates for the government as a whole. As the table shows, the sum of the two costs, direct and indirect, equals the total cost of the service.

Since cost drivers play such an important role in the overall ABC process, it is important that one pays attention to how the drivers and their corresponding levels are selected. Once the cost drivers have been selected and their corresponding levels specified, it becomes relatively easy to determine the cost driver rate. The following expression is generally used to determine the cost driver rate:

$$CDR = CACP / CDL \qquad [4.22]$$

where CDR is the cost driver rate, $CACP$ is the cost in the activity cost pool, and CDL is the cost driver level. For instance, to obtain the cost driver rate for insurance coverage, we divided the cost of insurance coverage by the cost driver level, which produced a rate of $2,083.33; that is, $25,000/12=$2,083.33/month. The remaining rates were obtained in a similar way.

While the example presented above is simple, the process can be easily expanded to deal with larger problems in that there will be multiple layers of drivers and levels, but the essential steps will remain the same.

A Note on Activity-Based Management

Since the introduction of ABC, a number of methods and approaches have been developed that use it as the basis for cost measurement, the most notable among them being activity-based management (ABM). Growing out of the work of the Texas-based Consortium for Advanced Manufacturing-International (CAM-I), ABM extends the use of ABC as a management tool to reduce costs and improve the decision process (Miller, 1996). The way ABM works is that it breaks down all activities into two distinct categories: those that add value to the goods or services an organization provides such as extending local library hours or upgrading a community recreation center; and those that do not add any value,

Table 4.7 Activity-Based Costing: Street Repair*

I Direct Costs				
Direct Labor		*No. Hours*	*Rate/Hour*	*Cost ($)*
Six-member crew		3,125	$40.00/hour	125,000.00
Direct Materials				
Material Types		*Amount*	*Rate*	*Cost ($)*
Asphalt		6,250 SF	$4.00/SF	25,000.00
Other materials		6,250 SF	$1.60/SF	10,000.00
Fuel		4,000 Gallons	$2.50/Gallon	10,000.00
II Indirect Costs				
Activity Pool	*Cost Driver*	*Cost Driver Level*	*Cost Driver Rate***	*Cost ($)*
Insurance coverage	No. months	12 months	$2,083.33/month	25,000.00
Support personnel	No. hours	400 hours	$25.00/hour	10,000.00
Utility expense	No. months	12 months	$166.67/month	2,000.00
Vehicle (4) insurance	No. miles	40,000 miles	$0.20/mile	8,000.00
Vehicle (4) maintenance	No. miles	40,000 miles	$0.125/mile	5,000.00
			Total cost	$220,000.00

Notes

* Year ended: December 31, 2017.

** Based on the overall rate for the government.

such as moving a cost item before installing it, which serves no purpose. In theory, ABM does not directly reduce costs but allows the decision makers to better understand costs that are worth undertaking and those that are not. However, like ABC, the ABM process can consume significant resources, including manpower.

Interestingly, the development of ABM and the host of other related methods are parts of a process known as continuous improvement. Continuous improvement deals with ongoing effort by firms and businesses to continuously improve goods, services, and processes. It does so by removing unnecessary activities, reducing costs, and improving organizational capability, efficiency, and quality over time (Albright & Lam, 2006). The underlying rationale is that improvements in general take time, but continuous improvement can bring about those improvements incrementally (i.e., a little at a time).

Chapter Summary

Cost accounting is integral to the effective functioning of an organization. Its purpose is to examine the cost structure of an organization; it does so by gathering, analyzing, and summarizing cost and related information, and assigning them to various cost objects to evaluate the efficiency of an operation. The process involves the use of various tools, methods, and techniques that have increased in number over the years. This chapter has focused on four of the most frequently used tools in cost accounting: job costing, process costing, cost allocation, and ABC. In job costing, costs are traced to a specific job, service, or activity. By tracing costs to a specific job, an organization can estimate the costs associated with each job accurately. In contrast, in process costing, costs are assigned equally to all units of a good or service. The units in process costing are generally homogeneous, which makes it possible to assign costs equally to these units.

Cost allocation, the third element of the cost accounting system discussed here, deals with the allocation of indirect costs to a responsibility center of an organization, which could be an agency, a department, or a unit of a department. In doing so, it not only assigns costs that are indirect, say, to a department, but also examines how costs that jointly benefit several departments can be assigned. Finally, as an extension of cost allocation, the chapter concluded with a brief discussion of ABC. In ABC, costs flow directly from an activity pool to associated cost objects, which make cost allocation much more precise than conventional costing methods where costs flow from a cost pool to cost objects.

Notes

1 J. R. Martin (2016) provides a lengthy discussion of a cost accounting system. According to Martin, the system should be based on five basic components: (1) It should have an input measurement focus based on pure, historical, or standard costing. (2) It should include an inventory evaluation method based on the amount of cost that could be

traced to an inventory. (3) It should include a cost accumulation method based on the manner in which cost data are collected and identified with respect to jobs (job-order costing), process (process costing), just-in-time, or some hybrid methods. (4) It should have a cost flow assumption that refers to how cost flows through an inventory, such as weighted average, first-in first-out, or last-in first-out. (5) It should have the capability of recording inventory cost flows at certain intervals, either periodically or permanently.

2 To balance out the effect of cost increase for inflation and other factors, firms using LIFO often give "tax holidays" to consumers, which lowers their income but also the tax they have to pay on the income.

3 For a 3×3 matrix, the determinant can be obtained by the expression (Gere & Weaver, 1965) shown below:

$$|A| = A_{11}A_{22}A_{33} + A_{12}A_{23}A_{31} + A_{13}A_{21}A_{32} - A_{13}A_{22}A_{31} - A_{11}A_{23}A_{32} - A_{12}A_{21}A_{33}$$

where the terms of the expression represent the coefficients of the original matrix, A. Substituting the respective values into the terms of the expression, taking their products, and adding them will produce the determinant of A, which turns out to be 0.6587.

References

Albright, T., & Lam, M. (2006). Managerial accounting and continuous improvement initiatives: A retrospective and framework. *Journal of Managerial Issues, 18*(2), 157–174.

Brimson, J. A., & Antos, J. (1994). *Activity-based management.* New York: Wiley & Sons.

Gere, J. M., & Weaver, W., Jr. (1965). *Matrix algebra for engineers.* New York: Van Nostrand Company.

Ingram, R. W. (1991). *Accounting and financial reporting for governmental and nonprofit organizations: Basic concepts.* New York: McGraw-Hill.

Kelly, J. T. (1984). *Costing government services: A guide to decision making.* Washington, DC: Government Finance Officers Association.

Lynn, E. S., & Norville, J. W. (1984). *Introduction to fund accounting.* Reston, VA: Reston Publishing Company.

Martin, J. R. (2016). *Management accounting: Concepts, techniques, and controversial issues.* Management and Accounting Web. Retrieved March 2016, from http://maaw.info/MAAWTextbookMain.htm.

Michel, R. G. (2004). *Cost analysis and activity-based costing for government.* Chicago, IL: Government Finance Officers Association.

Miller, J. A. (1996). *Implementing activity-based management in daily operations.* New York: Wiley & Sons.

Oliver, L. (2000). *The cost management toolbox: A manager's guide to controlling costs and boosting profits.* New York: American Management Association.

Shim, J. K., & Siegel, J. G. (1991). *Modern cost management and analysis.* Hauppauge, NY: Barron's Foundation Series.

5 Cost Control

Perhaps the single most critical element for an organization, especially from a cost management perspective, is cost control. Cost control is the process of ensuring that the cost of operation for an organization does not exceed its target cost and that there are no cost overruns. Ideally, the more an organization can control its costs without lowering the quality or quantity of goods and services it provides, the more efficient the organization. While occasional cost overruns are not unlikely because of events and circumstances over which an organization may not have enough control, it is important that all organizations, large or small, adopt measures that would ensure cost control as part of their normal operation. There is a diverse array of these measures, ranging from cybernetics, which deals with the study of all forms of control, to simple variance and ratio analysis, to cost internal control, to decision techniques such as those discussed in the next few chapters that one can use for cost control.

This chapter presents a brief discussion of several measures that serve as the foundation for cost control for an organization. They are standard costing, variance analysis, ratio analysis, cost internal control, performance auditing, information management, and planning and budgeting.

Need for Cost Control

For an organization to operate efficiently, it must perform two vitally important functions: (1) it must plan in that it must make a systematic effort to undertake those activities that would help it achieve the goals of the organization within a defined time frame; (2) it must undertake measures that would help control costs without significantly affecting the quality or quantity of goods and services it provides. The greatest advantage of using control measures is that they can help an organization evaluate its performance by comparing actual performance against planned or target performance, thereby creating opportunities for determining where the problems occur and how to correct them. The end result of the process would be reduced cost of operation and greater efficiency.

According to Clark and Lorenzoni (1998), a well-developed cost control system serves several important purposes for an organization. At a minimum, it can help focus an organization's attention on potential trouble areas by determining how its performance compares against planned or target performance. It can also help create a cost-conscious work environment that would allow the individuals working in an organization to appreciate the impact their work has on the cost of operations and make the efforts necessary to minimize these costs, thereby producing considerable savings for the organization.

Conceptually, one can think of cost control as a system with input, process, and output. Inputs typically are materials, labor, and overheads that go into the production or delivery of goods and services, and their associated costs (Ziegenfuss & Bentley, 2000). Outputs are the actual costs of production and delivery (Figure 5.1). The advantage of using a systems approach is that it would allow an organization to develop a set of procedures to monitor the costs of operation, from the initial stage (input) until they are produced and delivered (output). More important, it would allow an organization to measure the variation from planned or target cost and make the necessary adjustments in the process to keep the costs of operation under control. The systems approach has been extensively used in cost control, but often as a subsystem of cost management rather than a complete system per se (Kerzner, 2009).

Measures of Cost Control

Managing costs not only means cost control, but also an entire range of activities that are part of this process. This would typically include activities such as cost measurement, cost estimation, cost analysis, cost accounting, and other related activities such as productivity analysis, quality control, and so forth. While these activities are necessary to minimize the cost of operation in an organization, it is equally important, once a job has been completed, to apply

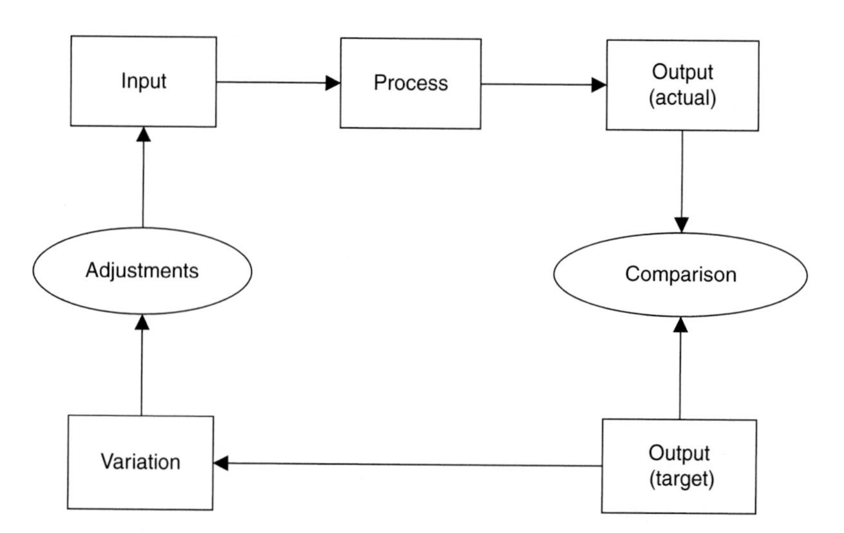

Figure 5.1 Input–Output Model in Cost Control

measures to determine how much the observed costs have deviated from planned or target costs. The analysis would allow the decision makers to determine the extent of the deviation and to take appropriate measures to correct the problem. The seven topics to be discussed next (i.e., standard costing, variance analysis, ratio analysis, cost internal control, performance auditing, information management, and planning and budgeting) are part of this exercise.

Standard Costing

Most discussions on cost control start with a measure that provides the basis that an organization can use to compare the actual costs of its operation against those planned or targeted, called standard costing. The term "standard" means benchmark or predetermined. When applied to cost accounting, it means a predetermined cost based on quantitative and other measures to determine how much it would cost to produce a good or deliver a service under a given circumstance. Three steps are commonly used in this process: (1) determine the standard costs of operation; (2) compare the actual cost against the standard cost; and (3) perform a variance analysis.

Determining Standard Costs

Determining standard costs follows a number of steps. The first in the process is to establish cost centers that would define the line of responsibility, followed by a classification of accounts, identifying the types of standards one would use, and setting the standards. Cost centers, as defined earlier, are anything for which the costs are ascertained; they could be anything – a department, an agency, a cost item, a group of items, and even the personnel in an organization. Accounts are usually classified under different headings using symbols and codes, based on different items of expenses. A detailed list of these accounts can be easily found in the financial statements organizations prepare throughout the accounting cycle and present as a formal report at the end of the year, called the comprehensive annual financial report. For cost accounting, they will be expenses related to labor, materials, and overheads.

Types of Standards

There are no hard-and-fast rules for determining the level of standard one would use. Different types of standards are used depending on the purpose of cost measurement but, for convenience, we will look at five different types that are frequently discussed in standard costing: basic, current, normal, expected, and ideal.

The following provides a brief summary of each:

> **Basic Standards.** These are standards that remain unchanged for a long time in that they do not reflect the changes in cost, price, and other variables that affect the quality of a good or service; as such, they do not have much value in cost control.

Current Standards. These standards are periodically adjusted to reflect the actual change in a good or service resulting from changes in costs of labor, materials, and overhead; hence, they are considered more useful in cost control.

Normal Standards. Unlike the basic or current standards, these standards are based on averages that are tied to a business cycle, so periodic adjustments are necessary. Since they are tied to the business cycle, they are more useful for long-term planning.

Expected Standards. Also called target standards, these standards are based on expectations of what an organization would like to achieve at some future points in time. The targets are usually set in such a way that they can be achieved under normal conditions; as such, they are considered more realistic than most other standards.

Ideal Standards. These standards are based on what can be achieved under the most desirable conditions; for instance, if all the factors of production are working at their fullest capacity with no underperformance of labor, equipment breakdown, or power failure, which are difficult to attain; hence, the term ideal.

There is a further category of standard that one can add to the above, called historical standards, which are based on the averages of the past. While more realistic than the ideal standard, there is an inherent weakness in historical standards since they can include past inefficiencies which can be easily passed on to new standards. As such, they are not considered as useful as current or expected standards; nevertheless, they can serve as a good starting point.

Setting Standards

Once the type of standard an organization wants to use for cost control has been determined, the next step in the process is to set the standards to be used. The responsibility for setting standards usually falls on a committee or group of individuals with good knowledge of accounting processes, planning and budgeting, performance auditing, and organizational responsibility, among others. Standards are generally set for individual elements of a cost structure, such as direct material, direct labor, and overheads.

Direct Material Cost (DMC). DMC has two components – standard material quantity (SMQ) and standard material price (SMP), and is obtained by multiplying the standard quantity by the standard price; that is, $DMC = SMQ \times SMP$. For instance, if the standard quantity to be used in the production of a good is 10 units and the standard price for each unit is \$5, the standard material cost for the good will be \$50 [$\$5 \times 10 = \$50$].

Direct Labor Cost (DLC). Like DMC, DLC also has two components: standard time (ST) to complete a job and standard rate (SR) at which labor is paid, and is obtained by multiplying the standard time by the standard rate; that is, $DLC = ST \times SR$. Thus, if it costs 25 hours to complete a job and the price of labor is $20 per hour, both of which are standards for the job, the standard labor cost will be $250 [$20 × 25 = $250].

Standard Overhead Cost (SOC). SOC, like normal overhead, is based on two components – a fixed overhead rate (FOR) and variable overhead rate (VOR). The FOR is obtained by dividing the standard overhead for a given period, usually the budgeted period, by budgeted units or budgeted hours, while the VOR is obtained by dividing the variable overhead by budgeted units or budgeted hours.

Revising Standards

An interesting aspect of standard costing is that the standards are flexible in that they can be established for any amount of time and there are no set rules for setting a time frame. The rule of thumb is that if they are set for a short period, frequent revisions may be necessary which could be time consuming and costly. On the other hand, if they are set for a long period they may not be useful because of changes in conditions such as changes in inflation, technology, material, and labor rates, which could make the standards obsolete or redundant. This means that standards have to be revised from time to time to take into consideration the changing circumstances that have a direct bearing on the goods and services an organization provides.

Standards may also need to be revised to reflect any permanent changes in the methods of production and delivery process, changes in physical capacity to deliver the goods and services, and whether there exists a large variation between standard and actual. The point is: revision in standards is necessary to facilitate control over costs and improve the overall functioning of the organization.

Variance Analysis

Once the standards to be used in cost control have been determined, the next step in the process is to compare them against the actual cost; the method that is used for this purpose is variance analysis. As the name implies, variance analysis deals with the difference between actual and standard costs. Standard costs, as defined earlier, are predetermined costs based on what production or delivery costs should be under efficient conditions. Efficient conditions generally means that all the factors that go into the production or delivery of a service for an organization are working at their fullest (optimum) capacity. It is a theoretical construct and, as such, serves as a standard for comparing observations on costs and performance. Therefore, standards can be construed as reference points around which one would expect the actual costs to vary. If variances are a normal occurrence, which they are for most organizations, then performance should be acceptable as long as the variance is kept to a minimum, or as long as it is within some intervals around the standard (as in an interval estimation, discussed previously).

Two things are necessary to perform a successful variance analysis: (1) It must be tied to a specific activity so that one is able to collect relevant data on it;

and (2) the data collected must be reliable – that is, they must be consistent so that the same results can be obtained from repeated applications. The latter is necessary to ensure that any deviation a variance analysis produces is not due to unreliability of data. Theoretically, one could calculate as many types of variances as data would permit but, for the purpose of illustration here, we will focus on five: price variance, usage variance, efficiency variance, overhead variance, and budget variance.

Since a variance is the difference between actual and standard or target cost, we can formally express it using a simple algebraic expression:

$$Var_i = P_i^A Q_i - P_i^S Q_i \quad [5.1]$$

where Var_i is the variance of the ith item or good, P_i^A is the actual price for the ith item, P_i^S is the standard price of the ith item, and Q_i is the quantity of the ith item (for $i = 1, 2, 3, \ldots, n$).

The expression used in Equation 5.1 is called the price variance – a standard measure used in cost control. This section discusses five different types of variances: price variance, usage variance, efficiency variance, overhead variance, and budget variance. In practice, one could construct as many variances as one would deem useful for analyzing cost data.

Price Variance

Price variance is the difference between actual and market price. Two types of price variance are commonly used in variance analysis, one related to materials used in the production or delivery of a good or service, called materials price variance, and the other related to the price an organization charges for the goods and services it provides, called the service price variance. Cost accounting is primarily concerned with the former (i.e., the materials price variance).

Materials price variance is obtained by comparing the actual costs of purchase of materials with their market costs. The assumption here is that the market price of materials reflects a competitive price absent any distortions in the marketplace. If such distortions are present in a marketplace, the convention is to use the average price for materials as a surrogate or proxy for the market price. To give an example, consider a federal agency that purchased 500 units of a certain item, say, screwdrivers, at a cost of \$7,625. The price the agency paid for each unit, on average, is \$15.25, while the market price is \$12.50 per unit. At this price, the market cost of the units purchased will be \$6,250. Therefore, the materials price variance for the agency is \$1,375, obtained by taking the difference between the actual and the market price. That is,

$$\begin{aligned} MPV &= AP - MP \\ &= \$7{,}625 - \$6{,}250 \\ &= \$1{,}375 \end{aligned} \quad [5.2]$$

where MPV is the materials price variance, AP is the actual price paid for materials, and MP is the market price of materials. Assume for the sake of argument that the market price is the standard price; the difference between the two would indicate the amount the agency has overpaid, which is $1,375.

Usage Variance

Usage variance is the difference between the actual and the standard usage of materials. When the usage variance is multiplied by the price of the materials used, it produces the cost of overutilization (or underutilization) of materials. To give an example, suppose that the public works department of a government recently purchased 100 units of an item for a job that actually requires 80 units. The department used 95 units to complete the job, thereby producing a usage variance of 95–80 = 15. Note that the five units that remain unused become part of the inventory to be used at a later date (assuming that it will not become obsolete or lose its use value).

Assume further that the department spent $45 for each unit, with a total cost of $4,500 for the 100 units purchased. Since the department spent 15 units more than is considered standard for the job, the cost of overutilization to the department would be $675.00. That is,

$$
\begin{aligned}
CO &= (c \times UU) - (c \times UR) \\
&= (\$45 \times 95) - (\$45 \times 80) \\
&= \$4{,}275 - \$3{,}600 \\
&= \$675
\end{aligned}
\qquad [5.3]
$$

where CO is the cost of overutilization, UR is the units required, UU is the units used, and c is the cost per unit. Interestingly, we could have also obtained the same result by simply multiplying the cost per unit by the number of units overutilized; that is, $\$45 \times 15 = \$675$.

Efficiency Variance

This is really an extension of the usage variance involving the usage of labor time. It is called the efficiency variance because it compares the actual labor time with the planned or standard labor time. As with the usage variance, multiplying the observed variance by the price of labor (i.e., the wage rate) will produce the cost of over- or underutilization of labor. To give an example, consider a local government that takes an average of 25 minutes to dispose of one ton of solid waste. Suppose that the government spent 2,000 hours of labor time last year to dispose of 25,000 tons of waste which, based on the available standard time, would have taken 1,875 hours. The result produces an efficiency variance of 2,000 – 1,875 = 125 hours.

Assume now that the government paid $14.50 per hour for labor time, with a total labor cost of $29,000. Therefore, the total loss to the government due to overutilization of labor would be $1,812.50. That is,

$$
\begin{aligned}
LGO &= TCL - SCL \\
&= (\$14.50 \times 2{,}000) - (\$14.50 \times 1{,}875) \\
&= \$29{,}000.00 - \$27{,}187.50 \\
&= \$1{,}812.50 \qquad [5.4]
\end{aligned}
$$

where *LGO* is the loss to the government due to overutilization, *TCL* is the total cost of labor, and *SCL* is the standard cost of labor.

Overhead Variance

Overhead variance is the difference between actual and standard overhead costs. The computation of this variance is somewhat more involved than any of the other variances in that it includes both fixed and variable overhead costs. Take, for instance, a government that spent $25,500 in variable overhead costs and $74,500 in fixed overhead costs on a particular service, with a total overhead cost of $100,000. Let us say that it cost the government 750 hours in indirect labor to provide the service. The standard cost of labor is $18.50 per hour. The overhead variance, based on the above information, therefore, will be $11,625. That is,

$$
\begin{aligned}
OV &= AOC - SOC \\
&= \$100{,}000.00 - [(\$18.50 \times 750) + \$74{,}500] \\
&= \$100{,}000 - \$88{,}375 \\
&= \$11{,}625 \qquad [5.5]
\end{aligned}
$$

where *OV* is the overhead variance, *AOC* is the actual overhead cost, and *SOC* is the standard overhead cost. It is the amount the government has overpaid in overhead costs for the service.

Note that we left the fixed overhead cost unchanged because it is likely that it will be the same under both conditions. In the event that it is different, we can always adjust it by following the same procedure as the one we used for variable overhead.

Budget Variance

Budget variance is the difference between budgeted and actual cost of an operation. Budget variance is a common occurrence in most organizations, especially in government, where budgeting is a recurring exercise. Several factors contribute to this variance, such as poor cost estimates, fluctuations in the economy, cost overruns due to unforeseen events, and so on. To give an example, suppose that it cost a government $1.925 million to run a social program last year that was initially budgeted for $1.682 million. The budget variance for the government, obtained by taking the difference between the two, therefore, will be $0.243 million; that is, $1.925 million−$1.682 million=$0.243 million.

Although budget variance in cost accounting generally focuses on the cost side of an operation, it can also deal with the revenue or income side, in which case it is called an income variance. Income variance occurs when there is a deviation between budgeted (expected) income and actual income. For instance, consider a government that had expected to earn $8.375 million in income from a service facility last year, but instead earned $7.986 million. The resultant variance in income for the government would be $0.389 million; that is, $8.375 million−$7.986 million=$0.389 million.

A Note on Decisions Involving Variance Analysis

All variance analyses serve two important purposes. One, they help us determine whether a problem has occurred in the course of providing a good or service. Two, they help us trace the cost effect the problem will have on an organization. Therefore, a properly conducted variance analysis can not only identify where in the process a problem has occurred, but can also save money for an organization by determining the cost effect the problem will have on its operations.

Performing variance analysis involving real-world data costs time and money. However, most of these costs entail the costs of labor and capital one would incur in collecting, refining, and analyzing cost and related data for these analyses. Although data requirements for individual analysis are usually low, they can be quite substantial as the number of analyses to be done increases. This raises an interesting question: at what point should an organization decide to do a variance analysis? Since cost is the primary factor in this case, the answer will depend on the cost of doing an analysis, as opposed to not doing an analysis. In general, if the expected cost of doing an analysis is less than the cost of not doing an analysis, it should be worth undertaking.

We can formally express this relationship as

$$pC+(1-p)(C+V)<(1-p)D \quad [5.6]$$

where p is the probability that a serious problem has occurred, C is the cost of doing an analysis, V is the cost of correcting a problem (if it is discovered from variance information that a serious problem has occurred that needs to be fixed), D is the cost of not correcting the problem, and $(1-p)$ is the probability that a serious problem has not occurred. Note that $(1-p)$ is the complementary probability of p; therefore, the sum of p and $(1-p)$ must be equal to 1; that is, $p+(1-p)=1$.

Suppose now that we have the following information for an agency that wants to do a variance analysis: C=$750, V=$2,500, and D=$5,000. Assume that we do not have any information on p. Since we do not have any information on p, we can make a conservative assumption that it is 0.50 (i.e., there is only a 50 percent chance that no problem has occurred). This is an acceptable assumption in statistical analysis if it can be argued that the agency concerned is basically doing a good job. Therefore, substituting these values into the expression in Equation 5.6, we can determine whether the expected costs of doing the analysis will be lower than not doing the analysis. That is,

$$pC+(1-p)(C+V)<(1-p)D$$

$$\text{or } (0.5)(\$750)+(1-0.50)(\$750+\$2{,}500)<(1-0.50)(\$5{,}000)$$

$$\text{or } \$375+\$1{,}625<\$2{,}500$$

$$\text{or } \$2{,}000<\$2{,}500$$

where \$2,000 is the cost of doing the analysis and \$2,500 is the cost of not doing the analysis. Since the expected cost of doing the analysis is less than the cost of not doing the analysis, the analysis is justified.

Ratio Analysis

Like variance analysis, ratio analysis can be used to determine the discrepancy between actual and standard performance in order to determine the cost effect this deviation will have on an organization. A ratio is a simple mathematical relationship between two quantities. Ratio analysis has been extensively used in finance, going back to the 1920s, when it was first applied to DuPont Corporation and has since been used in other areas, including accounting. As in variance analysis, there is a wide range of ratios that one can use but, for convenience, this section discusses only three: efficiency ratio, capacity ratio, and activity ratio.

Efficiency Ratio

Efficiency ratios are used for both labor and materials usage. An efficiency ratio for labor usage shows the relationship between actual labor hours and standard labor hours, and is given by the expression

$$ER=\left(\frac{Standard\ labour\ hours}{Actual\ labor\ hours}\right)\times 100 \qquad [5.7]$$

where ER is the efficiency ratio.

To give an example, consider a government that has used 3,500 hours of labor for an activity, say Z, where the standard for the activity is 3,000 hours. Therefore, the efficiency ratio for the activity, based on this standard, will be 0.8571, or $(3{,}000/3{,}500)\times 100=85.71$ percent. In general, the ratio must be greater than or equal to 1, or 100 percent, to be efficient. Since the observed ratio in our example is about 14 percent [100–85.71 = 14.29] less than the acceptable ratio of efficiency, we can say that there has been inefficiency in labor use by the government by that amount.

Assume now that the standard cost of labor for the activity is \$10.25 per hour, then based on the total labor hours used the government has spent \$5,125 more than it should have; that is, (3,500 − 3,000) × \$10.25 = \$5,125.

Capacity Ratio

Capacity ratio, on the other hand, shows the relationship between the actual time spent on an activity and the budget time. Since the comparison is based on labor hours, it is also called capacity labor ratio, and is given by the expression

$$CR = \left(\frac{Actual\ time}{Budgeted\ time} \right) \times 100 \qquad [5.8]$$

where *CR* is the capacity ratio.

Unlike the efficiency ratio, to be efficient the capacity ratio must be less than or equal to 1, or 100 percent. This is because we are using actual time or hours spent in the numerator, which means we expect the actual time not to exceed the budgeted time. To give an example, let us go back to the problem we have just discussed. Assume that the budgeted time for the activity is 3,300 hours. Budgeted times are usually based on past experiences of an organization and are often used in efficiency calculations when standard times are not available or when an organization wants to set its own targets.

Given this information, it seems that the government has used 200 hours more labor than budgeted, thereby producing a capacity ratio of 106.06 percent; that is, (3,500 / 3,300) × 100 = 106.06. In other words, it has used 6.06 percent more labor to complete the activity than allocated. Translated into dollar terms, it means that the government has actually spent \$2,050 more in labor cost than it should have if it had strictly gone by the budgeted time; that is, (3,500 − 3,300) × 10.25 = \$2,050.

Activity Ratio

When an organization plans to undertake an activity, it is quite likely that its budgeted time or cost will not correspond to standard time or cost. This is due to the fact that organizations vary in their capacity when utilizing resources; in particular, labor. The activity ratio shows this relationship by comparing the standard time (or cost) with the budgeted time (or cost). It is given by the expression

$$ER = \left(\frac{Standard\ labour\ hours}{Actual\ labor\ hours} \right) \times 100 \qquad [5.9]$$

where *AR* is the activity ratio. Since it does not directly deal with actual time or cost, the activity ratio merely sets the limit for the budgeted time (or cost), which is used for calculating capacity and other ratios.

To give an example, let us go back to the same problem and use the same information we have used for calculating the capacity ratio. The result produces an activity ratio of 0.9091, or 90.91 percent; that is, $(3{,}000/3{,}300)\times 100=90.91$. What this means is that the government's allocation of time for the activity is 9.09 percent (i.e., $100.00-90.91=9.09$) above the standard time, which does not speak well for efficiency of the government. In cost terms, it means that the government's estimate of the budgeted cost exceeds the standard cost by \$3,075; that is, $(3{,}300-3{,}000)\times \$10.25=\$3{,}075$.

Both variance and ratio analyses have been extensively used, especially in business. As can be seen from the preceding examples, they are simple to construct yet extremely useful in helping an organization control its costs, in particular those related to materials, labor, and overheads. Perhaps the greatest advantage of using these measures is their low data requirements. Both variance and ratio analysis require limited amounts of information, meaning that the costs of doing these analyses are relatively low, as long as the number of analyses to be done remains limited.

Cost Internal Control

Cost internal control deals with a set of measures and guidelines that are important for an organization to ensure that its resources are used for the intended purposes. These measures are essentially the same for all organizations, although their application may vary. Some of these measures are simple, requiring very little effort and few resources, while others are exhaustive, requiring detailed and careful analysis throughout the process. We present here six such measures that are frequently used in cost control: responsibility assignment, expenditure control, cash control, fraud and waste control, performance auditing, and information management.

Responsibility Assignment

Responsibility assignment deals with the allocation of responsibilities to individuals in order to carry out the diverse tasks of an organization. Assigning responsibilities allows an organization to keep track of its activities consistent with its goals and objectives. This, in turn, makes it possible to hold individuals, in particular those who have been entrusted with such responsibilities, accountable when deviations occur. In a typical organization, the responsibilities follow the chain of authority with appropriate separation of tasks. For instance, the accounting and finance functions of a government are vested in the finance department, usually headed by a director. Under the director, one would typically find a treasurer, a financial planner, and an accountant or controller, plus a number of support personnel, each with a set of clearly defined responsibilities.

In general, the tasks individuals perform in an organization must be nonoverlapping to avoid any potential problem of duplication. For instance, the responsibility of a treasurer in a government is to manage its cash and other credit activities, including cash flow analysis, credit rating, and portfolio management. Similarly, the responsibility of a financial planner is to monitor financial conditions, raising funds needed for capital projects, and forecasting costs and technological changes, among others. Finally, the responsibility of an accountant or a controller is to maintain records of financial transactions in a government and control its financial activities. The latter includes responsibilities such as identifying deviations from planned and efficient performance, managing inventories, including fixed assets, payroll, and matters related to revenue, as well as information system and computer operations. The separation of responsibilities in this manner not only reduces redundancy in operation, but also cuts costs and makes coordination among various activities much easier.

Expenditure Control

Expenditure control is the first and foremost financial measure an organization uses to bring its expenditure within the limits of its resource capabilities. Expenditure control, especially in government, takes place in two ways: (1) through regulating timing of obligations and expenditures; and (2) publishing a number of financial and accounting reports on a regular basis (Lynch, 1995). Regulating timing ensures that funds are obligated for intended purposes and that their misuse is kept to a minimum. Most organizations use variance analyses, similar to the ones discussed earlier, to show the amount of funds an agency has received and the actual amount it has spent on a specific activity within a given period. Similarly, publishing financial and accounting reports is essential to show what kinds of expenditures an organization has incurred and how they compare with those intended. These reports, which should be published on a quarterly, if not weekly or monthly, basis, can also serve as principal source documents for the annual financial report that most financial organizations, including government, are required to publish at the end of each fiscal year.

Cash Control

Cash control is the means by which an organization ensures control of its cash (i.e., its cash receipts and cash disbursements). Cash control is extremely important in government, especially at the local level, where a government receives the bulk of its revenue during certain times of the year, while its expenditure takes place consistently throughout the year. This creates two operational problems: (1) a cash surplus when more revenues are received than expenditure incurred; and (2) a cash shortage when expenditures exceed revenues received. Cash control helps a government maintain a balance between the two by ensuring that it has enough cash to carry out its normal day-to-day operations, while making sure that any surplus that is generated in the process does not remain idle. When surpluses are accumulated, they are usually invested in securities, earning additional revenue for the government.

The simplest way to control cash is to do a cash flow analysis. As discussed previously, a cash flow analysis is a fairly detailed and time-consuming exercise. It provides three vital pieces information that are useful for financial planning of a government: cash inflow, outflow, and net flow, with the latter being the difference between the first two flows. Inflows include the flow of funds from all sources of revenue, while outflows include all expenditures that have been incurred or will be incurred in the future. The analysis serves an important purpose by identifying in advance the cash position an organization will be in, and how to deal with problems that it may encounter in the future, such as cash shortage (Khan, 1996).

Table 5.1 presents the cash flows for a local government based on information on monthly receipts and disbursements over a three-month period. According to the table, the net flow for the government at the end of the first period is –$147,000, which then becomes the beginning balance for the next period. In general, if the ending balance (i.e., the balance at the end of a period) is positive,

Table 5.1 A Sample Cash Flow Analysis

Cash Flows	*Monthly Flow ($1,000)*		
	January	*February*	*March*
Beginning balance (BB)	–	(147)	573
Cash receipts (inflows)			
Tax revenue			
Property tax	27,238	28,165	28,956
(Local) sales tax	6,285	5,987	5,823
(Local) income tax	7,347	7,549	7,450
Non-tax revenue			
Charges and fees	2,455	2,138	2,107
Fines and forfeitures	1,570	1,296	1,325
Intergovernmental revenue	5,216	6,295	6,794
Miscellaneous	1,079	1,512	1,928
Total inflows (excl. BB)	51,190	53,142	54,183
Cash disbursements (outflows)			
Salaries and wages	31,514	32,925	32,975
Materials and supplies	3,295	2,764	2,516
Debt service	5,760	5,760	5,760
Benefit payments	2,127	2,479	2,863
Capital outlay	7,295	7,387	7,956
Miscellaneous	1,346	1,085	1,215
(–) Total outflows	51,337	52,410	53,285
Changes in cash balance			
= Net cash flow	(147)	720	1,098
+ Beginning balance	–	(147)	573
= Ending balance (EB)	(147)	573	1,671

it adds to the total revenue for the next period; the converse is true when it is negative. The process can be repeated for as many periods as one wants. In reality, however, a cash flow analysis for more than six or eight months may not be useful because of the likelihood of error that increases with time in predicting future flows, although annual cash flow analysis has become quite common in recent years.

Fraud and Waste Control

Every year government loses billions of dollars from fraud, waste, and abuse at all three levels of government, but more so at the federal level. In fact, the Government Accountability Office, in its annual report, provides a detailed listing of government programs and activities where waste, fraud, and abuse have taken place during a fiscal year, costing the government billions of dollars. Although the government has undertaken measures to control the problem, unfortunately it remains a serious concern.

From the perspective of cost management, an organization can control fraud, waste, and abuse by taking measures that would eliminate waste, such as by clearly identifying where the problem has occurred, increasing the speed of response to minimize time, improving quality, and reducing costs. These measures constitute the essence of a procedure called lean operating system. Over time, a number of lean tools have been developed to deal with the problem; these include, among others, value-stream mapping, five sigmas, six sigmas and continuous improvement, and technology.

The following provides a brief description of each:

Value-Stream Mapping (VSM). Also called value mapping, this deals with a series of activities that can improve the delivery of a good or service to achieve a desired goal or outcome. The mapping allows an organization to carefully work out the flow of goods and services, from the point of origin to the point of destination (i.e., the delivery point), as well as the decisions that are necessary at each stage of the process to achieve the desired goal.

Five Sigmas (5σs). Originally used in Japanese industries, five sigmas assume that employees in an organization cannot carry out their job efficiently if the workplace is unkempt and disorganized. Sigmas, in statistical terms, mean standard deviations from a standard such as an arithmetic average. The assumption is that five standard deviations would cover most of the deviations under the curve, assuming a normal distribution. Operationally, an organization can minimize the deviations from the standard if it follows a number of steps, which, for five sigmas, is a five-step process: sort (each item is properly placed or identified), order (arrange materials and equipment in a way that is easy to locate), shine (maintain a clean workplace at all times), standardize (develop formal procedures that are consistent), and sustain (keep the process in place through proper training, effective communication, and organizational structure).

Six Sigmas (6σs) and Continuous Improvement. Like five sigmas, six sigmas has become an integral part of lean operations management to ensure continuous improvement. The lean system assumes a quality output (i.e., quality good or service), but to ensure its continuity it is often combined with six sigmas. Based on Japanese Kaizan philosophy, it goes beyond productivity improvement by suggesting measures that would humanize a work environment, so the employees will have the incentive to detect waste and help minimize costs.

Technology. Although the principle behind lean operations management is to keep the work environment simple, uncomplicated, and well structured, technology can ease that process as it becomes increasingly accessible and affordable, especially that which has become commonplace in industries. For instance, automation used in industries for reasoning, learning, and control can provide useful guidance for improving organization performance (*Business week Online*, 2000).

Performance Auditing

Auditing is a process of investigating the accuracy of the information a government provides on its financial, accounting, and other activities for a fiscal or accounting year. There are three specific objectives an auditing process serves for an organization: (1) determine the extent to which it has been fair and accurate in presenting its financial position; (2) determine whether or not it has complied with the stated laws, regulations, and mandates that influence its financial transactions; and (3) determine the extent to which it has been able to achieve its stated goals and objectives. The first two objectives together are called financial auditing, while the third is called performance auditing.

Financial auditing is based on a set of standards by which one can measure and evaluate the financial position of a government, while performance auditing is much broader in scope. There are no set standards that can accurately guide and measure the performance of an organization. However, by focusing on goals and objectives, performance auditing can address the broader question of efficiency and effectiveness; in particular, how an organization carries out its operations and the manner in which it utilizes its resources in achieving those goals and objectives (Brown, Gallagher, & Williams, 1982). In other words, by placing emphasis on goals and objectives and their achievement, performance auditing attempts to measure an organization's bottom-line performance.

Since performance auditing is primarily concerned with efficiency, it tends to pay more attention to the internal operations of an organization. This is where cost accounting becomes useful. As noted earlier, cost accounting produces cost and related information that is essential for an organization to carry out its day-to-day operations. Therefore, by producing accurate, objective, and reliable information, cost accounting can significantly ease the tasks of performance auditing.

Information Management

All organizations need to have a good information system in place to maintain quality internal control. Without good, reliable, and quality information and their management, most organizations will find it difficult to make effective decisions. Information management deals with the management of data that have been processed and analyzed to meet the specific needs of an organization. There is a difference between information management and a term interchangeably used with it, data management. Data management refers to the management of raw, unprocessed facts and figures that are gathered by an organization as part of its normal, everyday operation. As organizations increase in size and complexity, their information requirements also increase, thereby placing additional demands on information management. With an abundance of information and computing facilities these days, more organizations are focusing on what is known as intelligent use of information. By intelligent use of information, we mean cutting through redundant and useless information to obtain those that have a direct bearing on the functioning of an organization. To make intelligent use of information, one requires a clear understanding of the critical needs of an organization and development of an information base to meet those needs.

Planning and Budgeting

There is a direct relationship between cost accounting, planning, and control. Planning deals with the specification of goals, objectives, and missions of an organization and the development of plans to achieve them. An important component of a planning process is the development and execution of a budget. A budget is a plan for carrying out the routine and nonroutine activities of an organization and a commitment of resources to accomplish them. As a plan, it helps an organization coordinate these activities to ensure that they are achieved in an orderly and timely manner.

Two types of budget are generally prepared by most financial organizations: an operating budget and a capital budget. An operating budget, which is usually prepared on an annual basis, deals with routine, day-to-day operations of an organization. As such, it is concerned with materials, labor, overheads, expenditure, revenue, cash, and other routine activities. In contrast, a capital budget deals with nonroutine, long-term activities of an organization, such as long-term assets, liabilities, capital structure, liquidity, and so on. The two, however, are interlinked by a process in which the operating budget includes a capital component, called capital outlay, that deals exclusively with the capital expenditure for a budget year plus any other expenditure that is carried over from previous years. What makes this integration possible is the fact that capital budgets, although prepared for several years into the future, are divided into annual segments so that capital outlays for any given year can be easily incorporated into the operating budget for that year.

In addition to being a planning instrument, budgets are extremely useful as a control device. For instance, governments are often required either by law or by some established procedure to prepare their budgets according to a formal structure called the budget format. The format, which frequently varies between governments, provides specific guidelines regarding how a budget should be prepared, including setting goals, objectives, programs, and activities.

An important characteristic of all budgets, in particular an operating budget, is that once it has been approved and funds have been allocated for its execution, it must be followed through. This may appear somewhat rigid to those who may not be familiar with public budgeting, but there is a rationale for this apparent inflexibility. In the private sector, competitive market forces serve as a vehicle to ensure that the resources of an organization are efficiently managed and utilized. In other words, when a private firm or business becomes inefficient, it will not make enough profit to stay in business and will eventually be forced out of the market. Since no such forces exist in government, nonmarket mechanisms such as laws, regulations, mandates, and various compliance requirements are necessary to achieve the same objective.

A Note on Flexible Budgeting

For both operating and capital budgets, it is important that in the event the actual budget deviates from the planned or target budget resulting from unforeseen circumstances, the decision makers have some flexibility in fully and efficiently utilizing the resources of the organization. Although the end result of a budget process is a single budget that an organization uses to carry out its routine and nonroutine operations, it is useful to have several alternative budget proposals under varying conditions to deal with unforeseen changes in the budget variables. Flexible budgeting allows the decision makers flexibility in making efficient allocation decisions under changed circumstances.

As a concept, flexible budgeting is nothing new. It has been extensively used in the private sector for a long time to construct various cost scenarios affecting both variable and fixed costs; in particular, overhead costs. Since it deals with alternative budgets, there is a similarity between scenario construction and flexible budgeting, except that it applies to costs. To give an example, let us go back to the garbage collection example we used in Chapter 1 under cost-plus pricing. Let us say that it costs a government \$5.25 to collect a ton of garbage (variable cost), with a fixed cost of \$750,000. Assume that \$750,000 is the fixed overhead cost and the variable cost includes both direct and indirect variable costs. Assume further that the estimated garbage collection for next year is 1.2 million tons, which will produce the total cost of garbage collection for next year, $t+1$, of \$7,050,000; that is, $TC=FC+VC(Q)=\$750{,}000+\$5.25\ (1{,}200{,}000)=\$750{,}000+\$6{,}300{,}000=\$7{,}050{,}000$. Table 5.2 shows a breakdown of the costs under different scenarios. It shows the flexible budget for the department under three different conditions of collection: 1 million tons, 1.2 million tons, and 1.5 million tons.

According to the table, the department has two alternative budget proposals in the event the actual collection falls below or goes above the projected amount

Table 5.2 Flexible Budget: Solid Waste Collection

Cost Category	*Amount to Collect (Tons)*			*Variable Cost ($)/Ton*
	1,000,000	*1,200,000*	*1,500,000*	
Direct labor	$3,000,000	$3,600,000	$4,500,000	3.00
Direct materials	750,000	900,000	1,125,000	0.75
Variable overheads				
Indirect labor	600,000	720,000	900,000	0.60
Indirect materials	400,000	480,000	600,000	0.40
Utilities	350,000	420,000	525,000	0.35
Miscellaneous	100,000	120,000	150,000	0.10
Total variable cost	$5,200,000	$6,300,000	$7,800,000	5.20
Fixed overhead				
Personnel	$495,000	$490,000	$490,000	
Depreciation	80,000	80,000	80,000	
Utilities	50,000	50,000	50,000	
Miscellaneous	130,000	130,000	130,000	
Total fixed overhead cost	$750,000	$750,000	$750,000	
Total cost	$5,950,000	$7,050,000	$8,550,000	

of collection, with corresponding budgeted amounts. It is important to note that the variations in the proposed budgets are due entirely to the variations in the amount of solid waste the department is estimated to collect next year. Although the example looks at two alternative proposals, in reality one could construct as many proposals as feasible, as long as the amounts of waste the department plans to collect in the coming years are based on reliable estimates.

Chapter Summary

Cost control plays an important role in containing the costs of operation for an organization. In order to contain these costs, organizations use various management and accounting measures that would improve performance and reduce costs. The primary objective underlying the use of these measures is to monitor and evaluate cost performance by identifying where the problem has occurred, what contributed to the problem and, eventually, to control cost. The range of methods that can be used for this purpose has increased considerably over the years. This chapter has focused on five such measures – standard costing, variance analysis, ratio analysis, cost internal control, performance auditing, information management, and planning and budgeting – that are simple and relatively easy to use.

References

Brown, R. E., Gallagher, T., & Williams, M. (1982). *Auditing performance in government*. New York: Wiley & Sons.

Business Week. (2000). Thinking machines. *Business Week Online*. August 7. Retrieved August 7, 2014, from www.businessweek.com/archives/2000/b3693096.arc.htm.

Clark, F., & Lorenzoni, A. B. (1998). *Applied cost engineering*. New York: CRC Press.
Kerzner, H. (2009). *Project management: A systems approach to planning, scheduling and control*. New York: Wiley & Sons.
Khan, A. (1996). Cash management: Basic principles and guidelines. In J. Rabin, W. Bartley Hildreth, & G. Miller (Eds.), *Budgeting: Formulation and execution* (pp. 313–322). Athens, GA: Carl Vinson Institute of Government, University of Georgia.
Lynch, T. D. (1995). *Public budgeting in America*. Englewood-Cliffs, NJ: Prentice Hall.
Ziegenfuss, Jr., J. T., & Bentley, J. M. (2000). Implementing cost control in health care: Strategies driven by organizational systems approach. *Practice and Systems Research, 13*(4), 453–474.

Part II

Optimization in Government

6 Classical Optimization

Optimization is a natural human tendency. When we encounter a problem, our mind automatically dissects it, evaluates alternatives for possible courses of action, and lets us choose the best option. This sequence of events (dissection, evaluation, and action) constitutes the basis of optimization. Optimization can be qualitative or quantitative. Qualitative optimization involves individual judgments and preferences, such as a sociologist predicting social unrest in a community resulting from a political decision, or a financial analyst predicting the most likely return a government could earn from a security portfolio based on their own personal experience, knowledge, and skill. Quantitative optimization, on the other hand, requires precise mathematical rules to produce the best result.

As a concept, optimization is nothing new. The ancient Greeks, Romans, and Egyptians are known to have used it extensively in many of their works of art, architecture, and astronomy. Optimization plays just as important a role in government today as it did centuries ago. The use of optimization as a means to improve performance can be found in almost every sphere of governmental activity. As the public demand for lean and efficient government continues to increase, so does the need for better and more efficient optimization tools to deal with the complex tasks of everyday operations in government. This chapter provides a brief discussion of the relationship between costs, operations management, and optimization; in particular, it focuses on the nature of optimization and areas of operations and cost management where optimization plays a critical role, such as inventory management, queuing, and simulation, which, for lack of better terms, we will call subsets in optimization.

Optimization in Cost and Operations Management

All government activities fall into three broad categories: operations, finance and accounting, and delivery. Operations include day-to-day activities of an organization that are often wide and varied. They can range from purchasing and acquisition, to inventory, to maintenance, and other support functions that frequently consume substantial resources – financial, as well as human resources. Managing these functions, known otherwise as operations management, is critical to effective functioning of an organization. Operations management deals with planning, control, logistics, and other measures that are necessary to ensure that these everyday activities are carried out efficiently. Put simply, a well thought-out operations management that pays careful attention to these activities can not only improve organizational performance, but can also produce substantial cost savings for an organization.

Conceptually, one can think of operations management as consisting of activities that convert inputs such as labor, capital, materials, entrepreneurship, and other resources into outputs (i.e., goods and services). From a cost perspective, the more efficient the process in converting the input resources into outputs, the more savings it will produce for an organization. Over time, there has been a concerted effort by organizations to undertake measures and use tools that would improve their performance. These range from simple qualitative methods such as lean management, to basic descriptive statistics such as measures of central tendency and dispersion, to complex analytical tools such as linear programming, CPM and PERT, queuing theory, inventory management, and game theory, among others. The underlying objective behind these and other techniques used in operations management, regardless of their analytical sophistication, is to optimize the management process that would lead to greater efficiency, productivity, and cost savings for an organization. In that regard, optimization in costs and operations management remains at the heart of all organizational activities – public and private.

At a more specific level, optimization plays an equally important role in finance, accounting, and delivery. Finance deals with the management of financial resources; in particular, money (i.e., revenue for government) such that it is available at all times to carry out the activities of an organization without having to incur debt, raise taxes, or run a shortfall. Optimization provides the tools necessary to manage those resources efficiently. Accounting is defined as the process of collecting, analyzing, summarizing, and disseminating cost and related information to those who would directly benefit from them. Optimization can help an organization carry out these and other activities efficiently, including those related to cost allocation, management, and control. The ultimate objective of an organization, including government, is to provide goods and services efficiently. Optimization can help to achieve that goal by finding ways to deliver goods and services at the lowest cost without significantly affecting the quality or quantity of provision.

Nature of Optimization

In conventional parlance, optimization means maximizing or minimizing an objective function such as maximizing revenue for a government or minimizing the cost of operation for a service agency. To a mathematician, optimization does not necessarily mean maximization or minimization, since there may be more than one mathematical optimum in a given problem. In general, a mathematical optimum occurs, for a single variable, at a point on a curve where the slope (defined as the change in a variable due to a change in another variable [i.e., $\Delta X/\Delta Y$] where Δ represents the change) is zero. This is a necessary but not a sufficient condition for mathematical optimum, as it may be an inflection point (i.e., the point at which the curve takes a sudden turn).

There is an important criterion that all mathematical optima must satisfy in that the variables to be optimized must be continuous and differentiable at every point on a curve. For instance, the points A, C, and F on the curve in Figure 6.1 represent mathematical maxima and the points B and E represent mathematical minima. Of these points, only C is considered a true maximum and E a true minimum since they have respectively the highest and lowest points on the curve. None of the other points represent a true optimum since $C>A$ and F and $E<B$. These points are called local optima. The remaining point on the curve, D, is neither a mathematical maxima nor a mathematical minima; it is a point of inflection.

In dealing with optimization problems it is important that the function to be optimized is clearly specified so that no error can result from incorrect specifications. Once the function is properly specified, the result or results can be carefully analyzed for their accuracy and reliability. For instance, an answer may be mathematically correct but may not be considered acceptable in the sense that it may fail to produce the intended results. This can pose a difficulty for the analysts dealing with the problem, but it is an important consideration that must not be overlooked by those who will use the results. Where the results derived from optimization seem unlikely, it may be necessary to do some sensitivity analysis.

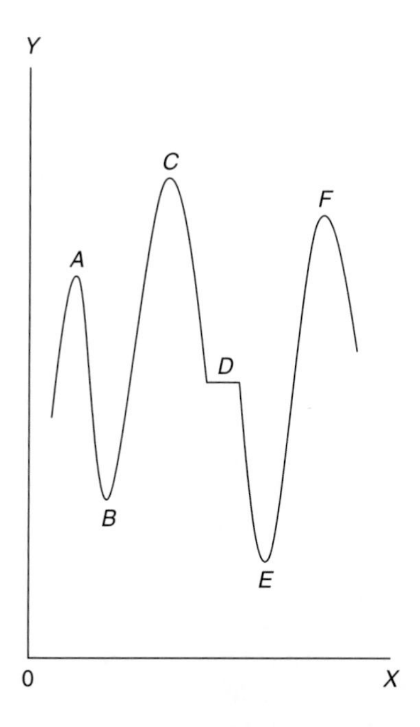

Figure 6.1 Multiple Mathematical Optimal Points

This means changing all or some of the values of the parameters of the model to see what kinds of results they will produce for the optimal solution. Sensitivity analysis is important in situations in which one needs to have some flexibility in terms of the solution the changes in model parameters produce for a problem.

Classical Optimization

All optimization problems begin with a clearly defined goal or objective called the objective function. An objective function is a statement of a mathematical relationship between a dependent and one or more independent variables. When more than one independent variable is involved in an optimizing function, it is called multivariate optimization. The purpose of optimization is to determine the value or values of the independent variable(s) that will maximize or minimize the value of a dependent variable. In traditional or classical optimization, the method commonly used to find the set of values of the independent variables that will yield the optimal value for the dependent variable is differential calculus. Differential calculus is particularly suitable for optimization problems where the objective function is at least twice differentiable within a feasible region (i.e., the space containing the optimum value of the objective function).

Frequently an optimization problem will include constraints that will determine the boundary of the feasible region. Simply put, constraints impose limitations on the feasibility of an objective function. An optimization problem with restrictions or constraints is called constrained optimization, and without constraints is called unconstrained optimization. When differential calculus is applied to optimization problems without constraints, it entails direct differentiation of the objective function. On the other hand, when it is applied to problems with constraints, the constraint equations are solved first for the independent variables, which are then substituted back into the objective function before differentiation is performed.

Unconstrained Optimization

Unconstrained optimization problems are much easier to deal with than constrained optimization because they do not involve any restrictions that can impose limitations on the realization of an objective function. This is generally true when one is dealing with single rather than multiple independent variables. To give an example, consider a situation in which the federal government plans to introduce a monthly job-training program for those currently unemployed to lessen their dependence on welfare. The training is also expected to produce a direct benefit to the government by saving money when an individual is taken off the welfare roll (assuming the individuals receiving training will be able to find gainful employment once they have gone through the program). Obviously, the more individuals that receive training, the greater will be the savings to the government. Let us assume that we have a savings function for the program that can be represented by a linear equation of the form:

$$S = 1{,}000 + 5{,}200Q \quad [6.1]$$

where S is the savings in dollars and Q is the number of individuals to receive training.

There is, however, a cost to the government of running the program. Assume further that it is given by a quadratic function:

$$C = 1.5Q^2 - 2{,}300Q + 5{,}000 \quad [6.2]$$

where C is the program cost in dollar terms and Q is the number of individuals to receive training, as before.

The objective of the government is to determine the number of individuals receiving training at which the net return (defined as the difference between savings and costs) will be the maximum. That is,

$$\begin{aligned} \pi &= S - C \\ &= 1{,}000 + 5{,}200Q - 5{,}000 + 2{,}300Q - 1.5Q^2 \\ &= -\ 1.5Q^2 + 7{,}500Q - 4{,}000 \quad [6.3] \end{aligned}$$

where π is the net return (net savings).

To find the quantity at which the net savings will be maximum, we first rewrite the function in Equation 6.3, and then take the first derivative of the function, called the necessary condition, and set it equal to 0. That is,

$$\begin{aligned} \pi &= -1.5Q^2 + 7{,}500Q - 4{,}000 \\ d\pi / dQ &= -3Q + 7{,}500 = 0 \\ \therefore\ Q &= 2{,}500 \quad [6.4] \end{aligned}$$

Thus, if the government is to maximize the net savings from operation of the program, it must train 2,500 individuals per month which, in turn, will generate a net saving of \$9.371 million, obtained by substituting the value of Q into Equation 6.4, as shown below:

$$\begin{aligned} \pi &= -1.5Q^2 + 7{,}500Q - 4{,}000 \\ &= -1.5(2{,}500)^2 + 7{,}500(2{,}500) - 4{,}000 \\ &= -9{,}375{,}000 + 18{,}750{,}000 - 4{,}000 \\ &= \$9{,}371{,}000 \end{aligned}$$

To further determine if the net saving is a true maximum, called the sufficient condition, we take the second derivative ($d^2\pi/dQ^2$) of the result obtained from the first derivative of the return function. There are three basic conclusions that can be drawn from the second derivative of the differentiated function: (1) if the second derivative is negative, it is a maxima; (2) if it is positive, it is a minima; and (3) if it is zero, it is a point of inflection. Thus, if we take the second derivative of the differentiated function, it produces a value of –3. That is,

$$d^2\pi/dQ^2 = -3$$

Since $-3 < 0$, we can say that the net savings is a maximum.

The example we used to solve the optimization problem deals with a single independent variable, but the process can be applied to multiple independent variables. The only difference is that it becomes operationally more complex when more than one independent variable is involved. Let us look at another example, this time with two independent variables. Suppose that a state correctional agency is concerned about the rising cost of running its juvenile crime prevention program. Although most taxpayers believe that juvenile crime is a serious problem that needs to be addressed, they also share the concern of the agency that the current trend in cost increase must be contained before it gets out of hand. Assume, for the sake of argument, that the agency attributes the problem to two major factors: the unemployment rate in the state (which, let us say, has been above the national average), and an increase in single-parent homes in recent years. We can operationalize the problem by defining a simple cost function for the program in terms of these two basic variables. That is:

$$c = 42.5 + 10U + 5P - 10UP + 2.5U^2 + P^2 \qquad [6.5]$$

where C is the cost to the agency of running the program, U is the unemployment rate, P is the percentage of single-parent homes in the state, and UP is an interactive term that includes both unemployment rate and single-parent homes. Assume that the coefficient of the terms in the equation are given in millions of dollars.

The objective of the agency is to find the value of U and P at which the cost to the government will be a minimum. To do so, we take the first derivative of C with respect to U and P, set them equal to zero and then solve them simultaneously to obtain their respective values. That is:

$$\partial C/\partial U = 10 - 10P + 5U = 0$$

$$\partial C/\partial P = 5 - 10U + 2P = 0$$

Now, solving the two equations simultaneously we get $U = 0.778$ and $P = 1.389$.

Next, we substitute these values into Equation 6.5 to obtain the value of C, which produces a total of \$49.86 million. That is:

$$C = 42.5 + 10U + 5P - 10UP + 2.5U^2 + P^2$$
$$= 42.5 + 10(0.778) + 5(1.389) - 10(0.778)(1.389) + 2.5(0.778)^2 + (1.389)^2$$
$$= 42.5 + 7.78 + 6.945 - 10.806 + 1.513 + 1.929$$
$$= \$49.86 \text{ million}$$

This is the minimum cost to the government of running the program.

To ensure that the cost is minimum at $U = 0.778$ and $P = 1.389$, we apply the second-order condition to the results we obtained from the first derivative of the cost function, including the interactive term, *UP*. As shown below:

$$\partial^2 C / \partial U^2 = 5$$
$$\partial^2 C / \partial P^2 = 2$$
$$\partial^2 C / \partial U \partial P = 10$$

Of the three second partials, the last one, $\partial^2 C / \partial U \partial UP$, called the cross-partial derivative, is used here for the interactive term, *UP*. The cross-partial derivative measures the ratio of the first partial derivative of the dependent variable with respect to the product of the first partial derivatives of the independent variables. In general, the square of cross-partial derivatives must be less than the product of the second derivatives for the value of the objective function to be a maximum. Conversely, it must be greater than or equal to this product for it to be a minimum. Since the second partial derivatives are positive and the product of the second partial is less than the square of the cross-partial derivative, the cost is minimum: that is, $(5)(2) < 10)^2 = 10 < 100$.

Constrained Optimization

Most optimization problems are associated with constraints that determine the extent to which an objective function will be a maximum or a minimum. The approach commonly used to solve a constrained optimization is the well-known Lagrangian method. The method simply states that given a function $f(x,y)$, subject to a constraint, $g(x,y)$, a new function can be constructed by setting the constraint equal to zero, multiplying it by λ, called the Lagrangian multiplier, and then adding the product to the original function. That is,

$$F(x,y,\lambda) = f(x,y) + \lambda g(x,y) \qquad [6.6]$$

where $f(x,y,\lambda)$ is the Lagrangian function, $f(x,y)$ is the initial function to be optimized, and $\lambda g(x,y)$ is the constraining function.

Since the constraint is always set equal to zero, $\lambda g(x,y) = 0$, the addition of the term does not change the value of the objective function. The solution values of x, y, and λ at which the function is optimized are obtained by setting the first plural derivative equal to zero, and then solving the equations simultaneously. We can look at a simple example to illustrate this. Suppose that the US Treasury Department wants to introduce two new commemorative coins, X and Y, to raise revenue for the preservation of a national monument that needs some major improvements. However, there is a cost associated with the production of the coins that the department must bear initially. Let us assume that we have a cost function for the department for coin production, given by the following equation:

$$\mathrm{C}=X^2+0.5XY+Y^2 \qquad [6.7]$$

where C is the cost of minting the coins.

Assume further that the department is restricted to minting a total of 1,500 coins to preserve their value over time. We can express this restriction or constraint on production as

$$X+Y=1{,}500 \qquad [6.8]$$

Our objective is to find the value of X and Y that will minimize the cost function, subject to this production constraint. To solve the problem we first multiply the constraint by the Lagrangian multiplier and then add the product to Equation 6.7. The complete equation can be written as

$$C-X^2-0.5XY+Y^2+\lambda\,(X+Y-1500) \qquad [6.9]$$

As before, we take the first partial derivative of C with respect to X, Y, and λ, and set them equal to zero. That is,

$$\partial C/\partial X=2X-0.5Y+\lambda=0$$

$$\partial C/\partial Y=-0.5X+2Y+\lambda=0$$

$$\partial C/\partial \lambda=X+Y-1{,}500=0$$

Now, solving the equations simultaneously, we get $X=750$, $Y=750$, and $\lambda=-1{,}125$. Therefore, the minimum cost the department will incur to produce the coins is $843,750, obtained by substituting the values of X and Y into Equation 6.7, as given below:

$$C=X^2-0.5XV+Y^2=750^2-(0.5)(750)(750)+(750)^2=\$843{,}750$$

The Lagrangian multiplier, λ, has a special meaning in this context which can be interpreted as the effect the objective function will have due to a unit change in the constant of the constraint function. If, for instance, λ is positive, a one-unit increase (decrease) in the constant will result in a decrease (increase) in the objective function by a value approximately equal to the value of λ. If λ is negative, the opposite would be true. Thus, with $\lambda=-1{,}125$, a one-unit increase in the production quota will lead to a decrease in cost by approximately $1,125.

The second-order test for determining whether an observed value is at a minimum for a constrained optimization is different from the one we used for an unconstrained optimization. The test most commonly used for constrained optimization is the bordered Hessian. A Hessian is a determinant, $|H|$, consisting of second-order partial derivatives where the direct partials are represented on the main diagonal and the cross-partials on the off-diagonal. When a Hessian is bordered by the first derivatives of the constraint with a zero on the principal diagonal, it is called a bordered Hessian, as shown below:

$$|H| = \begin{vmatrix} F_{xx} & F_{xy} & g_x \\ F_{yx} & F_{yy} & G_y \\ G_x & g_y & 0 \end{vmatrix} \qquad [6.10]$$

where the terms F_{xx} and F_{yy} represent the second partials, F_{xy} and F_{yx} the second cross-partials, and g_x and g_y the first partials of the constraint from Equation 6.9.

According to the test, if the determinant is negative – that is, if $|H|<0$, called negative definite – the optimizing function has a minimum. On the other hand, if the determinant is positive – that is, if $|H|>0$, called positive definite – the function is at a maximum.

The second-order conditions for our constrained optimization problem produce the following results:

$$F_{xx}=\partial^2 C/\partial X^2=2$$

$$F_{yy}=\partial^2 C/\partial Y^2=2$$

$$F_{xy,yx}=\partial^2 C/\partial X \partial Y=-0.5$$

$$g_x=\partial\lambda/\partial X=1$$

$$g_y=\partial\lambda/\partial Y=1$$

The differentials can now be organized in a matrix form to obtain the determinant of the bordered Hessian. That is,

$$|H| = \begin{vmatrix} 2 & -0.5 & 1 \\ -0.5 & 2 & 1 \\ 1 & 1 & 0 \end{vmatrix}$$

To find the determinant, we use a simple procedure known as the Laplace expansion, although other more conventional procedures could have been used. Laplace expansion is a method for evaluating a determinant in terms of cofactors (submatrices with plus and minus signs). It has an advantage over the conventional methods in that it allows the evaluation of a determinant along any row or column, which greatly simplifies the amount of computation that is necessary to solve a problem. The Laplace solution to the determinant produces a negative value for the Hessian, as can be seen from the following expressions:

$$\|H| = \begin{vmatrix} 2 & -0.5 & 1 \\ -0.5 & 2 & 1 \\ 1 & 1 & 0 \end{vmatrix}$$

$$= 2(-1)^{1+1}\begin{vmatrix} 2 & 1 \\ 1 & 0 \end{vmatrix} - 0.5(-1)^{1+2}\begin{vmatrix} -0.5 & 1 \\ 1 & 0 \end{vmatrix} + 1(-1)^{1+3}\begin{vmatrix} -0.5 & 2 \\ 1 & 1 \end{vmatrix}$$

$$= 2(-1) + 0.5(-1) + 1(-2.5)$$

$$= -2 - 0.5 - 2.5$$

$$= -5$$

Since $|H| < 0$, it is a negative definite and fulfills the second-order condition for a minimum. Therefore, the cost function, C, has a minimum.

Sensitivity Analysis

In most optimization problems, it is important to know the final outcome, in particular how sensitive the outcome is to a specific change in the values of the parameters, as well as of the independent variables that make up the problem. To illustrate, let us go back to the net return function for the job-training program problem.

$$\pi = -1.5Q^2 + 7{,}500Q - 4{,}000$$

where π and Q are respectively the dependent and the independent variables, and 1.5 and 7,500 are the values of the parameters associated with the dependent variable Q, we will call $a = -1.5$ and $b = 7.500$.

To see how sensitive π_{opt} is to a change in any of the terms on the right-hand side of the return function, we begin by taking the derivative of π with respect to Q, so that

$$d\pi / dQ = -3Q + 7{,}500 = 0$$

$$\therefore\ Q_{opt} = 2{,}500$$

as shown previously.

Next, we find a change in this value for a change, say, in the parameters of the function. Let us say that we have two such values for a and b, given by $a = 2$ and $b = 8{,}000$. Now, substituting these values into the return function we can obtain the changes in the optimum value of π. That is,

$$\pi = -2Q^2 + 8{,}000Q - 4{,}000$$

$$d\pi / dQ = -4Q + 8{,}000 = 0$$

$$\therefore Q_{opt} = 2{,}000$$

which is the new optimum value of Q.

The result presented above indicates the amount by which the optimum value of π will change for a change in the respective values of the parameters, a and b. In other words, this is how sensitive the optimum value of π is to a change in the values of the parameters of our quadratic return function. We can now apply this value to obtain the net return for the government, as given below:

$$\begin{aligned}\pi &= -\,2Q^2 + 8{,}000Q - 4{,}000 \\ &= -\,2(2{,}000)^2 + 8{,}000(2{,}000) - 4{,}000 \\ &= -\,8{,}000{,}000 + 16{,}000{,}00 - 4{,}000 \\ &= \$7{,}996{,}000\end{aligned}$$

which is approximately $8 million.

As before, to determine if the return is a true maximum we could have taken the second derivative of the result obtained from the first derivative of the return function, which would have given us a value of –4. Since it is negative, by definition, then the net return is a maximum. Similarly, we could have changed the value of the independent variable, Q, to a different figure than the one we used in the original expression to see how sensitive the net return, π, is to this change, and then follow similar steps to determine if the net return is maximum.

Subsets in Optimization

Optimization problems often use language and symbols of mathematics that are deterministic in nature. By deterministic, we mean the values of the terms used in an equation that are predetermined or known with certainty, but there are circumstances when it is not possible to obtain the values of some or all of the terms with certainly. In other words, they are probabilistic rather than deterministic. A vast majority of problems in social and behavioral sciences fall into this category. Therefore, traditional optimization techniques with all their characteristic requirements cannot be readily applied in these cases. What we have instead are problems that can only be partially optimized. As such, they are often regarded as subsets in optimization and treated as suboptimization problems. This section presents two well-known suboptimization problems frequently used in operations management: inventory and queuing.

Inventory Models

Like any firm or business, a government needs to maintain an inventory of materials and supplies that would meet its normal operational needs. Inventories will not be required if materials and supplies are available instantaneously, but they are not. Thus, it becomes necessary for a government to maintain an appropriate level of inventory so that there will not be any over- or undersupply of inventory. In determining what will be the most appropriate or optimum level of inventory, a government must take into consideration three sets of costs associated with an inventory: ordering cost, carrying cost, and stockout cost.

Ordering costs are related to the acquisition of materials and supplies that will eventually become an inventory or part of an existing inventory. These costs are incurred each time an order is placed with a supplier (vendor). Ordering costs mostly include the costs of preparing specifications as to the types of materials to be purchased, obtaining bids (a standard practice in government), negotiating contracts, and receiving items. There is often confusion between ordering cost and a term used interchangeably with ordering, called set-up cost. The two are not exactly the same; the latter deals with actual placement or installation of an ordered item such as setting up a network system after it has been acquired.

Carrying costs, also known as holding costs, are incurred once an organization has decided that it must maintain a certain amount of inventory, but there are costs associated with holding inventories if they are not going to be immediately used or consumed. These costs largely include costs of insurance, as well as premiums paid on insurance, storage, utility, plus any other costs associated with technological obsolescence, depreciation from normal wear and tear, and unexplained losses. Unfortunately, because of the imprecise nature of many of these factors, it is often difficult to determine an accurate estimate of these costs. As a result, organizations often calculate them on an annual basis and express them as a percentage of the average inventory value.

Finally, when an organization fails to undertake or complete a job for lack of supplies, it produces a cost to that organization that is directly attributable to this lack of inventory. Stockout or shortage costs result when an organization fails to meet the demand of its clients, or when it fails to function effectively for shortage of stocks. The determination of stockout costs is often approximate because it is not always possible to measure the exact amount of inconvenience a delay can cause and how to assign a dollar value to it. However, this does not mean they are insignificant or should not be taken into consideration in calculating the total cost of an inventory.

Thus, in designing an optimum inventory policy, the objective of a government should be to keep all three costs to a minimum. In other words, the objective should be to minimize (1) the ordering costs, (2) the carrying costs, and (3) the frequency of stockout costs which, in turn, will minimize the total cost of inventory.

A Simple Inventory Model

The simplest case that can be made for an inventory policy is to assume that: (1) when stock runs out it will be instantaneously replenished; (2) usage takes place at a constant rate; (3) there is no safety stock, i.e., no additional inventory that could be held as a protection against any possible shortage or stockout; and (4) the cost of ordering is constant and independent of lot (batch) size. Given these assumptions, we can illustrate a simple inventory model with the help of a schematic diagram, as in Figure 6.2, called a sawtooth diagram because it resembles a saw.

As shown in Figure 6.2, when an order is received the inventory level is Q units, at point Q. Since the inventory is used at a constant rate, it decreases by exactly the same amount, given by the negative slope of the straight line. B is the reorder point, meaning that when the inventory reaches point B a new order should be placed. The points b, d, and f represent the levels at which new inventories are received (given by the vertical lines) to replenish the stock. Since replacement is instantaneous, there are no shortages (stockout costs) according to the model. In reality, however, stocks may not be instantaneously replaced, indicating that shortages will occur, producing stockout costs before they can be replenished. To ensure that no such shortages take place, a government can maintain a safety stock, shown by the vertical distance OB. Thus, whenever the inventory reaches the reorder point a, c, and e, new orders should be placed to replenish the stock. This will eliminate the possibility of stockout costs.

Economic Order Quantity

A major weakness of the simple inventory model such as the sawtooth diagram is that it does not guarantee an optimum quantity. An optimum inventory ensures that it can be obtained in such a way that the total cost of maintaining the inventory to

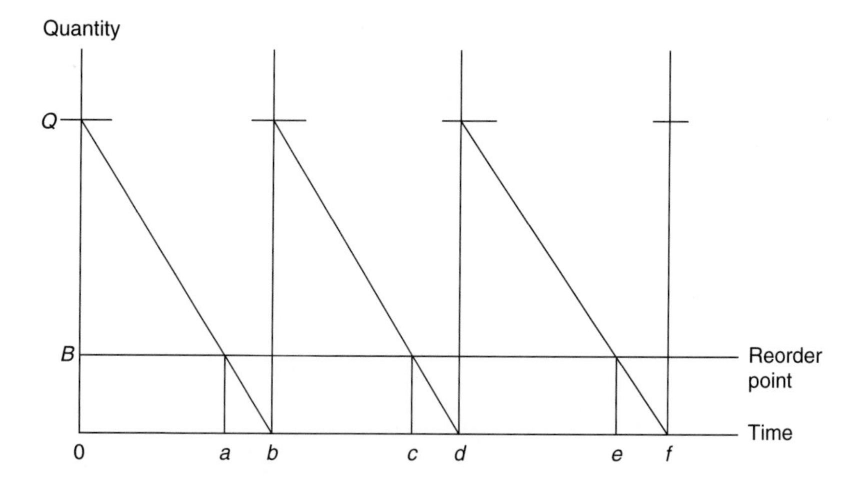

Figure 6.2 A Simple Inventory (Sawtooth) Diagram

the government will be minimum. In order to determine an optimum inventory, we introduce a model that was originally suggested by Baumol (1952), called an economic order quantity (EOQ). The model attempts to determine an order quantity (batch size) that minimizes the total cost of ordering plus carrying (holding) an inventory, based on the assumption that the demand for inventory is already known.

We can demonstrate the basic argument of the model with a simple example. Suppose that a government needs 250 units of a certain item each year. Theoretically, this could be any length of time (a week, a month, or a quarter) but it is convenient to express the requirement in annual terms. Let us say that the cost per unit, which is the purchase cost, is $20, with an ordering cost of $12.50 per order, and an annual carrying cost that is 20 percent of the value of the average inventory. The latter is frequently expressed as inventory carrying cost percentage (I_p).

Table 6.1 presents this information, along with the total cost of inventory for various order quantities. For instance, for a lot size of 250 with one order per year, the total cost of inventory is $512.50, obtained by adding the ordering and carrying costs. The table also shows the order or lot size at which the total cost will be minimum, which is 62.5 or 63 (after rounding off). It is important to note that this minimum cost occurs at a point where the carrying cost of inventory equals the ordering cost. In other words, this is the optimum point and the quantity corresponding to it is the optimum quantity.

This relationship between the various cost categories (total, carrying, and ordering) and the optimum quantity, Q^*, are also presented in Figure 6.3. As the figure shows, the total cost curve initially slopes downward with order quantity attaining the lowest point, where the ordering and the carrying cost curves intersect, and then increases as the order quantity increases. The lowest point is the optimum point and the cost corresponding to it is the optimum (minimum) cost. Note that the curve representing carrying cost slopes upward rather than downward, indicating that as the number of units held in inventory increases, the carrying cost also increases.

The optimum cost and quantity figures we obtained so far are based on a trial-and-error approach. In reality, we can determine these optimum values mathematically, provided we have information on cost per unit (assumed to be fixed),

Table 6.1 Order, Lot Size, and Cost Data for EOQ Model

Number of Orders	*Lot Size*	*Average Inventory*	*Carrying Cost ($)*	*Ordering Cost ($)*	*Total Cost ($)*
1	250	12	500.00	12.50	512.50
2	200	100	400.00	25.00	425.00
4	150	75	300.00	50.00	350.00
6	100	50	200.00	75.00	275.00
10	62.5	31.25	125.00	125.00	250.00
12	48	24	106.00	150.00	256.00
20	24	12	48.00	250.00	298.00
24	10	5	20.00	300.00	320.00
28	4	2	8.00	350.00	358.00

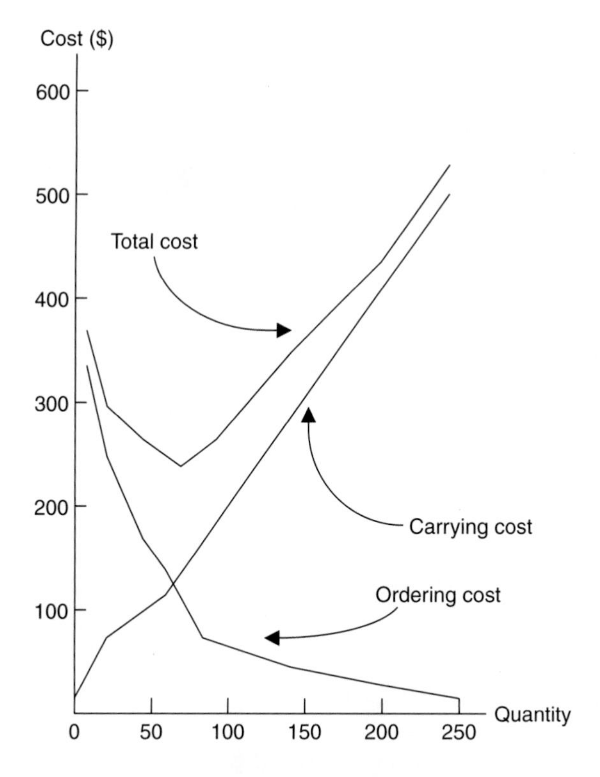

Figure 6.3 A Graphical Presentation of EOQ

carrying cost (assumed to vary with average inventory), ordering cost (assumed to be fixed per order), and total demand (known with certainty). As long as these conditions hold, we can have a total cost function that can be solved with simple differentiation to produce the optimum quantity.

Let us say that we have a total cost function, given by the expression

$$TC = (Q/2)h + (D/Q)\mathrm{O} \tag{6.11}$$

where TC is the total cost of inventory, Q is the order quantity, h is the per unit holding or carrying cost, O is the ordering cost, and D is the total demand.

The first part of Equation 6.11, $(Q/2)h$, represents the total carrying cost for a unit over a given period, where the term $Q/2$ represents the average inventory level for the same period. The second part, $(D/Q)O$, represents the total ordering cost for quantity Q, where the term D/Q represents the total number of orders The total cost of inventory per unit of time, therefore, is the sum of these two costs: total carrying and total ordering.

Theoretically we could have added a third component to the total cost function to take into account the cost of purchase. The cost of purchase is generally obtained by multiplying the total demand for a good or service, D, by the cost per unit, c; that is, $D \times c$. The total cost function, with the purchase cost included, can thus be written as:

$$TC = (Q/2)h + (D/Q)O + Dc \qquad [6.12]$$

In practice, however, the term Dc is left out of the cost function because the purchase cost is not considered a decision variable and, as such, it does not always enter into the determination of EOQ.

To determine the EOQ we take the first derivative of the total cost function in Equation 6.12 with respect to Q, and set it equal to 0. That is,

$$TC = (Q/2)h + (D/Q)O + Dc = 0.5Qh + DOQ^{-1}$$

$$d(TC)/dQ = 0.5h - DOQ^{-2} = 0$$

$$\text{or } h/2 - DO/Q^2 = 0$$

$$\text{or } Q^2h = 2DO$$

$$\therefore Q^* = \sqrt{\frac{2DO}{h}} \qquad [6.13]$$

where Q^* is the optimal EOQ.

The optimal order quantity, Q^*, means that at this level of inventory the total cost will be minimum. However, to ensure that it is a true minimum, we take the second derivative of the total cost function (obtained from the first derivative with respect to Q), as shown below:

$$d(TC)/d(Q) = h/2 - DO/Q^2 = 0$$

$$\frac{d^2(TC)}{dQ^2} = 2OQ^{-3}D = \frac{2DO}{Q^3}$$

Since the second derivative is positive (because D, O, and Q are positive), the total cost must be minimum for the EOQ, Q^*.

We can go back to the example again to see if the quantity we obtained is really an EOQ. To do so, we apply Equation 6.13 directly to the problem; the result produces an optimal quantity of 40, which is considerably lower than the value of 62.5 we obtained earlier through trial and error. That is,

$$Q^* = \frac{\sqrt{(2)(250)(\$12.50)}}{(20)(0.2)}$$

$$= \sqrt{\frac{\$6,250}{\$4}}$$

$$= 39.53 \approx 40$$

where Q^* is the optimal quantity, h is the carrying cost obtained by multiplying c by I_p; that is, ($20)(0.2) = $4, and the rest of the terms are the same as before.

Similarly, the total cost of inventory per unit to the government is also lower than the cost obtained previously by the same trial-and-error process (Figure 6.2). That is:

$$
\begin{aligned}
TC &= (Q^*/2)h + (D/Q^*)O \\
&= [40/2)(\$4)] + [(250/40)(\$12.50)] \\
&= \$80.00 + \$78.13 \\
&= \$158.13
\end{aligned}
$$

Since the government needs 250 units each year, the number of orders, based on the optimal quantity Q^*, will be $D/Q^* = 250/40 = 6.25$. In other words, there will be 6.25 orders per year or one order for every 58.4 days, each containing an order of 40 units $[40 \times 6.25 = 250]$. It should be worth noting that since one cannot have fractional orders, it should be rounded off to 6, which will lower the number to 240 $[40 \times 6 = 240]$ because of rounding-off error.

Volume Discounts

Volume discount means that suppliers often offer discounts on items when purchased in large quantities, but purchasing in large quantities has its pluses and minuses. On the plus side, it reduces unit costs, lowers ordering and stockout costs, and saves in transportation or shipping costs. On the minus side, it increases holding or carrying costs and also increases the chances of depreciation and obsolescence. This section presents a brief account of the EOQ model with volume discounts.

We start with the total cost function in Equation 6.12 and the same inventory problem where a government needs 250 units of a certain item each year. The cost per unit is $20, with an ordering cost of $12.5 and a carrying cost of $4 per unit. Assume now that the government receives a 7 percent discount when it purchases a minimum of 100 units of the item. The question is what would be the amount of savings under these conditions for the government?

To determine the savings, we first calculate the EOQ with the discount. We already know the EOQ without the discount, Q^*, which is 40. The EOQ with the discount can be obtained using the same procedure as we used for Q^*. That is:

$$
\begin{aligned}
Q' &= \sqrt{\frac{2DO}{h_d}} \\
&= \sqrt{\frac{(2)(250)(\$12.50)}{\$4 - (\$4)(0.07)}} \\
&= \sqrt{\frac{(2)(250)(\$12.50)}{\$3.72}} \\
&= 40.99 \approx 41 \qquad [6.14]
\end{aligned}
$$

where Q' is the EOQ with the discount and h_d is the carrying cost, adjusted for volume discount. The result produces an order quantity of 41.

Next, we compute the total cost with the discount, but to do so we need to modify the total cost function in Equation 6.12 so that we can see what effect the discount will have on the total cost and eventually on the savings for the government. Equation 6.15 shows the modified cost function, with the discount included:

$$TC' = (D/Q^*)(O) + (Q_d'/2)h(1-v) + Dc(1-v) \quad [6.15]$$

where TC' is the total cost of inventory with volume discount, v is the percentage of discount for volume purchase, and the rest of the terms are the same as before. Note that the ordering cost is not adjusted for discount since it is not affected by the quantity purchased.

The following presents the total cost of inventory to the government with and without the volume discount.

[A] Total cost (*TC) without the discount:**

$$\begin{aligned} TC^* &= (D/Q^*)(O) + (Q^*/2)h + Dc \\ &= (250/40)(\$12.50) + (40/2)(\$4) + (250)(\$20) \\ &= \$78.13 + \$80.00 + \$5{,}000 \\ &= \$5{,}168.13 \end{aligned}$$

[B] Total Cost (*TC*) with the discount:

$$\begin{aligned} TC &= (D/Q^*)(O) + [(Q'/2)h(1-v)] + Dc(1-v) \\ &= (250/40)(\$12.50) + [(41/2)(\$4)(1-0.07)] + (250)(\$20)(1-0.07) \\ &= \$78.13 + \$76.26 + \$4{,}650 \\ &= \$4{,}804.39 \end{aligned}$$

Subtracting B from A will produce a cost difference of $363.74. This is the amount of savings the discount will generate for the government. As should be clear by now, much of this reduction is coming from the reduction in carrying or holding costs as well as from savings in purchase costs. However, there are instances where a supplier may offer several discounts for varying levels of purchases, in which case the government should accept the one that reduces the total cost the most.

The EOQ Model with Stockout Cost

Earlier in the EOQ model, we assumed that there is no shortage or stockout cost. In general, if it can be argued that there is no cost effect of delay, then stockout can be assumed away but, in practice, it does have an effect that should be taken into consideration when considering an EOQ model.

To introduce shortages into the model, assume now that we have an initial inventory or stock, S, at the beginning of a cycle (defined as the time between orders), with w units required or withdrawn per unit time. The per unit time can be a week, a month, a year, or any time length. The inventory level is positive, with

S/w. The average inventory level during this period is $S/2$ and the corresponding cost is $h(S/2)$, where h is the carrying or holding cost, as before. The total carrying cost over the time inventory level is positive, and can thus be written as:

$$\left(\frac{hS}{2}\right)\left(\frac{S}{w}\right)=\frac{hS^2}{2w} \qquad [6.16]$$

where the terms of the equation are the same as those described earlier.

Assume further that shortages occur for a time given by $(Q-S)/w$, where Q is the units ordered in equal numbers or quantity. The overage shortages during this period is $(Q-S)/2$ units and the corresponding cost is $k(Q-S)/2$, where k is the shortage or stockout cost. Therefore, the total shortage cost over the time shortages can be expressed as

$$\left[\frac{k(Q-S)}{2}\right]\left[\frac{(Q-S)}{w}\right]=\frac{k(Q-S)^2}{2w} \qquad [6.17]$$

Similarly, the total cost of inventory per cycle, with shortages included, can be written as:

$$TC=\frac{DO}{Q}+\frac{hS^2}{2w}+\frac{k(Q-S)^2}{2w} \qquad [6.18]$$

where the first part of the equation represents the ordering cost, the second part the carrying cost, and the third part the stockout cost.

It should be pointed out that the total cost of inventory in Equation 6.18 is based on cost per cycle. As noted above, a cycle is the time between orders. Thus, if 40 units are ordered each time an order is placed, Q, and 15 units are used or withdrawn per unit of time, say, each month, the length of cycle, Q/w, called the time length of order, will be $40/15=2.67$ months.

To obtain the total cost of inventory per unit of time, we divide the total cost of inventory per cycle by the cycle length, Q/w. That is:

$$TC=\frac{DO}{Q}+\left[\frac{hS^2}{2w}+\frac{k(Q-S)^2}{2w}\right]/(Q/w)$$

$$=\frac{DO}{Q}+\frac{hS^2}{2Q}+\frac{k(Q-S)^2}{2Q} \qquad [6.19]$$

Next, to find the EOQ at which the total cost will be minimum, we do the following: first, revise the cost function, for ease of solution, by rearranging the terms of the expression in Equation 6.20, so that

$$TC=\frac{DO}{Q}+\frac{hS^2}{2Q}+\frac{k(Q-S)^2}{2Q}$$

$$=\frac{DO}{Q}+\frac{hS^2}{2Q}+\frac{k(Q^2-2QS-S^2)}{2Q}$$

$$=\frac{DO}{Q}-kS+\frac{kQ}{2}+\frac{kS^2+hS^2}{2Q} \qquad [6.20]$$

Second, take the partial derivatives of the revised cost function (Equation 6.20) with respect to S and Q, respectively, and set them equal to 0, so that

$$\frac{\partial(TC)}{\partial S} = -k + (k+h)\frac{S}{Q} = 0 \qquad [6.21]$$

$$\frac{\partial(TC)}{\partial Q} = -\frac{DO}{Q^2} + \frac{k}{2} - \frac{kS^2}{2Q^2} - \frac{kS^2}{2Q^2} = 0 \qquad [6.22]$$

Third, and finally, solve the equations simultaneously to obtain two optimum quantities, as shown below:

$$S^* = \sqrt{\frac{2DO}{h}}\sqrt{\frac{k}{k+h}} \qquad [6.23]$$

$$Q^* = \sqrt{\frac{2DO}{h}}\sqrt{\frac{k+h}{k}} \qquad [6.24]$$

where S^* is the optimum stock and Q^* is the new EOQ.[1]

To illustrate this further, let us add a dollar value to the cost of shortage, k, while keeping the values of the rest of the terms the same as before. Let $k=\$5$, the optimum stock (inventory) for the period and the corresponding order quantity will be:

$$S^* = \sqrt{\frac{2DO}{h}}\sqrt{\frac{k}{k+h}} = \frac{\sqrt{(2)(250)(\$12.50)}}{\$4}\sqrt{\frac{\$5}{\$5+\$4}} =$$
$$\sqrt{1,562.5}\sqrt{0.556} = 29.488 \approx 30$$

$$Q^* = \sqrt{\frac{2DO}{h}}\sqrt{\frac{k+h}{k}} = \sqrt{\frac{(2)(250)(\$12.50)}{\$4}}\sqrt{\frac{\$5+\$4}{\$5}} =$$
$$\sqrt{1,562.5}\sqrt{1.8} = 53.047 \approx 53$$

30 and 53 units, respectively.

Taking the difference between the two optimum quantities would produce an optimum shortage of 23 units for the period; that is, $S^* - Q^* = 30 - 53 = -23$. What this means is that the government should allow a shortage of 23 units to accumulate before ordering 53 units of the item, thereby raising the inventory level to 30 units. Furthermore, if we divide Q^* by a fixed amount of use or withdrawal, w, it will produce the optimum (time) length of order, Q^*/w. Let 15 be the number of units used per unit of time, then the optimal length of time for which the order must be placed is $53/15 = 3.53$ months. In other words, with shortage permitted, the government should set up the order every 3.53 months.

Sensitivity Analysis

The optimal quantities we obtained for the EOQ model are basically estimates; that is, they are subject to change if the model conditions change (i.e., the values of the terms used in the equation change). We can do a sensitivity analysis to see, for instance, how sensitive the optimal quantity or cost will be to a change in demand or in the cost categories. We can change the value of any one, or all, of the terms of the optimum inventory model simultaneously. Let us look at a situation in which the ordering cost changes from \$12.50 to \$10.50. The optimum inventory without the stockout cost (Equation 6.13) will be:

$$Q^* = \sqrt{\frac{2DO}{h}} = \sqrt{\frac{(2)(250)(\$10.50)}{4}} = 36.228 \approx 36$$

which is lower than the optimum quantity of 40 we had earlier. This is understandable since the ordering cost was lower and none of the other terms in the equation were changed.

If, for instance, we lower the ordering cost by \$4 from \$10.50 to \$6.50, the new optimum order quantity will further decrease to $28.504 \approx 29$ units. Let us say that the demand also changes from 250 to 300. This will increase the EOQ by 2, from 29 to 31; that is, $\sqrt{2DO/h} = \sqrt{(2)(300)(\$6.50)/\$4} = 31.225 \approx 31$. It is important to note that even if there is not a significant change in some or all of the value of the terms in our model, the quantity changed will be smaller because it changes by the square root of the ratio of ordering to carrying cost.

A Note on Simulation

So far our discussion has concentrated on inventory models that are deterministic in nature, in that we know for certain the exact values of the variables used in these models, but inventory problems may involve situations in which these values may not be known with certainty or a priori. In other words, they may be random (i.e., probabilistic) rather than deterministic. A random variable is a special variable that has its own probability distribution. The probability distribution may be discrete or continuous, depending on the characteristics of the random variable. If the random variable is discrete, the probability distribution will be discrete; if it is continuous, the distribution will be continuous.

Inventory problems with uncertainty are far more complicated than their deterministic counterparts, since solutions to these problems by direct analytical methods are not always possible. One alternative is to use simulations. A simulation is an experimental model of a system designed to generate a solution that cannot be achieved through trial and error or using conventional analytical methods. Simulations generally require a great deal of computation to arrive at a solution, but with easy availability of mainframe, high-speed desktop and portable computers it should not pose a problem. Monte Carlo simulation is a good example in which one uses a computer to generate the uncertainty that may exist in a problem.

Monte Carlo Simulation. To understand how a Monte Carlo simulation works, one has to have some familiarity with random numbers. Random numbers are numbers that are generated by a computer or collected from a random number table in a manner that is free of any pattern or bias. This unique characteristic of random numbers makes it possible to use them to simulate any type of distribution – empirical, theoretical, discrete, or continuous.

Let us think of a simple inventory problem with uncertainty. Suppose that it costs a government $25 to order an item that it uses regularly, with a carrying cost of $5 per unit per week, and a shortage (stockout) cost of $10 per unit per week. Also suppose that orders are placed at the beginning and arrive at the end of the week, and that charges are based on the end-of-week inventory, but before the arrival of a new batch. Our objective is to find the optimum stock and reorder level for the government.

However, before we can apply the simulation, we need to make a couple of assumptions about the distribution of demand and the time between when an order is placed and when the delivery is made, called the lead time. Assume that the mean demand for the item in question is five units per week and the mean lead time is four weeks. The distribution of the two variables along with their respective probabilities is presented in Table 6.2. The last column in the table represents the assignment of random numbers based on a rectangular distribution, considered here for the sake of simplicity, although we could have considered other types of distributions.

Returning to the example, we start with an arbitrary reorder level, say, 20. It does not really make any difference what number we choose since in the long run the effect of any arbitrary starting value will be balanced out over a large number of use cycles. Table 6.3 shows the tabulations for randomly generated numbers for the two rectangular variates – demand and lead time. As can be seen from the table, we consider only two cycles, each consisting of six weeks. We assume the cycle begins on the first of the week after a new order has been

Table 6.2 Distribution of Demand and Lead Time

Demand (d)	*P*(d)	*Cumulative Relative Frequency*	*Random Numbers Assigned*
2	0.05	0.05	00–05
3	0.10	0.15	06–15
4	0.15	0.30	16–30
5	0.20	0.50	31–50
6	0.25	0.75	51–75
7	0.15	0.90	76–90
8	0.10	1.00	91–100
Lead Time (l)	*P(l)*	*Cumulative Relative Frequency*	*Random Numbers Assigned*
3	0.30	0.30	00–30
4	0.45	0.75	31–75
5	0.25	1.00	76–00

Table 6.3 Monte Carlo Simulation for Two Cycles with $Q=15$ and $R=20$

Cycle	*Cumultive Period*	RN[1]	d[2]	RN[1]	*Lead Time*	*Stock End of Period*	*Carrying Cost ($)*	*Ordering Cost ($)*	*Stockout Cost ($)*	*Cumulative Cost ($)*	*Ave. Cost per Period ($)*
1	0			68	4	20					
	1	65	6			14	70	0	0	70	70.00
	2	17	4	42	4	10	50	0	0	120	60.00
	3	98	8			2	10	0	0	130	43.33
	4	70	6			–4	0	0	40	170	42.50
	5	45	5			35 (26)	130	25	0	325	65.00
	6	78	7			19	95	0	0	420	70.00
2	7	20	4			15	75	0	0	495	70.71
	8	67	6			9	45	0	0	540	67.50
	9	69	6	83	5	3	15	0	0	555	61.67
	10	55	6			–3	0	0	30	585	58.5
	11	21	4			35 (28)	140	25	0	750	68.18
	12	98	8			20	100	0	0	850	70.83

Notes
1 *RN* = random number.
2 *d* = demand.

placed, which occurs whenever the inventory level reaches or falls below 20 at the end of the week. The table also shows a random number, 68, which represents a lead time of four weeks for the order placed at that time.

During the first week, the demand is 6 and the inventory level drops to 14 at the end of the week. The process continues until the fourth week, when the demand produces a shortage of 4, costing the government $40 [$10 × 4 = $40] in shortage cost. Let us say that at the end of the fourth week and the beginning of the fifth week, a new lot of, say, 35, arrives. However, during this period the actual level of inventory was 26, obtained by subtracting from the reorder level at 35 (the shortage of 4 we had the previous week and 5 used during the fifth week). The last two columns in Table 6.3 present the cumulative cost and the average cost per period. The latter is obtained by dividing the cumulative cost by the number of cumulative periods (weeks).

We can use the method further to calculate the various averages for the inventory problem, such as average cost, average shortage, and so forth. For instance, the average cost of inventory for the two cycles over a 12-week period is $70.83, as shown in the table. The implication here is that if we have a large number of cycles to consider, then over the range of these cycles the observed value, say the average, will be equal or close to the true average of the distribution of the variable(s). In statistics, this is known as the law of large numbers.

Other Methods

Besides Monte Carlo simulation there are other methods that one can use to deal with uncertainties in inventory analysis. One such method is the Newby problem. Newby is primarily used for single orders that do not require reorders. As such, it is mostly used for problems with perishable items or things that do not have enough opportunity for application in the public sector. Another method that also deals with uncertainty is incremental analysis. In incremental analysis, one tries to decide how much to order by comparing the cost or loss of ordering one additional unit with the cost of not ordering an additional unit. The difference between the two usually produces an optimal order for a single period.

Two additional developments in inventory problems are worth noting here. The first is an approach that has received considerable attention in recent years, called just-in-time or JIT (Wantuck, 1989). JIT is a philosophy that focuses on eliminating all kinds of waste and abuse in an organization, including inventory. The principle underlying the approach is the notion that materials and supplies must be acquired only when they are needed, thus either eliminating or minimizing the need to maintain an inventory. To utilize a JIT approach, one needs to coordinate a set of activities with the help of an information system, called Kanban, a Japanese word for card. The Kanban's are used to write down the activities in such a way that they can be synchronized from the first to the last stage of the activity process. The end result is an inventory system with small lot sizes and virtually no safety stocks. Since lot sizes are small and the safety stocks are maintained at a minimum, the cost of inventory is also very low under JIT.

The other development does not exclusively deal with the type of situation discussed here, but is frequently mentioned alongside inventory problems. It is called material requirements planning (MRP). Its purpose is to determine the optimal requirement for raw materials, components, and subassemblies in order to minimize the total cost of an operation. Designed for industrial production, the method can also be used in government for those activities where it is possible to identify a clear need for MRP. Construction of roads, bridges, and highways are good examples of where some variations of MRP can be used.

Queuing Models

A queue is a waiting line or delay. Anyone who has ever called an agency or a service department with a problem is familiar with the situation when the system was either busy or failed to respond on time. The performance of a service system can have a significant impact on how citizens view the levels of services they receive from their government. Queuing theory, a branch of operations research, explores such relationships between the demand on a service system and their result, such as delays or inconveniences experienced by the users of the system. A large number of services a government provides such as fire, police protection, hospital emergency, and so forth falls into this category, where queuing theory plays a key role in the analysis and planning of these services. This section presents a broad overview of the queuing theory and its application in government.

Elements of a Queuing System

The physical structure of a queuing system consists of three components: input source, queues, and a service facility. The input source is the pool of potential consumers, called the population that will require the service. The size of this population relative to the individuals in a queue can be inexhaustible, in which case it is called an infinite population. On the other hand, if it is limited compared to the size of the queue being served, it is called a finite population. The number of telephone calls being served by a telephone exchange is a good example of an infinite population (assuming that the circuit is not entirely tied up).

The queue is also a function of size; that is, it can be infinite if the size is allowed to grow to any length, or finite if it is restricted by the amount it can grow. A queue can be single, such as waiting in a line to board a bus, or multiple, such as when several groups of a population are being simultaneously served from different operations. An example of the latter will be when individuals are waiting in line at different counters at a local post office. The queues, the service facility, and the services together are called the queuing system. Figure 6.4 shows a simple queuing system.

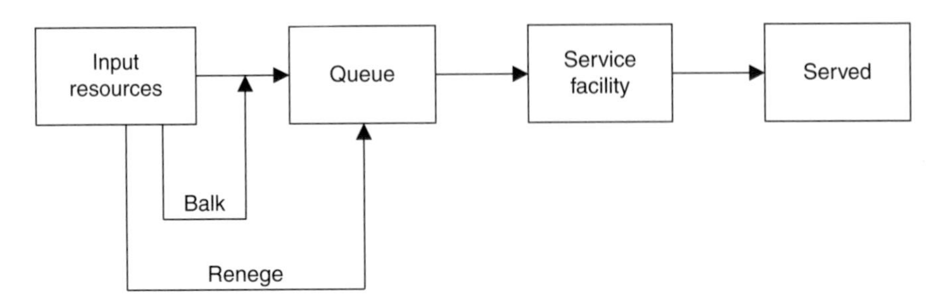

Figure 6.4 Elements of a Queuing System

On the other hand, the manner in which the individuals or units of a calling population are taken from a queue is called the queue discipline. The latter may be on a first-come, first-served basis, or random, or subject to priorities such as in an emergency clinic. It may also be possible for individuals to switch queues, and the choice of which queues to join may depend on arrival time. Finally, individuals attempting to join a queue may balk if the queue length is excessive and may decide not to join, or they may join the queue and subsequently renege (i.e., change their mind and leave before being served). In either case, they are not included as part of the system.

A Simple Queuing Model

The easiest way to construct a queuing model is to assume a single queue, infinite population, with constant arrival and service time. In other words, individuals arrive at equally spaced intervals and receive service at a constant rate. It is safe to assume that if the service time is less than the time periods between arrivals, we do not have an infinite queue (i.e., it is finite). Similarly, if the waiting time for an individual is equal to the service time, then there is no queue except for waiting at the service channel.

We can think of a situation in which an individual, say, a customer, enters a service channel at an interval t_a, and remains there for t_a time period. Therefore, the waiting cost for the individual is

$$c_w t_s \qquad [6.25]$$

where c_w is the cost of waiting for an individual for one time period, and t_s is the service time for the individual.

There is also a cost associated with the service. To a large extent, the cost of service as well as the time it takes to provide the service depends on the size of the service channel. In general, the time for serving an individual is inversely proportional to the size of the channel, while the cost of operating a service channel is proportional to its size. What this means is that as the service channel gets larger, it will take less time to serve an individual. However, the cost of operation for large channels will also be more than the cost of operation for small channels.

Since the time it takes to serve an individual is t_s, the cost of operating a service channel for the time period can be written as

$$C_o = c_p / t_s \qquad [6.26]$$

where C_o is the cost of operating a service channel for one time period, c_p is the cost for a service channel that can serve one individual in one time period, and t_s is the service time for serving an individual, as before.

Thus, the cost of a time period, t_s, will be the sum of the cost of waiting (also called lost time) for the individual being served and the cost of operating the service channel for t_a time periods. That is:

$$C = c_w t_s + t_a(c_p / t_s) \qquad [6.27]$$

where Ct_s is the cost of a time interval, t_a, and the rest of the terms are the same as before.

From Equation 6.27 we can obtain the total cost for one time period, TC_t, by dividing it by t_a, so that

$$TC_t = c_w(t_s / t_a) + c_p / t_s \qquad [6.28]$$

where TC_t is the total cost for one time period.

Now, if we take the derivative of TC_t with respect to t_s, set it equal to 0, and then solve for t_s, we get the optimum (minimum) service time:

$$\frac{\partial(TC)}{\partial t_s} = \frac{c_w}{t_a} - \frac{c_p}{(t_s)^2} = 0$$

$$t_{S,opt} = \sqrt{\frac{c_p t_a}{c_w}} \qquad [6.29]$$

Since t_s is optimum, substituting this value into Equation 6.28 will produce the optimum (minimum) cost for a single period.[2] That is:

$$TC_{t,opt} = 2\sqrt{\frac{c_p c_w}{t_a}} \qquad [6.30]$$

Let us look at a simple example to illustrate this. Suppose that the central administration of a community hospital is planning to improve its service time by making the best use of its available time and resources. Let us say that it costs the hospital $3,000 per hour to provide and maintain the facility. The hospital currently receives an average of 12 patients per hour, although it can effectively serve, say, 15 patients during the same time. If the hospital loses a patient, it costs the hospital $8,500 per day. The administration is interested in determining the optimum service time the hospital must spend on a patient and the cost associated with it.

Since the hospital can effectively serve 15 patients at the current cost of $3,000 per hour, the potential cost for one patient per hour at this rate is:

$$C_p = \$3{,}000/15 = \$200$$

The time interval between arrivals is:

$$t_a = 1/12 = 0.0833 \text{ per hour}$$

The cost of waiting is:

$$C_w = \$8{,}500/24 = \$354.17 \text{ per hour per patient}$$

where 24 represents the number of hours in a day. Therefore, the optimum service time per hour, based on Equation 6.29, is:

$$t_{s,opt} = \sqrt{\frac{(\$200)(0.0833)}{\$354.17}} = 0.2169$$

Similarly, the total cost corresponding to the optimum service per hour, based on Equation 6.30, is:

$$TC_{t,opt} = 2\sqrt{\frac{(\$200)(\$354.17)}{0.0833}} = \$1{,}844.29$$

Since the hospital currently serves 12 patients per hour, the total cost of serving one patient per hour at the optimum level is $153.69; that is, $\$1{,}844.29/12 = \$153.69$. If the hospital were to serve 15 patients instead of 12, assuming that no additional costs are involved, the cost per patient per hour would be $122.95; that is, $\$1{,}844.29/15 = \$122.95$.

Queuing Models with Uncertainty

In the example presented above, we assumed that the arrival rate and service time were fixed (i.e., they were known with certainty). In reality, they are seldom known with certainty. When the arrival time and service time are not known a priori, one can generate them with the help of a probability distribution that best approximates it. The probability distribution that is frequently used in this case is the Poisson distribution.

Poisson distribution, also known as Poisson process, assumes that arrivals occur at random rather than at fixed intervals, represented by a parameter λ. The parameter λ is the average (mean) arrivals per unit of time. Thus, if n denotes the number of arrivals at time t, the probability function for a Poisson process, $P[t_a(n)]$, is given by the expression

$$P[t_a(n)] = \frac{\lambda^n e^{-\lambda}}{n!} \quad [6.31]$$

where t_a is the arrival time described before, λ is the mean arrival rate, e is the base of the natural logarithm (2.71828), $n=0,1,2,3,\ldots$, and $n!=n(n-1)(n-2)\ldots(3)(2)(1)$.

To illustrate how the process works in reality, consider a situation in which a local 911 Emergency System receives on average five calls per minute. Our objective is to find the probability of the system receiving exactly zero, one, two and three calls, say, between 7:30 p.m. and 7:31 p.m. on a given day. If we assume the arrival rates for these calls follow a Poisson distribution, we can calculate the respective probabilities as follows:

$$P[t_a(0)] = \frac{5^0 e^{-5}}{0!} = 0.0067$$

$$P[t_a(1)] = \frac{5^1 e^{-5}}{1!} = 0.0337$$

$$P[t_a(2)] = \frac{5^2 e^{-5}}{2!} = 0.0842$$

$$P[t_a(3)] = \frac{5^3 e^{-5}}{3!} = 0.1396$$

and so on.

An important characteristic of the Poisson process is that if the number of arrivals per unit time is Poisson distributed with a mean λ, then the length of time between arrivals is exponentially distributed with a mean of $1/\lambda$, which, in this case, will be $1/5=0.20$ minutes. Similarly, if the time between arrivals has an exponential distribution, the number of arrivals for each time period must have a Poisson distribution. Furthermore, it has generally been found that if the arrival time follows a Poisson distribution, the service time follows a negative exponential distribution, given by the expression

$$f(t_s) = ue^{-ut} \quad [6.32]$$

where $f(t_s)$ is the probability of service time, u is the mean service rate, $1/u$ is the mean service time, and e is the base of the natural logarithm. What this means is that as the service time increases, the probability function decreases exponentially toward a probability of zero. Figure 6.5 shows a typical negative exponential distribution.

Since the negative exponential distribution is a continuous distribution, the area under the curve represents the probability of service time, $P(t_s \leq T)$. We can easily find this time – that is, the area under the curve – by taking the integral of the negative exponential function in Equation 6.36. That is,

$$F(T_s) = \int_0^T ue^{-ut}\,dt = -e^{-ut}\big|_0^T = -e^{-uT} + e^0 = 1 - e^{-uT} \quad [6.33]$$

where $F(T_s)$ is the cumulative or total service time.

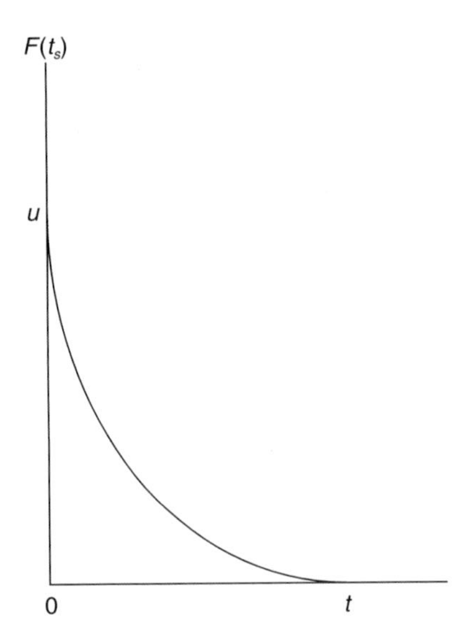

Figure 6.5 A Negative Exponential Probability Distribution

If the final expression in Equation 6.33 can be subtracted from 1, it will produce the probability the service time will take for a specific value of T, as shown below:

$$P(t_s) = 1 - F(T_s) = 1 - (1 - e^{-uT}) = e^{-uT} \quad [6.34]$$

To give an example, let us go back to the 911 problem. Suppose that the mean time $(1/u)$ it takes to dispatch help for a 911 call is five minutes. The probability that the dispatching help will take T or more minutes for different values of T, where $T = 0, 1, 2, 3, \ldots$, can be obtained with the help of Equation 6.34, as shown below:

$$F[t_s(0)] = e^{(-1/5)(0)} = 1.0000$$

$$F[t_s(1)] = e^{(-1/5)(1)} = 0.8187$$

$$F[t_s(2)] = e^{(-1/5)(2)} = 0.6703$$

$$F[t_s(3)] = e^{(-1/5)(3)} = 0.5488$$

$$F[t_s(4)] = e^{(-1/5)(4)} = 0.4493$$

and so on.

Queuing Models with Multiple Channels

The queuing models we discussed up to this point focused on a single channel (queue). In practice, queuing problems with multiple channels are quite common, although they are much more complex than single-channel problems. This is due in part to the difficulty associated with the amount of computations involved when the number of channels increases from one to n, but there are standard procedures one can use to significantly change the problem.[3]

Let us look at another example. Suppose that a local motor vehicle registration office gets busy during certain hours of the day, say, between 11:30 a.m. and 3:30 p.m. Individuals visiting the office during this period arrive, on average (λ), at the rate of 90 per hour. It takes an average (u) of one minute to serve an individual, which means 60 individuals per hour. The service time is assumed to be exponentially distributed. As the ratio of arrival to service time is greater than 1 (i.e., $\lambda/n = 90/60 = 1.5 > 1$), it is not possible for a single person to serve all the individuals needing the service. A second service desk is opened. Our objective is to find the value of the three most frequently asked questions in a queuing problem: (1) the average length of a queue; (2) the average wait before being served; and (3) the average number of individuals waiting for service.

To find answers to these questions, we first need to find the average length of a queue, which will then be used to find the average waiting time. The average length of a queue can be obtained from a standard expression used for queuing problems with multiple channels:

$$\bar{l}_w = (f_0)\frac{c^c k^{c+1}}{c!(1-k)^2} \qquad [6.35]$$

where $\bar{l}_w$ is the average length of queue, f_0 is the probability that no one is waiting for service, c is the number of channels, and k is the ratio of mean arrival to service time.

Note that the probability function f_0 and the ratio k are given by the following expression:

$$f_0 = \left[\sum_{n=0}^{c-1}\frac{(c_k)^n}{n!} + \frac{(c_k)^c}{c!}\left(\frac{1}{1-k}\right)\right]^{-1} \qquad [6.36]$$

$$k = \frac{\lambda}{c(n)} \qquad [6.37]$$

where the terms c, n, k, λ, and u are the same as before.

Similarly, the average waiting time for an individual before being served, and the average number of individuals waiting for service, $\bar{n}$, can be obtained by solving the following equations, provided that we have some information on the average length of a queue:

$$\bar{t}_w = \frac{\bar{l}_w}{\lambda}(n) \qquad [6.38]$$

$$\bar{n} = \frac{\lambda}{n}(\bar{t}_w + n) \qquad [6.39]$$

Next, to determine the solution for each of the three questions, we substitute the values of the parameters from our motor vehicle registration office example into the appropriate equations and solve, respectively, for $\bar{l}_w$, $\bar{t}_w$, and $\bar{n}$. The results of the solution are presented below.

[I] Average length of a queue, $\bar{l}_w$:

$$\bar{l}_w = (f_0)\frac{c^c k^{c+1}}{c!(1-k)^2}$$

$$= (0.1429)\frac{(2)^2(0.75)^3}{2!(1-0.75)^2}$$

$$= (0.1429)\frac{(4)(0.4219)}{(2)(0.0625)}$$

$$= (0.1429)\frac{1.6876}{0.125}$$

$$= 1.9293 \approx 2 \text{ persons}$$

where $k = \dfrac{90}{(2)(60)} = 0.75$

and $f_0 = \left[\dfrac{(1.5)^0}{0!} + \dfrac{(1.5)^1}{1!} + \dfrac{(1.5)^2}{2!}\left(\dfrac{1}{1-0.75}\right)\right]^{-1}$

$$= [1 + 15 + (1.125)(4)]^{-1}$$

$$= 0.1429$$

[2] Average waiting time before being served, $\bar{t}_w$:

$$\bar{t}_w = \frac{\bar{l}_w}{\lambda}(n)$$

$$= \left(\frac{1.923}{90}\right)(60) = 1.282 \text{ minutes}$$

where 60 represents the 60 minutes in an hour.

[3] Average number of individuals in the motor vehicle office, $\bar{n}_w$:

$$\bar{n} = \frac{\lambda}{n}(\bar{t}_w + n)$$

$$= \frac{90}{60}(1.282 + 1)$$

$$= 3.4293 \approx 3 \text{ individuals}$$

As the results indicate, the average length of queue in the office consists of two persons, the average waiting time is 1.282 minutes, and the average number of individuals waiting for service is three.

Monte Carlo Simulation. As with the inventory problem, we can use simulations to deal with uncertainty in queuing. All queuing models have two uncertainty characteristics – one related to the arrival time of a population (finite or infinite) and the other related to the service time distribution. Once we have information on them, most queuing problems can be solved without much difficulty. This section presents an overview of how simulation models such as Monte Carlo, discussed earlier for inventory problems, can be used to determine the arrival time and the service time distribution. As noted before, Monte Carlo simulation does not necessarily produce an optimal solution; however, it is possible to obtain a near-optimal solution if the simulated trials can be repeated an infinite number of times, which is possible only with a high-speed computer. In reality, most simulation processes are truncated after a finite number of trials.

To illustrate the point, let us go back to the motor vehicle registration office problem and set a simple objective – say, finding the mean waiting time for the customers arriving between 8:00 a.m., when the office opens, and 9:30 a.m., before the rush hour sets in. We begin with a simple assumption that we have some knowledge of the interval service time for individuals along with their probability distributions. This is shown in Table 6.4. The table also shows the allocation of 100 two-digit random numbers from 00 to 99 (both inclusive) to these distributions, corresponding to the interval and service time. Note that we have used two-digit random numbers because the probabilities obtained were correct to two decimal points.

Next, we generate 20 random numbers, compare them with the information contained in the table, analyze the results, and calculate the mean waiting time.

Table 6.4 Random Number Allocation: Interarrival and Service-Time Distribution

Interarrival Time (Minutes)	*Probability*	*Cumulative Probability*	*RNs* Allocated*
2	0.15	0.15	00–15
3	0.20	0.35	16–35
4	0.30	0.65	36–65
5	0.25	0.90	66–90
6	0.10	1.00	91–00
Interarrival Time (Minutes)	*Probability*	*Cumulative Probability*	*RNs* Allocated*
2	0.10	0.10	00–10
3	0.15	0.25	11–25
4	0.30	0.55	26–55
5	0.30	0.85	56–85
6	0.15	1.00	86–99

Note
*RNs = random numbers.

Let us say that the first random number we draw is 12. Therefore, the first simulated arrival is 8.02 a.m. since the office opens at 8:00 a.m. All simulated arrival and service times can be constructed in this way. Since no other person was present before the first arrival, the person can be served right away. However, given that the person arrived at 8:02 a.m. and that there was no other person before him or her, the service counter will remain idle for the first two minutes. This means that the simulated service time for the first person will be two minutes and the time when the service ends will be four minutes, or at 8:04 a.m. The next arrival is, say, at 8:05 a.m., which indicates that the service has been idle for an additional minute; that is, the server had to wait for one minute before the arrival of the second person, and so forth. Table 6.5 presents the arrival, service, and waiting times for the problem.

Whenever a person has to wait upon arrival due to how busy the server is serving an earlier person, the waiting time is entered in the column for the length of queue. As can be seen from Table 6.5, at no point during the experiment were there more than two persons waiting in the queue. We can now obtain the mean time for our sample of 20 individuals. As the table shows, the total waiting time for all 20 individuals is 21 minutes (third column from the right). Therefore, the mean waiting time for a person is 1.05 minutes; that is, $21/20=1.05$

Monte Carlo simulation is a valuable tool for solving queuing problems with uncertainty. As we saw earlier, the simulation generates a set of random numbers that can be used to predict the outcomes of a problem and analyze its results. However, a simulation itself does not eliminate the need for developing formal models to deal with a queuing problem. Most simulation techniques, including Monte Carlo simulation, serve a very specific need in problem solving, especially where uncertainty conditions exist in a problem that requires continuous observation of outcomes over a large number of trials. Without the uncertainty, all trials will produce the same result and, as such, there will not be any need for simulation.

Chapter Summary

The primary goal of an organization is to provide goods and services efficiently. An organization is efficient if it can provide goods and services at the lowest cost without substantially reducing the quality or quantity of the provision. This chapter has briefly introduced how optimization, in particular classical optimization, can be used to improve the operations of an organization. Two areas where the concept has been frequently used are inventory management and queuing. Although used mostly in the private sector, inventory management has received increasing attention in government in recent years (Schwartz, 1987), in part because it is based on the same principles as those that underlie the cash management practices of an organization (Baumol, 1952; Schwartz, 1987).

Queuing models, on the other hand, are more complex than inventory models, but their application has considerable merits for government. This chapter has presented several examples of both deterministic and probabilistic queuing

Table 6.5 Monte Carlo Simulation: Arrival, Service, and Waiting Time

RN	*Interarrival Time (Min)*	*Arrival Time (a.m.)*	*Service Starts (a.m.)*	*RN*	*Service Time (Min)*	*Service Ends (Min)*	*Waiting Time Customer (Min)*	*Waiting Time Server (Min)*	*Length of Queue*
12	2	08:02	08:02	14	2	08:04	0	2	0
28	3	08:05	08:05	31	3	08:08	0	1	0
50	4	08:09	08:09	42	4	08:13	0	1	0
62	4	08:13	08:13	59	4	08:17	0	0	0
24	3	08:16	08:17	77	5	08:22	1	0	1
93	6	08:22	08:22	63	4	08:26	0	0	0
98	6	08:30	08:30	82	5	08:35	0	4	0
35	3	08:33	08:35	11	2	08:37	2	0	1
96	6	08:39	08:39	37	4	08:43	0	2	0
79	5	08:44	08:44	54	4	08:48	0	1	0
9	2	08:46	08:48	14	2	08:50	2	0	1
18	3	08:49	08:50	83	6	08:56	1	0	1
97	6	08:55	08:56	98	5	09:01	0	0	0
83	5	09:00	09:01	60	4	09:05	1	0	1
69	5	09:05	09:05	23	3	09:08	0	0	0
47	4	09:09	09:09	81	5	09:14	0	1	0
62	4	09:13	09:14	95	6	09:20	1	0	1
26	3	09:16	09:20	22	3	09:23	4	0	2
15	2	09:18	09:23	46	4	09:26	5	0	2
87	5	09:23	09:27	67	5	09:32	4	1	0

models involving single, as well as multiple, service channels. In particular, it has focused on models with infinite populations and queues because they are simple to understand and relatively easy to use. This does not mean that problems with finite populations and queues are not important; indeed, they are. Operationally, a finite population tends to produce results that are different from those produced by an infinite population, especially when the number of individuals to be served is also finite. A finite queue is the result of restrictions imposed by physical space. In other words, space limitation may permit only a limited number of individuals to be served at any given time.

In spite of the fact that queuing problems with finite populations are more complex, they can be solved with the help of finite queuing tables (Hazlewood, 1958). Although not considered here, they can also be solved by modifying the single- or multiple-channel problems with infinite population. However, the level of computation that is involved in the process can be enormous (Gupta & Cozzolino, 1975; Daellenbach & George, 1983).

Notes

1 The model, which was originally suggested by Baumol for determining demand for an inventory of cash, is given by the expression

$$\frac{bT}{C}+\frac{iC}{2} \tag{1}$$

where T is a steady stream of dollars (i.e., the value of the transaction that is predetermined), C is cash withdrawn in lots of C dollars spaced evenly throughout the year, b is a fee that must be paid each time cash is withdrawn, and i is the interest rate; both b and i are assumed to be constant.

The first part of the expression in Equation (1), bT/C, is the average cash holding, while the second part of the expression, $iC/2$, is the annual interest cost of holding cash. Now taking the derivative of Equation (1) with respect to C and setting it equal to zero produces

$$-\frac{bT}{C^2}+\frac{i}{2}=0$$

$$C=\sqrt{\frac{2bT}{i}} \tag{2}$$

which is the optimal inventory of cash.

2 The following equations show how these quantities were obtained. We start with the equation for S (Equation 6.21) and rewrite it such that:

$$-k+(k+h)\frac{S}{Q}=0 \tag{3}$$

$$(k+h)\frac{S}{Q}=k$$

$$S=\frac{kQ}{k+h} \tag{4}$$

Next, we take the modified equation for S, Equation (2), plug it in the solution obtained for the revised cost function in Equation 6.22, then solve for Q:

$$-\frac{DO}{Q^2}+\frac{k}{2}-\frac{kS^2}{2Q^2}-\frac{hS^2}{2Q^2}=0$$

$$-\frac{DO}{Q^2}+\frac{k}{2}-\frac{S^2}{2Q^2}(h+k)=0$$

$$-\frac{DO}{Q^2}+\frac{k}{2}-\frac{1}{2Q^2}\left(\frac{k^2Q^2}{(k+h)^2}(h+k)\right)=0$$

$$-\frac{DO}{Q^2}+\frac{k^2+hk-k^2}{2(k+h)}=0$$

$$-\frac{DO}{Q^2}+\frac{hk}{2(k+h)}=0$$

$$\frac{DO}{Q_2}=\frac{hk}{2(k+h)}$$

$$\frac{Q^2}{DO}=\frac{2(k+h)}{hk} \tag{5}$$

With slight algebraic manipulation, the last expression produces the optimum Q^*, as shown in Equation 6.24:

$$Q^2=\left(\frac{2DO}{h}\right)\left(\frac{k+h}{k}\right)$$

$$Q^*=\sqrt{\left(\frac{2DO}{h}\right)}\sqrt{\left(\frac{k+h}{k}\right)} \tag{6}$$

Finally, we substitute the value of Q back into Equation (4) for S to obtain the optimum S^*, as in Equation 6.23:

$$S^*=\frac{kQ}{(k+h)}$$

$$=\frac{k}{(k+h)}\sqrt{\frac{2DO}{h}}\sqrt{\frac{(k+h)}{k}}$$

$$=\sqrt{\frac{k}{(k+h)}}\sqrt{\frac{2DO}{h}}$$

To obtain the expression for Equation 6.30, we simply substituted Equation 6.29 into Equation 6.28:

$$TC_{t,opt}=(c_w)\sqrt{\left(\frac{c_pt_a}{c_w}\right)}/t_a+(c_p)\frac{1}{\sqrt{\frac{c_pt_a}{c_w}}}$$

$$= (c_w)\left(\frac{\sqrt{c_p}\sqrt{t_a}}{\sqrt{c_w}}\right) / t_a + (c_p)\left(\frac{\sqrt{c_w}}{\sqrt{c_p}\sqrt{t_a}}\right)$$

$$= \left(\sqrt{c_w}\sqrt{c_p}\sqrt{t_a}\right) / t_a + \frac{\sqrt{c_p}\sqrt{c_w}}{\sqrt{t_a}}$$

$$= 2\sqrt{\frac{c_p c_w}{t_a}}$$

3 We can extend the expression for the single channel to determine the expressions for multiple channels. Assume there are N number of service channels instead of one; we can rewrite Equation 6.28 to obtain the total cost for one time period as:

$$TC_t = c_w(t_s / t_a) + c_p(N / t_s) \tag{7}$$

Similarly, Equations 6.29 and 6.30 can be rewritten for optimum service and cost, with N service channels, respectively, as:

$$T_{s,opt} = \sqrt{(c_p t_a N / c_w)} \tag{8}$$

$$TC_{t,opt} = \sqrt[2]{(c_p c_w N / t_a)} \tag{9}$$

where $t_{s,opt}$ is the optimum service time, and $TC_{t,opt}$ is the optimum service cost.

References

Baumol, W. J. (1952). The transaction demand for cash: An inventory theoretic approach. *Quarterly· Journal of Economics, 66*, 546–556.

Daellenbach, H. G., & George, J. (1983). *Introduction to operations research techniques.* Boston, MA: Allyn and Bacon.

Gupta, S. K., & Cozzolino, J. M. (1975). *Foundations of operations research for management.* San Francisco, CA: Holden-Day.

Hazlewood, R. N. (1958). *Finite queuing tables.* New York: Wiley & Sons.

Schwartz, E. (1987). Inventory and cash management. In J. R. Aronson and E. Schwortz (Eds.), *Management policies in local government finance* (pp. 342–363). Washington, DC: ICMA.

Wantuck, K. A. (1989). *Just-in-time for America.* Milwaukee, WI: Forum Limited.

7 Mathematical Programming

Traditional optimization methods such as those based on differential calculus are useful and work well in decision environments that are not complex. When a decision environment is complex either due to resource limitations or the presence of interdependent goals and alternatives, traditional optimization methods do not always produce the best result. Given that most decision environments are complex, one needs to use more efficient tools like mathematical programming that can deal with these complexities more efficiently. Mathematical programming is a generic term used for a group of (programming) models where each model is based on a set of interrelated but well-defined structures, equations, and assumptions. Models are abstract representations of the real world. However, as abstractions of the real world, they can only deal with the most important elements of the world. It is not possible for any model to capture every single element of a system in all its intricate details; it is costly, time consuming, and even difficult to analyze. Mathematical programming models are no exception.

As quantitative tools, mathematical programming attempts to capture only the most important elements and relationships that exist in a real system by utilizing mathematical relations. In general, the more complex a system, the more complex are the relations. This chapter discusses several programming models – linear, integer, dynamic, heuristic, and goal programming. Because of its widespread use, the chapter provides more attention to linear programming than the other approaches.

Linear Programming

Linear programming is the most well-known among all mathematical programming models. Developed during World War II as a mathematical technique to deal with the problems of military logistics (Dantzig, 1963), it has been applied to a wide variety of problems and disciplines, ranging from industrial technology to education, to medicine, to economics and business, among others. It is the forerunner of all mathematical programming models and serves as the foundation for many of them.

Although it can be applied to almost any situation, linear programming is ideal for problems involving resource allocation, especially where resources such as labor, materials, equipment, and the like are scarce or in short supply. The primary objective of linear programming is to find an optimum solution for a function called the objective function, subject to resource and other constraints. The objective function of a linear programming model is to describe the goal or objective a decision maker wants to achieve, subject to a set of constraints where the constraints describe the resource limitation or any other restriction imposed by the environment within which the system operates.

Model Assumptions

Like most quantitative models, a linear programming model is based on a number of conditions. Success of the model depends on the extent to which one is able to successfully apply these conditions when dealing with a real-world problem. Important among these conditions are:

1. **Linearity.** Both the objective function and the constraint equations must be linear (i.e., each term in these equations can be expressed in only one variable raised to power one). In ordinary terms, it means that when a function involving, say, two decision variables is plotted on a two-dimensional plane, it forms a straight line.
2. **Divisibility.** The decision variables are continuous (i.e., they can take on any value). In other words, they cannot be restricted to integers for the decision variables.
3. **Certainty.** All the input parameters are known precisely and are not subject to any variation (i.e., they are deterministic, not stochastic).
4. **Homogeneity.** Each unit of the available resource is as productive as the rest, which means that if n units of a resource are being utilized to produce an output, then each of these n units will be as productive or efficient as the rest.
5. **Proportionality.** The relationship between the terms of all the equations in a model is directly proportional (i.e., any change in the value of a decision variable will result in exactly the same relative change in the functional value of that relationship). For instance, consider a decision variable X, with a value of 10, and a coefficient b that is functionally related to X with a value equal to 2 (i.e., $b=2$). The functional value of this relationship, bX, will be 20; that is, $bX=(2)(10)=20$. Therefore, if X increases by, say, 10 percent, the functional value will also increase by 10 percent to 22 (i.e., $20\times0.10+20=22$).
6. **Additivity.** The whole is equal to the sum of its components (i.e., the total measure of resource use must be equal to the sum of the resources used on each decision variable). For instance, the total cost of a service will be equal to the sum of all individual costs associated with that service.
7. **Independence.** There is no interconnectedness or relationship between the decision variables, resource availability, or operations performed.

When these conditions are fully specified in a linear programming problem, one can assume the model to be in good working form; that is, any solution the model produces can be accepted as valid and reliable. However, reality does not always reflect perfect model conditions, which means that some of these conditions may need to be relaxed in order for the model to work or, given a choice, alternative models other than linear programming should be used.

The Basic Structure

Structurally all linear programming models must contain three basic components: (1) an objective function component consisting of a set of decision variables (variables whose values are determined as part of achieving the optimal solution) and associated coefficients called the objective function coefficients; (2) a component containing resource and other constraints, given by the capacity for each decision variable called technical coefficients, and an upper or lower limit of resource availability for each constraint equation; and (3) a component containing a set of non-negativity constraints, specifying the range of values the decision variables must not exceed.

The following presents a typical linear programming model with n decision variables and m constraints:

$$\text{Maximize } Z = \sum_{j=1}^{m} C_j X_j \quad [7.1]$$

Subject to

$$\sum_{i=1}^{n} \sum_{j=1}^{m} a_{ij} X_j \leq B_i \quad [7.2]$$

$$X_j \geq 0 \quad [7.3]$$

where Z is the objective function to be optimized (maximized or minimized; maximized in this case), X_j is the jth decision variable (for $j=1,2,3,\ldots,m$), c_j is the jth objective function coefficient (for $j=1,2,3,\ldots,m$), a_{ij} is the ith technological coefficient associated with the jth decision variable (for $i=1,2,3,\ldots,n$; $j=1,2,3,\ldots,m$, and $i=j$), B_i is ith resource constraint (for $i=1,2,3,\ldots,n$). The terms c, a, and B are also known as input parameters of the model (the constants whose values are specified by the decision makers).

We can easily expand the model to include a complete set of relationships between the variables and the terms of the equations, as follows:

$$\text{Maximize } Z=c_1X_1+c_2X_2+c_3X_3+\ldots+c_mX_m \quad [7.4]$$

Subject to

$$a_{11}X_1+a_{12}X_2+a_{13}X_3+\ldots+a_{1m}X_m\leq B_1 \quad [7.5]$$

$$a_{21}X_1 + a_{22}X_2 + a_{23}X_3 + \ldots + a_{2m}X_m \leq B_2 \qquad [7.6]$$

. . . . .

. . . . .

. . . . .

$$a_{n1}X_1 + a_{n2}X_2 + a_{n3}X_3 + \ldots + a_{nm}X_m \leq B_n \qquad [7.7]$$

$$X_1, X_2, X_3, \ldots, X_m \geq 0 \qquad [7.8]$$

where Equation 7.4 is the objective function to be maximized, Equations 7.5–7.7 are the constraint equations, and Equation 7.8 is the non-negativity constraint, corresponding to Equations 7.1–7.3.

The approach to solving a linear programming problem is not as complex as it may seem at first glance, provided that one follows four basic steps: (1) formulate the problem by specifying the input parameters of the decision variables, the objective function, and all the relevant constraints; (2) solve the problem, using any of the following methods – a graphical approach, the conventional simplex method, or a computer-based solution algorithm; (3) interpret the results; and (4) perform a sensitivity analysis on the solution, known as post-optimality analysis.

The Graphical Solution

The simplest way to solve a linear programming problem is to use a graphical approach since it makes it easier to visualize the problem, especially when one is dealing with two decision variables and a limited number of constraints, and follow the solution process. But the method most frequently used is the simplex method, which uses an iterative process to obtain the optimal solution. In recent years, especially with the advancements in microcomputers, a variety of software packages are now available for solving linear and other programming problems. The most common among these packages is the Solver, a compendium of Microsoft Excel®. The software has proved to be quite useful not only in dealing with mathematical programming problems, but also a variety of other analytical problems, including forecasting.

To give an example of a graphical solution, suppose that a government wants to undertake two community development projects, X_1 and X_2, for young adults during summer months. The goal of the government is to get the young men and women involved in community services and to develop a sense of responsibility when there are no schools and they have plenty of time on their hands. The government plans to spend a maximum of $1.25 million on the projects, mostly in the form of hourly compensation. The government will pay $10 per hour for Project X_1 and $12.50 an hour for Project X_2. The objective of the government is for the students to provide as many as 75,000 hours of community services for X_1 and 65,000 hours for X_2 without exceeding the cost.

We can formulate the problem as a maximization problem, since the goal is to provide as many hours of services as possible within the stated constraints, as shown below:

$$\text{Maximize } Z = X_1 + X_2 \qquad [7.9]$$

Subject to

$$X_1 \leq 75{,}000 \qquad [7.10]$$

$$X_2 \leq 65{,}000 \qquad [7.11]$$

$$10X_1 + 12.50X_2 \leq 1{,}250{,}000 \qquad [7.12]$$

$$X_1, X_2 \geq 0 \qquad [7.13]$$

where Z is the objective function representing the total (optimal) hours of activities that can be undertaken by the government, subject to the constraints given by the constraint equations. For convenience, we will call the first two constraints (Equations 7.10 and 7.11) the service constraints, and the third constraint (Equation 7.12) the cost constraint. The last constraint (7.13) is the non-negativity constraint, indicating that fractional provision is possible.

Figure 7.1 shows the graphical solution of the problem. According to the figure, Project X_1 is presented on the vertical axis and Project X_2 on the horizontal axis. The three constraint equations are also shown on the graph. The vertical line shows the upper limit of the constraint for X_2. Thus, any point to the left will produce a solution, called the feasible (i.e., achievable) solution, and any point to the right will exceed the upper limit of the constraint set at 65,000. Similarly, the horizontal line shows the upper limit of the constraint for X_1, which is restricted to 75,000. As before, any point below this line will produce a feasible solution, since it will not exceed the upper limit of the constraint, and any point above the line will exceed the upper limit. The third constraint, which indicates the total amount the government can spend on the two services, and must not exceed \$1.25 million, can be drawn by setting $X_1 = X_2 = 0$, so that when $X_1 = 0$, $X_2 =$ \$0.1 million, or when $X_2 = 0$, $X_1 =$ \$0.125 million. The intersection of the three lines produces a region called the feasible region, which contains the solution that will be optimal. Therefore, any point inside the area *OABCDE* will produce a feasible solution for the first two constraints, while any point inside the area *OABDE* will produce a feasible solution for all three constraints. In other words, the optimal solution must lie within this region because in order to have an optimal solution it must be able to satisfy all the constraints in the problem.

In linear programming, the optimal solution always lies at the corner point, called the corner point solution, but the question is which corner: *B* or *D*? The

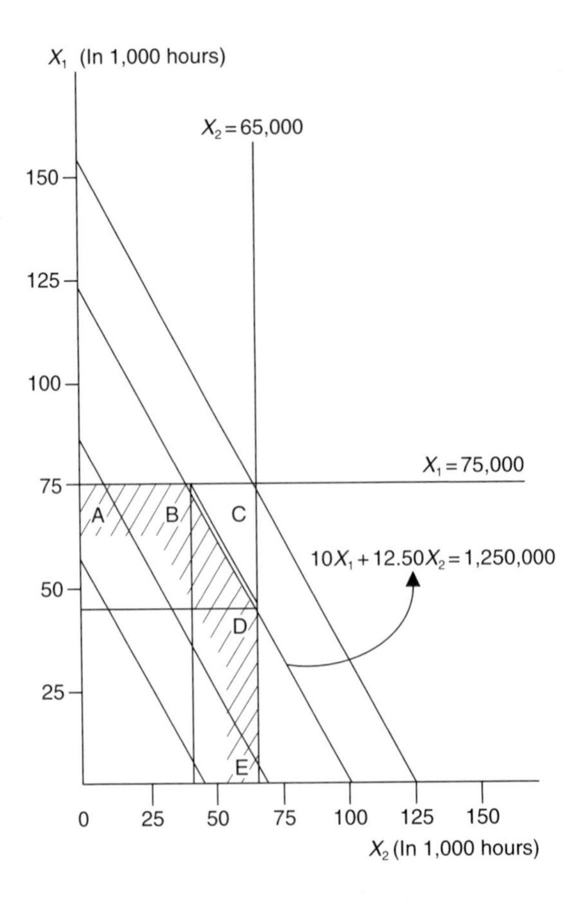

Figure 7.1 Graphical Solution of the LP Problem

problem can be easily solved by drawing a series of lines with a slope obtained from the equation of the objective function, as shown below:

$$Z = X_1 + X_2$$

$$\text{or } X_1 = Z - X_2;\ X_2 = Z - X_1$$

$$\text{or } X_1 = (-1)X_2 + Z;\ X_2 = (-1)X_1 + Z \qquad [7.14]$$

The result produces a slope of –1. What this means is that any point on these lines inside the feasible region will produce a feasible solution, but the point that reaches the farthest as the lines move outward will produce an optimal solution, and at that point the objective function will be optimized (the amount of services to be provided by the government will be maximum, given the constraints). This is point *B*, where $X_1 = 75,000$ and $X_2 = 40,000$.

Now, substituting these values into Equation 7.24 will produce:

$$Z = X_1 + X_2 = 75{,}000 + 40{,}000 = 115{,}000$$

which is the optimum value for Z. In other words, the optimum solution for the government will be 75,000 hours of service for X_1 and 40,000 hours of service for X_2, which will exhaust the available funds of \$1.25 million; that is, \$10(75,000) + \$12.5(40,000) = \$750,000 + \$500,000 = \$1,250,000 = \$1.25 million. Incidentally point D will also produce a perfectly feasible solution, as it will utilize the full \$1.25 million [\$10(43,750) + \$12.5(65,000) = \$437,500 + \$812,500], but it will not be optimal because it produces a maximum of 108,750 hours (43,750 + 65,000), which is 6,250 hours (115,000 − 108,750) less than the optimal quantity at point B.

We can use the above information to determine how many young people could be hired with the available funds and the optimal number of hours they must work. Assuming each young man or woman will work for 20 hours each week for a maximum of six weeks, the number of individuals the government can hire will be 625 for for project X_1 [75,000/(20 hrs/wk × 6 weeks)] and 333.3333 or 333 for project X_2 [40,000/(20 hrs/wk × 6 weeks)], with a total of 958 individuals for both projects. To see if the answer is correct, that is, it does not exceed the allocated amount, we can multiply each figure by the corresponding dollar value for that project, so that $\$10X_1 = [(\$10 \times 625) \times 120] = \$750{,}000$ and $\$12.50X_2 = [\$12.50 \times 333.3333) \times 120] = \$499{,}999.95 \approx \$500{,}000$. The result appears to be correct.

The Simplex Solution

It was easy to find an optimal solution for our problem as we had only two decision variables and three constraints to deal with, but if one has to deal with a large number of decision variables and constraints, it will not be possible to show the optimal solution on a two-dimensional graph. It will be too cumbersome, if not impossible. The alternative is to use a method of solution that is capable of dealing with multiple decision variables, as well as constraints, and has been a convention for solving standard linear programming problems for a long time. This is called the simplex method.

The simplex method provides an iterative (step-by-step) solution for the linear programming problem. It does so, first, by converting the constraints to equations; second, by defining a set of new variables called the basic variables; third, by finding an initial feasible solution for the problem; and, finally, by continuing the search for an improved solution (by presenting the results in each of the solutions in a table, called a tableau) until an optimal solution has been found. In common sense terms, it means that each time the method produces a solution for the objective function as large or larger (for a maximization problem) and as small or smaller (for a minimization problem) than the previous solution, it is reaching closer and closer to the optimal solution. The process continues until the optimal solution has been reached.

Convert the Constraints to Equations. The method begins by requiring that all inequality constraints (constraints with less than or equal to, ≤, or greater than or equal to, ≥) are converted to strict equalities. To do so, the method introduces two new variables into the system of equations – one slack (for less than or equal to) and one surplus (for greater than or equal to). The slack variables represent an underutilization of resources (below the upper limit of capacity), while the surplus variables represent an overutilization of resources (above the upper limit of capacity) within each constraint. By adding a slack variable to a less than or equal to constraint and subtracting a surplus variable from a greater than or equal to constraint, the inequalities are converted to strict equalities. This process of adding a slack or subtracting a surplus is known as augmentation.

Since all three constraint equations in our problem contain less than or equal to terms, we can add a slack variable to each of the three equations. The augmented model with the slack variables added is:

$$X_1 + S_1 = 75{,}000 \quad [7.15]$$

$$X_2 + S_2 = 65{,}000 \quad [7.16]$$

$$10X_1 + 12.50X_2 + 1S_3 = 1{,}250{,}000 \quad [7.17]$$

where X_1 and X_2 are the decision variables, and S_1, S_2, and S_3 are the slack variables corresponding to the three constraint equations.

The interpretation of the slack variables is quite simple. For instance, if we obtain a value of 72,000 hours of service for X_1 and 58,500 hours of service for X_2, it means a slack of 3,000 for X_1 ($S_1 = 3{,}000$) and 6,500 for X_2 ($S_2 = 6{,}500$). In other words, it is the amount by which the government will underprovide each service, given its original intention.

The simplex method requires that all variables (decision as well as slack) are included in each equation, meaning that the variables not appearing in an equation are added with a coefficient of 0. A variable with a zero coefficient essentially means that it has no influence on the equation in which it is included, but makes it possible for us to keep track of the variables as the solution continues. The revised augmented model with all the variables added to the constraint equations can be shown as follows:

$$1X_1 + 0X_2 + 1S_1 + 0S_2 + 0S_3 = 75{,}000 \quad [7.18]$$

$$0X_1 + 1X_2 + 0S_1 + 1S_2 + 0S_3 = 65{,}000 \quad [7.19]$$

$$10X_1 + 12.50X_2 + 0S_1 + 0S_2 + 0S_3 = 1{,}250{,}000 \quad [7.20]$$

$$X_1, X_2, S_1, S_2, S_3 \geq 0 \quad [7.21]$$

Since the slack variables have no influence on the equation in which they do not appear, they can be easily added to the objective function with 0 coefficients, in which case our Equation 7.9 becomes:

$$\text{Maximize } Z = 1X_1 + 1X_2 + 0S_1 + 0S_2 + 0S_3 \quad [7.22]$$

Find the Basic Variables. Next, we need to define the basic variables. The purpose of these variables is to determine the contribution each variable (decision, slack, and surplus) makes to the objective function. The solution starts by placing S_1, S_2, and S_3 in a table, called the simplex tableau, as the basic variables under a column called basis. The main body of the tableau consists of two parts: the body matrix consisting of the coefficients for the decision variables, and the identity matrix (with 1s in the main diagonal and 0s in the off-diagonal) representing the coefficients for the slack variables, as shown in Table 7.1. The column to the left of the basis, called C_i (for $i = 1, 2, 3 \ldots, n$), represents per unit contribution by each slack variable i (for $i = 1, 2, 3 \ldots, n$) to the solution. For instance, a value of 0 indicates that the slack variable does not make any contribution to the objective function at this stage of the solution.

In general, if a variable is in the basis or solution mix it is called a basic variable; if it is not, it is called a non-basic variable. Since neither of the decision variables is in the basis, they are non-basic variables.

The First Simplex Tableau. The first basic feasible solution can now be read from the initial simplex tableau (Table 7.1). As the table shows, the first three rows in the table correspond to the three constraint Equations 7.18–7.20. For instance, the numbers (1, 0, 1, 0, 0) in the first row represent the coefficients of the first equation (Equation 7.18); that is, $1X_1 + 0X_2 + 1S_1 + 0S_2 + 0S_3$. Similarly, the numbers in the second and third rows represent the coefficients of the second and third equations (Equations 7.19 and 7.20). The initial solution is produced at the point of origin (Figure 7.1) where $X_1 = 0$ and $X_2 = 0$, which means that the values of the slack variables in the initial solution must be non-zero; that is, $S_1 = 75{,}000$, $S_2 = 65{,}000$, and $S_3 = 1{,}250{,}000$. The three slack variables constitute the initial solution mix in the basis and their values are provided under the

Table 7.1 The LP Problem: The First (Initial) Simplex Tableau

C_i	*Basis*	X_1	X_2	S_1	S_2	S_3	B_i
0	S_1	[1]	0	1	0	0	[75,000]
0	S_2	0	1	0	1	0	65,000
0	S_3	10	12.5	0	0	1	1,250,000
	C_j	1	1	0	0	0	–
	Z_j	0	0	0	0	0	0
	$\Delta_j = Cj - Z_j$	[1]	1	0	0	0	–

Pivot column (the X_1 column); Pivot element (the boxed 1 in row S_1); Pivot row (the S_1 row)

column B_i. Since X_1 and X_2 are not in the solution mix, their values can be assumed to be 0.

We can show this initial solution with the help of a column vector:

$$\begin{bmatrix} X_1 \\ X_2 \\ S_1 \\ S_2 \\ S_3 \end{bmatrix} = \begin{bmatrix} 0 \\ 0 \\ 75,000 \\ 65,000 \\ 1,250,000 \end{bmatrix}$$

If the problem would produce an optimal solution, the decision variables will have positive values and the slack variables will have a value of 0 in the column vector. On the other hand, if the optimal solution has slack variables with positive values, it means that some resources have not been fully consumed, which is quite common in linear programming.

The last three rows in our initial (first) tableau need some elaboration (Table 7.1). The first of the three rows, C_j, represents the coefficients of the objective function, corresponding to Equation 7.22. The last two rows, Z_j and $\Delta_j = C_j - Z_j$, are used to determine whether or not an optimal solution has been reached or whether the solution would need further implementation. The Z_j row shows the contribution each variable makes toward the realization of the value of the objective function and is obtained by multiplying each value in the C_i column by the corresponding value in the simplex tableau, X_{ij}, and adding them up, as given by:

$$Z_j = \sum_{i=1}^{n} C_i X_{ij} \qquad [7.23]$$

(for $j = 1, 2, 3 \ldots, m$). For instance, the first value of the Z_j row is obtained by multiplying each value of the C_i column by the values of the first decision variable, X_1, such that:

$$Z_1 \text{ (for column } X_1) = (0 \times 1) + (0 \times 0) + (0 \times 10) = 0$$

The Z_j values for the other columns can be produced in the same way; that is, multiplying each value in the C_j column by the corresponding value in the body matrix and adding them, as shown below

$$Z_2 \text{ (for column } X_2) = (0)(0) + (0)(1) + (0)(12.5) = 0$$

$$Z_3 \text{ (for column } S_1) = (0)(1) + (0)(0) + (0)(0) = 0$$

$$Z_4 \text{ (for column } S_2) = (0)(0) + (0)(1) + (0)(0) = 0$$

Z_5 (for column S_3) = (0)(0) + (0)(0) + (0)(1) = 0

Z_6 (for column B_j) = (0)(75,000) + (0)(65,000) + (0)(1,250,000) = 0

The computation of the third row, $\Delta_j = C_j - Z_j$, is relatively straightforward in that it is obtained by taking the difference between the C_j and Z_j rows, that is, $1-0=1$, $1-0=1$, $0-0=0$, $0-0=0$, and $0-0=0$. However, the evaluation of the results is critical to understanding the solution of the linear programming problem. In general, when there are no more positive values left in the Δ_j row (for a maximization problem), we can assume that an optimal solution has been reached. Conversely, when no more negative values remain in the Δ_j row (for a minimization problem), the same conclusion can be reached. If neither situation occurs, we have not reached an optimal solution, indicating that a new basic solution may be needed. On the other hand, when the value of the objective function is 0, as was the case here, we can safely assume that we need a new basic solution.

The Second Simplex Tableau. Since the initial solution produced a value of 0 for the objective function Z_j, we need a new basic solution. The new solution can be obtained by adding a new basic variable and deleting an existing one currently in the basis. The process of selecting a new basis variable and deleting one in its place is called the change of basis. In general, the element corresponding to the decision variable in the Δ_j row with the largest positive value (for a maximization problem) and the largest negative value (for a minimization problem) determines the variable to be entered. Since both elements in the Δ_j row in the current example have an identical value of 1, we arbitrarily select the first variable, X_1, making it the entering basic variable (Table 7.1). The corresponding column then becomes the optimum column, called the pivot column. Pivoting is the process of moving from one column to the next.

To maintain the number of variables in the basis constant, for each variable that enters one must leave. To determine which variable must leave, we use a procedure called the displacement ratio, which is obtained by dividing the constant column, B_i, by the elements of the pivot column. The row with the smallest ratio (excluding the ratios that are ≤ 0) determines the variable leaving the basis. This happens to be the first slack variable under the basis, S_1, since it has the smallest displacement ratio of the three: 75,000 (Table 7.1):

S_1: 75,000 / 1 = 75,000

S_2: 65,000 / 0 = ∞

S_3: 1,250,000 / 10 = 125,000

The smallest displacement ratio indicates the maximum number of hours of service for project X_1 that can be produced without violating any of the original constraints. The row with the smallest displacement ratio is called the pivot row and the element corresponding to the leaving and entering basic variable is called the pivot element.

Having selected the new basic variable, the next step is to find the elements of the first row in the second tableau. This is accomplished by dividing each element of the replaced row by the pivot element. Thus, dividing the elements of the replaced row, S_1, by the pivot element of 1 produces the elements of the new row: $1/1=1$, $0/1=0$, $1/1=1$, $0/1=0$, $0/1=0$, and $75{,}000/1=75{,}000$ (Table 7.1). This is what we have in the first row of the second tableau (Table 7.2).

The derivation is different for the second and third row. It involves a three-step process. First, select the element to be entered from the non-pivot row of the initial tableau. Second, multiply the element of the remaining row corresponding to the pivot element in the initial tableau by the new corresponding element in the replacing row. Third, subtract the result from the element to be entered from the non-pivot row of the initial tableau in step 1. We can informally write it as:

> {Element to be entered from the non-pivot row of the initial tableau} − {(Element of the remaining row corresponding to pivot element in the initial tableau) × (new corresponding element in the replacing row)} = New value

Accordingly, we select the first element to be entered from the second (non-pivot) row of the first column in the initial tableau, which is 0 (Table 7.1). Next, we multiply the element of the remaining (second) row corresponding to the pivot element in the initial tableau, which is 0, by the new corresponding element in the replacing row (which is the first element of the first row of the second tableau) $1/1=1$ (Table 7.2). The result will be $(0)(1/1)=0$. Finally, we subtract this from the element in step 1, so that $0-(0)(1/1)=0$. This will be the entering value of the slack variable S_2 corresponding to X_1 in the second row of the second tableau, $S_{21}=0-(0)(1/1)=0$ (Table 7.2).

Table 7.2 The LP Problem: The Second Simplex Tableau

C_i	*Basis*	X_1	X_2	S_1	S_2	S_3	B_i
1	X_1	1	0	1	0	0	75,000
0	S_2	0	1	0	1	0	65,000
0	S_3	0	[12.5]	−10	0	1	[500,000]
	C_j	1	1	0	0	0	–
	Z_j	1	0	1	0	0	75,000
	$\Delta_j=C_j-Z_j$	0	[1]	−1	0	0	–

Pivot column (1); Pivot element (12.5); Pivot row (500,000)

We repeat the process to calculate the remaining elements of the second row of the second tableau:

$$S_{22} = 1 - (0)(0/1) = 1$$

$$S_{23} = 0 - (0)(1/1) = 0$$

$$S_{24} = 1 - (0)(0/1) = 1$$

$$S_{25} = 0 - (0)(0/1) = 0$$

$$S_{26} = 65{,}000 - (0)(75{,}000/1) = 65{,}000$$

where, as before, the first subscript of the variable S represents the row and the second subscript represents the column.

The elements of the third row can be obtained in the same way: first, we select the first element to be entered from the third (non-pivot) row of the first column in the initial tableau, which is 10 (Table 7.1). Next, we multiply the element of the remaining (third) row corresponding to the pivot element in the initial tableau, which is 10, by the new corresponding element in the replacing row (which is the first element of the first row of the second tableau) $1/1 = 1$ (Table 7.2). The result will be $(10)(1/1) = 10$. Finally, we subtract this from the element in step 1, so that $10 - (10)(1/1) = 10 - 10 = 0$. This will be the entering value of the slack variable S_3 corresponding to X_1 in the third row of the second tableau, $S_{31} = 10 - (10)(1/1) = 10 - 10 = 0$ (Table 7.2).

We repeat the process to calculate the remaining elements of the third row of the second tableau (Table 7.2):

$$S_{32} = 12.5 - (10)(0/1) = 12.5 - 0 = 12.5$$

$$S_{33} = 0 - (10)(1/1) = 0 - 10 = -10$$

$$S_{34} = 0 - (10)(0/1) = 0 - 0 = 0$$

$$S_{35} = 1 - (10)(0/1) = 1 - 0 = 1$$

$$S_{36} = 1{,}250{,}000 - (10)(75{,}000/1) = 1{,}250{,}000 - 750{,}000 = 500{,}000$$

where the first subscript of the variable S represents the row and the second subscript represents the column.

Once the coefficients of the row elements have been calculated, the method proceeds to calculate the values of Z_j and the Δ_j rows. The procedures for obtaining the values of Z_j and Δ_j rows are the same as before. Thus, to obtain the value of Z_1 in the second tableau, we multiply the elements of the C_j column by the elements of the first decision variable, X_1; that is, $Z_1 = (1)(1) + (0)(0) + (0)(0) = 1$.

We do the same for the remaining values of Z_j:

$$Z_2 = (1)(0) + (0)(1) + (0)(12.5) = 0$$

$$Z_3 = (1)(1) + (0)(0) + (0)(-10) = 1$$

$$Z_4 = (1)(0) + (0)(1) + (0)(0) = 0$$

$$Z_5 = (1)(0) + (0)(0) + (0)(1) = 0$$

$$Z_6 = (1)(75{,}000) + (0)(65{,}000) + (0)(500{,}000) = 75{,}000$$

As before, the values of the last row are obtained by taking the difference between the C_j and Z_j row. The result produces one negative value for the S_1 column and the rest are a combination of 0 and 1. As noted earlier, the rule for stopping is to make sure there are no positive values left in the Δ_jth row (and vice versa for a minimization problem). Since there is still a positive value left in the row corresponding to the X_2 column, we need to compute a third tableau using the same set of procedures as those in the second tableau.

The Third Simplex Tableau. Table 7.3 presents the third simplex tableau, obtained the same way as the second tableau. As can be seen from the tableau, all the elements in the Δ_j row are now less than or equal to 0, indicating that an optimum solution has been reached. No further calculations are necessary. The optimum solution for the problem, therefore, is: $X_1 = 75{,}000$ and $X_2 = 40{,}000$, with a combined total of 115,000 hours; that is, $Z = 75{,}000 + 40{,}000 = 115{,}000$. This is also what we obtained earlier with the graphical solution to the problem.

However, looking at the final solution we can see that S_2 is still in the basis, which indicates that the optimal solution did not fully satisfy the second constraint of all 65,000 hours of service for X_2. In fact, it is short by 25,000 hours, as shown under the B_i column. The absence of S_1 and S_3 in the basis indicates that these two constraints have been fully satisfied.

Shadow Prices

An important contribution of the simplex method, other than finding the optimal solution, is that it produces a variety of information that policy makers will find

Table 7.3 The LP Problem: The Third (Final) Simplex Tableau

C_i	*Basis*	X_1	X_2	S_1	S_2	S_3	B_i
1	X_1	1	0	1	0	0	75,000
0	S_2	0	0	0.8	1	–0.08	25,000
1	X_2	0	1	–0.8	0	0.08	40,000
	C_j	1	1	0	0	0.00	–
	Z_j	1	1	0.2	0	0.08	115,000
	$\Delta_j = C_j - Z_j$	0	0	–0.2	0	–0.08	–

useful for making allocations and other decisions. Of particular significance is shadow pricing. Shadow prices show the amount of latitude the decision makers have in utilizing the available resources. More specifically, they reflect the changes in the objective function as a result of marginal (one-unit) changes in the right-hand side (RHS) coefficients of the constraint equations, B_i.

Since the constraints are generally expressed in terms of equalities (=), less than or equal to (≤) for slack variables, and greater than or equal to (≥) for surplus variables, the changes are directly related to the nature of these constraints. For a less than or equal to constraint, the shadow prices represent the amount by which the value of the objective function will increase (for a maximization problem) or decrease (for a minimization problem) as a result of a one-unit increase (or decrease) in the input. For a greater than or equal to constraint, they indicate the amount by which the objective function will increase for a one-unit decrease in the input. In other words, the shadow price represents the maximum amount a decision maker will be willing to pay to acquire one additional unit of resource for a slack variable or to have a requirement relaxed by one unit for a surplus variable.

Interestingly, the information on shadow prices can be directly obtained from the final tableau of the simplex solution, especially the objective function (Z_j) row or from the dual of a primal problem. The values appearing under the slack and surplus variables corresponding to Z_j are really the shadow prices. Thus, in the current example, the shadow price for $S_1=0.2$, for $S_2=0$, and for $S_3=0.08$. What this means is that output will increase by 0.2 for a one-unit increase in the RHS coefficient for S_1, by 0 for a one-unit increase in the RHS coefficient in constraint for S_2, and by 0.8 for a one-unit increase in the RHS coefficient for S_3, as shown in Table 7.3. Since the second constraint, C_2, has a value of 0 for S_2, it means that its marginal contribution is 0. In other words, increasing the service for program X_2 by an additional unit will not change the optimal solution in the objective function. It is important to note, however, that shadow prices are valid within the specific range of additional units one can acquire. What this means is that one must be able to determine the amount of additional resources the government will need in order to realize the potential increases, as given by the shadow prices.

Sensitivity Analysis

As noted previously, prudent decision making requires that we make an effort to determine how sensitive an optimal solution is to discrete changes in the input parameters of a linear programming problem. For instance, what would happen if the objective function coefficients were different for any of the decision variables or what would happen if there was a change in the technical coefficients of the constraint equations? Similarly, what would be the nature of the optimal solution if some of the input resources (RHS coefficients) were different from those included in the model? Post-optimality or sensitivity analysis provides answers to these questions by offering the flexibility a model needs to determine the optimal solution under changed conditions. We can use the current example to examine how the questions raised above affect the optimal solution. We can consider them simultaneously or one at a time.

Changing the Objective Function Coefficients. An important aspect of sensitivity analysis is to deal with the effect the optimal solution will have as a result of a change or changes in the objective function coefficients. Assume that the government is going through some budget cuts and has imposed restrictions on the maximum number of hours it can afford to pay for the two programs. Let us suppose that it is 65 percent for program X_1 and 50 percent for program X_2 compared to what was initially planned. The new objective function will, therefore, be: maximize $Z=0.65X_1+0.5X_2$. Now applying the simplex algorithm to the problem with the same constraint equations as before would produce a solution of 64,000 hours, which is considerably lower than the previous solutions (Table 7.4a).

Alternatively, we could have substituted the optimal values for the decision variables into the objective function equation and solved it for Z. The result would have been the same as that produced by the simplex procedure: $Z=0.65X_1+0.5X_2=(0.65)(75{,}000)+(0.50)(40{,}000)=48{,}750+20{,}000=68{,}750$. Since the number of hours to be completed is much lower according to the revised estimate, this will leave a surplus or unused amount in the budget constraint to the tune of \$512,500; that is, \$1,250,000−[(\$10)(0.65 × 75,000)+ (\$12.50)(0.50)(40,000)] = \$1,250,000 − [(\$10)(48,750) + (\$12.50)(20,000)] = \$1,250,000−[\$487,500+\$250,000]=\$1,250,000−\$737,500=\$512,500.

Changing the Technological Coefficients. The optimal solution may also be affected by changes in the technological coefficients. Changes in technological coefficients can occur either due to measurement error or technological improvements, or some other factors that may actually reduce the requirement of a particular resource. Assume for a moment that our government has decided to lower the hourly payment to \$8.50 for program X_1 and to \$9.5 for program X_2, so that the new cost constraint will be $8.5X_1+9.5X_2\leq 1{,}250{,}000$. Assume further that nothing else has changed from the previous situation other than the hourly rates, the simplex solution for the revised problem would produce the target value of 139,474 hours (75,000+64,474), with no surplus in the budget constraint; that is, \$1,250,000−[(\$8.5)(75,000)+(\$9.5)(64,474)]=\$1,250,000−(\$637,500+ \$612,500)=\$1,250,000−\$1,250,000=\$0 (Table 7.4b).

Changing the RHS Coefficients. Finally, let us look at the effect on the optimal solution of changes in the availability of resources, vis-à-vis the RHS coefficients. The changes in the availability of resources imply that the resource constraints are either relaxed (increased) or restricted (decreased). Assume that the government is under budget pressure to lower the maximum amount it can spend on the two programs from \$1.25 million to \$1 million. Equation 7.12, which represents this constraint, will thus be $10X_1+12.5X_2 \leq 1{,}000{,}000$ to reflect this change. As before, we can solve the problem using the simplex procedure. The result would produce an optimal solution of 95,000 (75,000 hours for X_1 and only 20,000 hours for X_2), which is 20,000 hours less

Table 7.4 The LP Problem: Results of Sensitivity Analysis

a Changing the Objective Function Coefficients (Final Tableau)

C_j	*Basis*	X_1	X_2	S_1	S_2	S_3	B_i
0.65	X_1	1	0	1	0	0	75,000
0.50	X_2	0	1	–0.8	0	0.08	40,000
0	S_2	0	0	0.8	1	–0.08	25,000
	C_j	0.65	0.50	0	0	0	–
	Z_j	0.65	0.50	0.25	0	0.04	68,750
	$\Delta_j = C_j - Z_j$	0	0	–0.25	0	–0.04	–

b Changing the Technological Coefficients (Final Tableau)

C_j	*Basis*	X_1	X_2	S_1	S_2	S_3	B_i
1	X_1	1	0	1	0	0	75,000
1	X_2	0	1	–0.89	0	0.10	64,474
0	S_2	0	0	0.89	1	–0.10	562
	C_j	1	1	0	0	0	–
	Z_j	1	1	0.10	0	0.10	139,474
	$\Delta_j = C_j - Z_j$	0	0	–0.10	0	–0.10	–

c1 Changing the RHS Coefficients (Final Tableau)

C_j	*Basis*	X_1	X_2	S_1	S_2	S_3	B_i
1	X_1	1	0	1	0	0	75,000
0	S_2	0	0	0.8	1	–0.08	45,000
0	X_2	0	1	–0.8	0	0.08	20,000
	C_j	1	1	0	0	0	–
	Z_j	1	1	0.2	0	0.08	95,000
	$\Delta_j = C_j - Z_j$	0	0	–0.2	0	–0.08	–

c2 Changing the RHS Coefficients (Final Tableau)

C_j	*Basis*	X_1	X_2	S_1	S_2	S_3	B_i
1	X_1	1	0	1	0	0	60,000
0	S_2	0	0	0.8	1	–0.08	18,000
0	X_2	0	1	–0.8	0	0.08	32,000
	C_j	1	1	0	0	0	–
	Z_j	1	1	0.2	0	0.08	92,000
	$\Delta_j = C_j - Z_j$	0	0	–0.2	0	–0.08	–

than the original solution of 115,000 hours, but will use up all $1 million (Table 7.4c1).

We can do the same with the other two constraints. Let us say that we lower the first constraint (service requirement for program X_1) to 60,000 hours and the second constraint (service requirement for program X_2) to 50,000 hours, with a total of 110,000 hours. The result would produce an optimal solution of 92,000

(60,000 hours for X_1 and 32,000 hours for X_2), which is 3,000 hours lower than the previous solution of 95,000 and 18,000 hours short of the overall target of 110,000 hours, but, as before, it will use up all $1 million (Table 7.4c2). If our objective is to change the RHS coefficients marginally (say, by one unit) instead of by a large amount, the resultant effect on the objective function will be the shadow price, discussed above.

The Dual

Every maximization (minimization) problem in linear programming has a corresponding minimizing (maximizing) problem. The original problem is called the primal and the corresponding problem is called the dual. The primary purpose of using a dual is to determine the accuracy of the primal. Another reason why the dual is used is that it is sometimes easier to solve a linear programming problem first by solving its associated dual than solving the primal problem directly, especially when the primal has a large number of variables but few constraints. For instance, a primal problem with n decision variables and two constraints can be easily transformed into a two-variable problem that can be solved graphically. Finally, the duality in linear programming provides useful information that can lead to a better understanding of the problem solution.

Properties of the Dual

All dual problems have three important properties. First, the optimal value of the primal objective function is always equal to the value of the dual objective function, provided that an optimal feasible solution exists. Second, if a decision variable in the objective function has a value that is non-zero, the corresponding slack (or surplus) variable in the dual problem must have an optimal value of zero. Finally, if a slack (or surplus) variable in the primal has a value that is non-zero, the corresponding decision variable in the dual must have an optimal value of zero. Although all linear programming problems have their duals, interpreting the dual solution can be complicated unless the problem is properly converted into its dual first. The following rules are commonly used in converting a standard linear programming problem into its corresponding dual:

1. The direction of optimization is reversed in that maximization in the primal becomes the minimization in the dual, and vice versa.
2. The inequalities of the constraint equations are reversed in that the less than or equal to becomes greater than or equal to, and vice versa, while non-negativity constraints remain unchanged.
3. The constraints rows in the primal become the columns in the dual.
4. The RHS constraints in the primal become the objective function coefficients in the dual, while the objective function coefficients become the RHS constraints in the dual.

5 The decision variables in the primal, X_js, become the decision variables in the dual, Y_js.

To illustrate the procedure, let us go back to the community service problem and convert it to a dual problem without incorporating any changes in the original values or parameters, so that the new problem will look as follows:

$$\text{Primal: maximize } Z = X_1 + X_2 \quad [7.9]$$

Subject to

$$X_1 \leq 75{,}000 \quad [7.10]$$

$$X_2 \leq 65{,}000 \quad [7.11]$$

$$10X_1 + 12.50X_2 \leq 1{,}250{,}000 \quad [7.12]$$

$$X_1, X_2 \geq 0 \quad [7.13]$$

$$\text{Dual: minimize } C = 75{,}000Y_1 + 65{,}000Y_2 + 1{,}250{,}000Y_3 \quad [7.24]$$

Subject to:

$$Y_1 + 10Y_3 \geq 1.0 \quad [7.25]$$

$$Y_2 + 12.5Y_3 \geq 1.0 \quad [7.26]$$

$$Y_1, Y_2, Y_3 \geq 0 \quad [7.27]$$

where Y_js are the new decision variables, and C is the objective function to be minimized.

Surplus and Artificial Variables. Interestingly, the simplex solution to the dual of a linear programming problem is more complicated, especially when the objective is to find the minimum value for a problem with greater than or equal to constraints. It needs two sets of new variables instead of one: a surplus and an artificial. The surplus variable tells us how much the solution exceeds the constraint resource, B_i, as opposed to the slack variable, which tells us how much the solution produces unused or below the constraint resource, B_i. Since a surplus is the opposite of a slack variable, it is often called a negative slack. As a general rule, the surplus variables are subtracted from greater than or equal to constraints to convert them to strict equalities, just the way the slack variables are added to less than or equal to constraints to convert them to strict equalities.

To convert the first constraint to strict equality, we simply subtract a surplus variable, S_1, from it so that Equation 7.25 becomes:

$$Y_1 + 10Y_3 - S_1 = 1.0 \quad [7.28]$$

Assume that we have a solution for the linear programming problem that produces a value of 75,000 for Y_1 and 40,000 for Y_2, as in the original solution. We can use this information to find the amount of the surplus or unused resource for the first constraint as:

$$Y_1 + 10Y_3 - S_1 = 1.0$$

$$1(75{,}000) + (10)(40{,}000) - S_1 = 1.0$$

$$475{,}000 - S_1 = 1.0$$

$$-S_1 = 1.0 - 475{,}000$$

$$S_1 = 474{,}999$$

or 474,999 hours.

There is a minor problem in trying to use the first constraint in setting up the initial simplex tableau, in that the surplus variables take on a negative value when the "real" variables such as X_js are set to 0 in the initial tableau, as can be seen below:

$$Y_1 + 10Y_3 - S_1 = 1.0$$

$$1(0) + (10)(0) - S_1 = 1.0$$

$$-S_1 = 1.0$$

$$S_1 = -1.0$$

Since the non-negative constraint of linear programming requires that all variables – real, slack, and surplus – must be non-negative, the solution violates the condition. To correct the problem, we need to introduce a new variable called the artificial variable, A, to the constraint equation, so that the new equation will appear as:

$$Y_1 + 10Y_3 - S_1 + A_1 = 1.0 \qquad [7.29]$$

The new equation will allow us to set the decision, as well as the surplus variables, equal to 0 in the initial simplex tableau. The resultant solution will leave us with $A_1 = 1.0$. That is:

$$1(Y_1) + (10)(Y_3) - S_1 + A_1 = 1.0$$

$$1(0) + (10)(0) - 0 + A_1 = 1.0$$

$$A_1 = 1.0$$

Similarly, we can write the second constraint equation as:

$$Y_2 + 12.5Y_3 - S_2 + A_2 = 1.0 \qquad [7.30]$$

In general, whenever an artificial or surplus variable is added to one of the constraints, it must be added to other constraints, including the objective function, as we did with the slack variables. The complete linear programming problem (for the dual) can now be presented as:

$$\text{Minimize } C=75{,}000Y_1+65{,}000Y_2+1{,}250{,}000Y_3+S_1+S_2+A_1+A_2 \quad [7.31]$$

Subject to:

$$Y_1+0Y_2+10Y_3-S_1-0S_2+A_1+0A_2=1.0 \quad [7.32]$$

$$0Y_1+Y_2+12.5Y_3-0S_1-S_2+0A_1+A_2=1.0 \quad [7.33]$$

$$Y_1, Y_2, S_1, S_2, A_1, A_2 \geq 0 \quad [7.34]$$

The Big *M* Method. Artificial variables do not have any real economic meaning in linear programming; they are introduced essentially to generate the initial linear programming solution without which we would have been left with negative surplus values. Since they do not have any real economic meaning, we need to force them out of the solution. One way to do this is to assign a very high dollar value to each artificial variable. The low-dollar-value variables are usually more desirable in minimization problems and, as such, are the first to enter the solution. The high-dollar value variables, on the other hand, hardly enter the solution, or leave it as quickly as possible. Rather than set an unspecified high-dollar value for each artificial variable, the convention is to use the letter M to represent an unspecified large value called the big M method.[1] Surplus variables, on the other hand, carry a zero dollar value.

The revised problem with all the surplus, artificial, and big M variables can thus be expressed as:

$$\text{Minimize } C=75{,}000Y_1+65{,}000Y_2+1{,}250{,}000Y_3+0S_1+0S_2+MA_1+MA_2 \quad [7.35]$$

Subject to:

$$Y_1+0Y_2+10Y_3-1S_1-0S_2+1A_1+0A_2=1.0 \quad [7.36]$$

$$0Y_1+Y_2+12.5Y_3-0S_1-1S_2+0A_1+1A_2=1.0 \quad [7.37]$$

$$Y_1, Y_2, S_1, S_2, A_1, A_2 \geq 0 \quad [7.38]$$

The Simplex Solution

The solution of the dual minimization problem is very similar to the maximization problem, described earlier. The major difference is in the C_j-Z_j row. Unlike the maximization problem, the new variable to enter the solution in each simplex tableau (the pivot column variable) is the one with a negative C_j-Z_j value. This means that the value of the objective function, C, will decrease if that variable is selected to enter the solution. In other words, each variable to enter the solution is the one with a negative C_j-Z_j that will produce the greatest improvement in the objective function. Other than that, the steps are very much the same as in the maximization problem. The final solution is obtained when the values of the decision, as well as artificial variables in the C_j-Z_j row, are 0 or positive, which is the exact opposite of the maximization problem. Table 7.5 shows the final solution to the dual problem.

As shown in Table 7.5, the solution to the dual problem is $X_1=0$, $Y_3=0$, and $Z_j=115{,}000$, which corresponds to the optimal solution obtained for the primal (maximization) problem. However, the presence of $Y_2=-25{,}000$ indicates that the optimal solution did not fully satisfy the second constraint of all 65,000 hours of labor for X_2. In fact, it is short by 25,000 hours.

Special Problems in Linear Programming

Conventional wisdom suggests that as long as the basic conditions of the model requirement are satisfied, linear programming will produce an optimal solution in every instance. In reality, that is not always the case. There are circumstances in which it will be difficult to obtain an optimal solution. The following section provides a brief description of three situations in which an optimal solution is not possible: infeasible solution, unbounded solution, and multiple optimal solutions.

Infeasible Solution. There are cases in linear programming when it is not possible to obtain a solution that will satisfy all the model constraints, in which case it is known as an infeasible solution. In general, in an infeasible solution there is no common or feasible region such that it is possible to identify a single point that can satisfy all the model constraints, as we saw in our two project investment programs problem. Therefore, regardless of the nature of the objective function, it will not produce a feasible solution. On the other hand, an infeasible solution does not necessarily mean that the problem is completely insolvable; it simply means that additional resources may be needed to produce an optimal solution.

Unbounded Solution. In some linear programming problems, a situation may exist that will produce an infinitely large-valued solution for the objective function without necessarily violating any of the objective function constraints in a problem. This is generally known as the unbounded solution. In most instances, an unbounded solution occurs when the problem has not been correctly specified, such as unlimited cost, unlimited return, or output, but if a problem is correctly formulated it is unlikely that it will produce an infinitely large solution.

Multiple Optimal Solutions. Occasionally, one may come across a situation in which a linear programming problem will produce multiple optimal solutions rather than a single optimal solution. This usually occurs when the slope of the

Table 7.5 The Dual: The Final Simplex Solution

C_j	*Basis*	Y_1	Y_2	Y_3	S_1	S_2	A_1	A_2	B_i
75,000	Y_1	1	–0.80	0	–1	0.80	1	0.80	0.20
1,250,000	Y_3	0	0.80	1	0	–0.08	0	–0.08	0.08
	C_j	75,000	65,000	1,250,000	0	0	0	0	–
	Z_j	75,000	90,000	1,250,000	75,000	40,000	–75,000	–40,000	115,000
	$\Delta_j=C_j-Z_j$	0	–25,000	0	–75,000	–40,000	75,000	40,000	

objective function and one of the constraints are identical, forcing the optimal point to lie on the line segment generated by the constraints, unlike our investment problem. However, multiple optimal solutions are not always undesirable, because they increase the range of decision options which may allow for greater flexibility, especially when a decision maker is required to choose from several different optimal solutions.

In addition to the above, there are at least three other technical problems that are worth noting here. First, there may be a situation in which one of the basic variables in the final solution may contain a value of zero, which can lead to a degenerate solution known as degeneracy. Degeneracy occurs when there are ties in the rows during pivoting. Second, it is quite possible that one may deal with a problem in which one of the constraints does not have any effect on the feasible solution. This usually occurs when the constraints lie outside the boundaries of the feasible region. These types of constraints are called redundant constraints. Like ordinary outliers in statistical analysis, the simplest way to deal with this type of problem is to throw out these variables from the model since their contribution to the optimal solution is zero. Third, which is not really a technical problem, is a situation in which it is not possible to apply linear programming. This occurs when the objective function and/or any of the constraints are nonlinear, in which case one should use nonlinear programming. Although quite common, nonlinear programming models are often complex, requiring complex algorithms to solve because of the inherent computational difficulties.

Integer Programming

In many real-world situations, solutions to linear programming problems make sense only if they have integer values. Optimal solutions such as 79,920.64 meals during the summer obtained for the linear programming problem does not seem very realistic, although it is acceptable when rounded off to 79,921. That is what we basically did, but there are situations where simply rounding off values to the nearest whole number (integer) may not necessarily produce an optimal solution. Also, forcing an integer solution in some cases may even violate the constraints. The alternative to dealing with this kind of problem is to use a closely related programming method, known as integer programming

Integer programming is a natural extension of linear programming with an added constraint in that the decision variable, X, must have integer values. When the decision variables in a programming problem are all integer valued, it is called a pure integer programming problem; otherwise, it is called a mixed-integer programming problem. Mixed-integer programming problems are not uncommon. In fact, for many real-world problems, it is rather common to find continuous decision variables where a mixed-integer programming solution would be more appropriate than either a linear or a pure integer programming solution.

Integer Programming Solution

Since integer programming is an extension of linear programming, its solution follows very much the same procedure as the one used in linear programming. In other words, one can treat it as a linear programming problem by ignoring the integer requirement, then solving it using the simplex method as in linear programming. If the solution produces all integer values, we have an optimal solution for the problem; if not, there are alternatives one can use to deal with the problem. We will discuss here one such method that has become an integral part of integer programming solution procedure, called the branch-and-bound method.

Developed primarily by Land and Doig (1960) and subsequently improved by Little, Murty, Sweeney, and Karel (1963) and others, the branch-and-bound method proceeds by branching off a feasible non-integer solution into a set of subproblems in such a way that continuous solutions that do not satisfy the integer requirement can be easily eliminated in favor of discrete solutions. The branching process is accomplished by introducing mutually exclusive constraints so as to satisfy the integer requirement of the problem, while guaranteeing the inclusion of a feasible integer solution.

Let us look at a simple problem to illustrate the method. Suppose that a County Commission wants to conduct a public-opinion survey using both door-to-door and telephone interviews. Let us assume that it takes an interviewer an hour to complete five telephone and three door-to-door interviews. The survey requires at least 15 interviews per hour. The Commission currently has three summer interns with some experience in door-to-door interviews and will need additional interviewers to complete the survey. Assume further that it costs $7.25 per hour for telephone interviews and $8.50 per hour for door-to-door interviews. The Commission's objective is to find the number of interviewers that will produce the maximum number of interviews per hour at the lowest cost to the Commission.

We can formally state the problem as follows:

$$\text{Minimize } Z=7.25X_1+8.5X_2 \quad [7.39]$$

Subject to:

$$5X_1+3X_2\geq 15 \quad [7.40]$$

$$X_2\geq 3 \quad [7.41]$$

$$X_1,X_2\geq 0 \quad [7.42]$$

where X_1 represents telephone interviews, and X_2 represents door-to-door interviews.

To solve the problem we begin with the simplex method, as usual. The optimal simplex solution for the problem is: $X_1=1.2$, $X_2=3$, and $Z=\$34.20$.

What this means is that in addition to the three interviewers it already has, the Commission will need 1.2 interviewers to conduct the telephone surveys. This is not a realistic solution since one cannot have fractional interviewers. However, the solution represents the initial (non-integer) upper bound, which can be rounded down for a lower bound (integer) solution at $X_1=1$, $X_2=3$, and $Z=34$, but this will not be a normal integer solution. Therefore, to obtain a proper solution to the problem we use the branch-and-bound method.

We start by dividing the initial real solution into two parts, P_1 and P_2, for the variable with the highest fractional value, then eliminating the fractional part in favor of an integer solution. In the current example, the variable with the highest (only) fractional value is X_1. Next, we create two mutually exclusive constraints closest to the fractional value of X_1. By creating two new constraints, $X_1 \leq 1$ and $X_1 \geq 2$, we obtain two new problems. The new constraints, in effect, eliminate all possible fractional values of X_1 between 1 and 2, leaving us with fewer integer solutions to decide from.

The integer problem with new constraints can thus be written as P1:

$$\text{P1: minimize } Z = 7.25X_1 + 8.5X_2 \qquad [7.43]$$

Subject to:

$$5X_1 + 3X_2 \geq 15 \qquad [7.44]$$

$$X_1 \leq 1 \qquad [7.45]$$

$$X_2 \geq 3 \qquad [7.46]$$

$$X_1, X_2 \geq 0 \qquad [7.47]$$

$$\text{P2: minimize } Z = 7.25X_1 + 8.5X_2 \qquad [7.48]$$

Subject to:

$$5X_1 + 3X_2 \geq 15 \qquad [7.49]$$

$$X_1 \geq 2 \qquad [7.50]$$

$$X_2 \geq 3 \qquad [7.51]$$

$$X_1, X_2 \geq 0 \qquad [7.52]$$

The simplex solutions to the new problems are as follows: $X_1=1.00$, $X_2=3.33$, and $Z=-\$35.58$ for P1 and $X_1=2.00$, $X_2=3.00$, and $Z=\$40.00$ for P2. Thus, P2 produces an all-integer solution with an optimal number of interviews that can be conducted in an hour for both telephone and door-to-door surveys. The optimal number is 19; that is, $5X_1+3X_2=5(2)+3(3)=19$. The corresponding Z value or cost per hour for five interviews, two telephone and three door-to-door, is \$40; that is, $7.25X_1+8.SX_2=\$7.25(2)+\$8.5(3)=\$40$. On the other hand, P1 yields a non-integer solution for X_2, meaning that it needs to be branched off further into two additional subproblems, P1A and P1B, with constraints $X_2 \leq 3$ and $X_2 \geq 4$, respectively. Hopefully, this will eliminate all possible fractional values between 3 and 4.

The integer problem with revised constraints can, therefore, be written as:

$$\text{P1A: minimize } Z=7.25X_1+8.5X_2 \quad [7.53]$$

Subject to:

$$5X_1+3X_2 \geq 15 \quad [7.54]$$

$$X_1 \leq 1 \quad [7.55]$$

$$X_2 \leq 3 \quad [7.56]$$

$$X_1, X_2 \geq 0 \quad [7.57]$$

$$\text{P1B: minimize } Z=7.25X_1+8.5X_2 \quad [7.58]$$

Subject to:

$$5X_1+3X_2 \geq 15 \quad [7.59]$$

$$X_1 \leq 1 \quad [7.60]$$

$$X_2 \geq 4 \quad [7.61]$$

$$X_1, X_2 \geq 0 \quad [7.62]$$

Interestingly, the simplex method produces an infeasible solution for P1A (since an artificial variable still remains in the final tableau, not shown here), and a non-integer solution for P1B; that is, $X_1=0.60$, $X_2=4.00$, and $Z=\$38.35$. Since P1B produces a non-integer solution we need to repeat the process by introducing two subproblems for X_1: P1Ba, with a constraint of ≤ 0, and P1Bb, with a constraint of ≥ 1.

As before, the integer problem with revised constraints can be written as:

$$\text{P1Ba: minimize } Z=7.25X_1+8.5X_2 \quad [7.63]$$

Subject to:

$$5X_1+3X_2 \geq 15 \quad [7.64]$$

$$X_1 \leq 0 \quad [7.65]$$

$$X_2 \geq 4 \quad [7.66]$$

$$X_1, X_2 \geq 0 \quad [7.67]$$

P1Bb: minimize $Z = 7.25X_1 + 8.5X_2$ [7.68]

Subject to:

$$5X_1 + 3X_2 \geq 15 \quad [7.69]$$

$$X_1 \geq 1 \quad [7.70]$$

$$X_2 \geq 4 \quad [7.71]$$

$$X_1, X_2 \geq 0 \quad [7.72]$$

The result of the last branching effort produces an integer solution for the new subproblems, indicating that no additional branching will be needed. The optimal simplex solutions for the two subproblems, therefore, are: $X_1 = 0$, $X_2 = 5.00$, $Z = \$42.50$ for P1Ba, and $X1 = 1.00$, $X_2 = 4.00$, $Z = \$41.25$ for P1Bb. Figure 7.2 shows all the branches for the problem in sequence, including the optimal solution obtained at each. According to the figure, the last two solutions (although they produce integer values) are inferior to the optimal solution produced initially for P2. The cost of interviews is also higher in both cases. To provide some statistics, the number of interviews to be conducted came out to be lower for both P1Ba (15) and P1Bb (17), compared to the initial solution of 19 for P2. Therefore, the final optimal solution for the problem will be 19 interviews per hour, with two individuals doing the telephone interviews and three doing the door-to-door interviews.

The interesting thing about the problem here is that it deals with a situation in which the solution values for the decision variables can be any number as long as it is an integer, thus making it possible to take advantage of methods such as branch-and-bound to obtain an integer solution. However, there are situations in which the solution values can be restricted to binary integers such as 0 or 1, in which case alternative methods such as the implicit enumerative procedure (Balas, 1965) can be used to produce an optimal solution. Capital budgeting, project scheduling, and location–allocation decisions are among the best-known examples for which the solutions require binary values or integers.

Dynamic Programming

Another optimization technique frequently used in decision problems is dynamic programming. Like linear and integer programming, dynamic programming

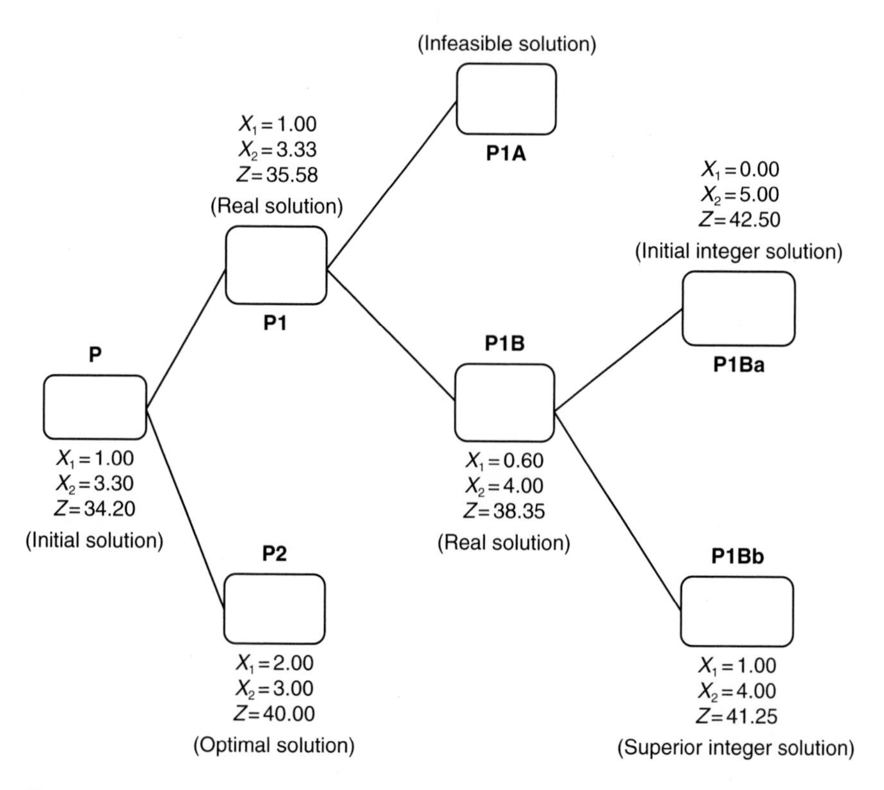

Figure 7.2 A Branch-and-Bound Integer Solution

attempts to optimize an objective function subject to a set of constraints but, unlike the first two, it divides a problem into sets of interrelated components called stages, where each stage produces an optimal solution. Because of the way in which a problem is structured into interrelated components, the decision made at one stage has a direct bearing on the decision made at the next stage. Furthermore, the decision made at a given stage must not only take into consideration its effects on the next stage, but also on all subsequent stages in a recursive manner. The term recursive means what happens in one stage has a direct consequence on all future stages, but has no effect on what took place in previous stages. The latter is called recycling. The ultimate objective of dynamic programming is to find an optimum combination of decisions that will, over time, optimize the overall outcome.

Dynamic programming was introduced by Bellman (1957) based on a concept known as the principle of optimality. The principle states that an optimal decision (policy) has the property that whatever the initial state or initial decisions, the remaining decisions must constitute an optimal decision (policy) with regard to the state resulting from the first decision. This simply means that no matter what the initial states or decisions, the remaining decisions will constitute an optimal decision (policy) with regard to the information derived from the first decision. Another way of stating the principle is to suggest that if an incorrect decision has been made in the first or second stage, it does not prevent a decision maker from making the correct decision in future stages.

Dynamic Programming Solution

Like linear programming, the procedure for solving a dynamic programming problem involves several basic steps. They are (1) determining what decision variables to include and setting up the objective function to be optimized subject to a set of constraints; (2) specifying the stages of the problem and determining the variables called states, whose values constitute the basis of decisions at each stage; (3) identifying the recursive relationship between different stages and finding the optimal solution; and (4) presenting the results in a table.

A simple example will help illustrate the procedure. Suppose a state Treasury department wants to invest $3 million in three different securities, *X*, *Y*, and *Z*. Assume that the return each security can produce ranges from $10,000 to $60,000, depending on the amount invested. The objective of the department is to find the combination of securities that will produce the maximum return from the investment of all $3 million. We can formally present the problem as follows:

$$\text{Maximize } \Sigma R_{X,Y,Z} = R_X + R_Y + R_Z \qquad [7.73]$$

Subject to:

$$V_X + V_Y + V_Z \leq K \qquad [7.74]$$

$$R, V, K \geq 0 \qquad [7.75]$$

where *R* is the return on investment, *V* is the amount invested in each security, and *K* is the total amount available for investment.

Table 7.6 presents this basic information as well as the results of the optimal combination of investments. The first part of the table shows the return on investment by security type and amount. The second part of the table shows all possible combinations for investing $3 million or less between two of the high-yielding securities, *Y* and *Z*. For instance, when all $3 million are invested in *Y* and *Z*, the maximum return that could be earned is $40,000 by investing $2 million in *Y* and $1 million in *Z*. Similarly, when the same $3 million are invested in *Y* and *Z*, the maximum return that could be earned is $50,000 by investing $1 million in *Y* and $2 million in *Z*, and so on. The second part of the table presents a summary of investment returns for combinations of the two securities.

The third part of the table shows that when the securities *Y* and *Z* are considered simultaneously, with $1 million invested in *Y* and $2 million in *Z*, the

Table 7.6 The DP Solution: Investment and Return*

V_x	R_x	V_y	R_y	V_z	R_z
0	0	0	0	0	0
1	10	1	15	1	15
2	15	2	25	2	35
3	25	3	35	3	45

V	V_y	R_y	V_z	R_z	*Optimal* R_{y+z}
3	3	35	0	0	35
	2	25	1	15	40
	1	15	2	35	50
	0	0	3	40	40
2	2	25	0	0	25
	1	15	1	15	30
	0	0	2	35	35
1	1	15	0	0	15
	0	0	1	15	15
0	0	0	0	0	0

$V_{y,z}$	V_y	V_z	*Optimal* R_{y+z}
0	0	0	0
1	1	0	15
2	0	2	35
3	1	2	50

$V_{x,y,z}$	V_x	R_x	*Optimal* V_{y+z}	R_{y+z}	*Optimal* R_{x+y+z}
3	0	0	3	50	50
	1	10	2	35	45
	2	15	1	15	30
	3	25	0	0	25

Note
* *V* = Securities Investment ($ Million); *R* = Return on Investment ($1,000).

optimal return from the combination will be $50,000. The fourth and final part of the table shows the tabulation for all three securities taken together, but this requires the consideration of security *X* in conjunction with the optimal policy for *Y* and *Z*. As the table shows, the optimal policy will be to invest $0 in *X* and $3 million in *Y* and *Z* (using a combination of $1 million in *Y* and $2 million in *Z*). This, in turn, will produce a maximum return of $50,000.

Although the optimum decision in the current example is to invest $1 million in *Y* and $2 million in *Z*, the department had a choice to stop at any of the earlier stages during this process and select the combination that would have best fitted its objectives. This unique characteristic is what makes dynamic programming interesting and perhaps more useful than most programming techniques. In fact, one can think of numerous situations in which a dynamic programming model would be appropriate, such as finding an optimal route at a minimum cost or determining the optimum output plan by adjusting seasonal fluctuations in demand for services such as water or electricity, as in production smoothing. Even government inventories could be treated as a dynamic programming problem if the objective could be couched in terms of determining a policy that would minimize the expected cost resulting from situations such as shortage or stockout. All of these problems have one thing in common: they are all multi-stage problems that can be solved through a sequence of decisions similar to the problem discussed here.

Heuristic Programming

The fourth method discussed here is heuristic programming. Heuristic programming is useful in decision problems where it is not required to have an optimal solution based on standard methods of solution such as those discussed under earlier programming methods. This does not mean that heuristic programming does not yield an optimum solution; it simply means that it is not a prerequisite for problem solving. In fact, good heuristics or rules of thumb, as they are often called, are capable of producing optimum solutions, but there are cases where the most one could expect is a solution that comes as close to being optimal as possible. These types of solutions are called semi-optimal or near-optimal solutions.

There is an advantage of using heuristic programming in that it takes less time to find the solution to a problem than most conventional programming techniques. Heuristic programming is particularly suitable for problems which are poorly defined or ill-structured (i.e., those, for example, that cannot be expressed in precise mathematical terms and lack the attributes essential for formal problem structuring that other programming techniques do. Many social and administrative problems fall into this category. When faced with these kinds of situations, a decision maker would be satisfied with a near-optimal solution rather than going through the rigorous process of obtaining an optimum solution that is difficult to establish.

Heuristic Programming Solution

Solving a heuristic programming problem is more of an art than a science. Much depends on how the problem is structured and the extent to which it is amenable to an optimal solution. There are several different methods currently available that one can use to solve a heuristic programming problem, all of which are capable of producing near-optimal to optimal solutions. We present here two such methods that have become quite common in heuristic solution methodologies: the northwest corner rule and the greedy heuristic.

Northwest Corner Rule. As the name implies, the northwest corner rule produces solutions to a heuristic programming problem that lie at the extreme northwest corner of a matrix or a table containing all the relevant information on the problem. The matrix is set up in such a way as to ensure that the northwest corner always has the highest value. The solution starts by allocating all the resources to the cell in the northwest corner, then moving on to the corner with the next highest value and allocating the maximum available resources to the cell corresponding to that corner, and continuing with the process until all the resources have been fully allocated. It is important to note that since the allocation process begins by assigning all or most of the available resources to the cell with the highest corner value, it fails to take into account the cells with lower corner values, which may have a direct bearing on the solution. This is especially true in situations in which the objective is to minimize costs or reduce services. Consequently, it tends to produce results that are not always optimal.

Let us look at a simple example to illustrate the procedure. Suppose that the transportation department of a large metropolitan government wants to determine how best it could provide transit services to three different parts of the city that will produce the greatest amount of savings for the government. Table 7.7 presents information on the origin and destination points along with information on the costs of operation for each trip (transit service), actual demand for service, and the number of daily trips that could potentially be offered from each point of origin to their points of destination.

According to the table, it costs the department $50 per trip from point *A* to point *X*, $40 from point *B* to point *Y*, and $50 from point *C* to point *Z*, and so on. The number of trips that could potentially be made from points *A*, *B*, and *C* to points *X*, *Y*, and *Z* are 10, 15, and 25, respectively, which may be different from the actual demand for services. The top of the table shows the northwest corner solution for the problem. Since the solution requires that we start with the northwest corner, we allocate all 10 trips to the cell corresponding to this corner, *AX*, at a total cost of $500 or $50 per trip. The next cell with the highest corner value is *CZ*. Theoretically we should be allocating all the available trips to this cell since it has the highest corner value, but in reality we can allocate only 15 trips out of a total of 25. To allocate more will force us to exceed the maximum demand for service to the destination, *Z*. Of the remaining 10 trips,

Table 7.7 HP: Northwest Corner Solution

From \ *To*		*X*		*Y*		*Z*	*Service Potential*
A	10	$50		$40		$25	10 ($500)
B		$30	15	$40		$20	15 ($600)
C	5	$35	5	$30	15	$50	25 ($1,075)
Actual demand	15		20		15		50 ($2,175)

we allocate five to *CX* and five to *CY*, in each case making sure we do not exceed the limit.

We continue the process until all 50 trips have been fully allocated to the respective destinations. It should be worth noting that if a destination failed to receive a service allocation, it is more than likely that it was not operationally viable unless, of course, the government is willing to subsidize the service in its entirety. As the table shows, the total cost of operation for all 50 trips produced by the northwest corner rule *is* $2,175. It is obvious that the result will not be optimal since the allocations were made on the basis of the highest cost corners. However, it provides a basis for comparing the results produced by other solution methods. To determine the optimum number of trips at which the savings will be the greatest for the department, let us turn to the next solution method, the greedy heuristic.

The Greedy Heuristic. Like the northwest corner rule, the greedy heuristic also produces a quick solution. However, the difference between the two is that instead of starting with the highest northwest corner value, the greedy heuristic begins with the lowest corner value in the matrix (Table 7.8). It allocates all or most of the resources to the cell corresponding to that corner, and then moves on to the cell corresponding to the second lowest corner value by allocating the maximum permissible amount of resources to that cell, and so forth until all the resources to be allocated have been exhausted.

Looking back at the problem in the table, we can see that the cell with the lowest corner value is *BZ*, with $20 per trip. Therefore, we allocate all 15 trips to this cell. The cell with the next lowest cost is *AZ*, with $25, but no allocation was made to this cell since destination *Z* has already reached its limit of 15. We continue the process until all 50 trips have been fully allocated to their respective destinations. The total cost of operation resulting from the greedy solution amounts to $1,575, which is significantly lower than the results obtained by the northwest corner rule.

Interestingly, although the greedy heuristic showed a substantial improvement over the solution produced by the northwest corner rule, it has a weakness in that it does not have the ability to absorb more than one type or category of information

Table 7.8 HP: Greedy Heuristic Solution

From \ *To*	*X*		*Y*		*Z*		*Service Potential*
A	10	$50		$40		$25	10 ($500)
B		$30		$40	15	$20	15 ($300)
C	5	$35	20	$30		$50	25 ($775)
Actual demand	15		20		15		50 ($1,575)

at any given time. This may be critical in situations in which one would need information on multiple categories to produce an optimal solution. There are methods such as Vogel Approximation that are better equipped to deal with problems that require a diverse range of information, but for most ordinary problems the northwest corner rule or the greedy heuristic should be adequate.

Goal Programming

An interesting characteristic of the models discussed up to this point is that they all deal with a single goal or objective. In reality, one often encounters problems when one has to deal with not one but several different and frequently conflicting goals and objectives. The conventional programming models, discussed earlier, are less than sufficient to deal with these kinds of problems. Thus, when faced with an optimization problem with multiple goals and objectives, the most appropriate method to use is goal programming.

Introduced by Charnes and Cooper (1961) and subsequently refined by Lee (1972), Ignizio (1976), and Steuer (1986), and others, goal programming is treated as an extension of the linear programming model. However, instead of maximizing or minimizing the objective function directly, as in linear programming, goal programming attempts to minimize the deviations from goals sequentially. The deviations, represented by a set of variables, called *d*s, take on major significance in the goal programming formulation as the objective function becomes the minimization of these variables based on a set of weights and priorities assigned to them.

The following is a typical structure of a goal programming model:

$$\text{Minimize } Z = \sum_{j=1}^{m} P_j d_j \qquad [7.76]$$

Subject to:

$$\sum_{i=1}^{n} \sum_{j=1}^{m} a_{ij} X_j + d_i^+ - d_i^- = B_j \qquad [7.77]$$

$$a, X, d, B \geq 0 \qquad [7.78]$$

where m goals are expressed by some estimated targets B, P represents the priorities assigned to the deviational variable d, and Xs represent the decision variables. The as represent the coefficients associated with goal and other constraints (for $i = 1, 2, 3 \ldots, n$ and $j = 1, 2, 3 \ldots, m$) and $n > m$.

Operationally, each of the m goals is analyzed in terms of whether overachievement or underachievement of a goal is satisfactory. For instance, if the objective is to achieve a minimum level of a goal, negative deviation (d^-) should be left in the objective function; if the objective is to exceed a minimum level of a goal, positive deviations (d^+) should be left in the objective function. On the other hand, if the objective function is to come as close as possible to a specified level of a goal, both positive and negative deviations $(d^{+}+d^{-})$ should be included in the objective function. Likewise, if the objective is to maximize the value relative to a given level of goal, both negative and positive deviations $(d^{-}-d^{+})$ should appear in the objective function. Finally, if the objective is to minimize the value achieved relative to a goal level, both positive and negative deviation $(d^{+}-d^{-})$ should be included in the objective function.

The constraints of the model are also worth noting. There are two types of constraints one would find in a typical goal programming problem: environmental and goal. The environmental constraints represent resource limitations or restrictions imposed by the decision environment, whereas the goal constraints represent policies of the decision makers. The environmental constraints are similar to the resource constraints in linear programming and, as such, they require the usual slack or surplus variables. On the other hand, goal constraints are specified as strict equalities that contain two deviational variables, d_i^{+} and d_i^{-}, to indicate overachievement or underachievement of a goal.

It is worth noting that in formulating a goal programming model, each goal is set at a level that may not necessarily be the most attainable, but one that the decision makers would be satisfied to achieve, given the multidimensional nature of the goals. Operationally, then, the objective becomes one of finding a set of solutions for these goals in terms of some stated targets along with their associated constraints. In general, goal achievement starts with the most important (highest priority) goal and continues until the achievement of the least important (lowest priority) goal. This is known as the process of sequential solution. The solution gives the decision makers an option to see the consistency of their decision by allowing them to observe the extent to which their priorities have been achieved and whether or not they need to be refined further.

An Illustrative Example. To illustrate the model and how it works, let us look at a simple example. Suppose that a local government wants to privatize two of its sports facilities – an employee gymnasium and a public swimming pool, both of which have been losing money for some time. The private firm that wants to take over the facilities will introduce an annual membership fee that they do not currently have, say, \$50 for the gymnasium and \$45 for the swimming pool, for unrestricted use of the facilities throughout the year.

Two types of costs are involved in running the operation: labor cost and maintenance cost. Assume that the current operating cost for labor is \$20 per user for the gymnasium and \$30 for the swimming pool. The future management wants to make sure the total labor cost for the year does not exceed \$17,500. The management also wants the maintenance cost not to exceed \$12,500 for the year, which currently stands at \$15 per user for the gymnasium and \$20 for the swimming pool. The number of individuals presently using both facilities is 500, but the management wants to increase the number substantially to maximize the return from operation.

Since goal programming is an extension of linear programming, we can start by formulating the problem as a linear programming problem, then convert it into a goal programming problem. The linear programming formulation of the problem is presented below:

$$\text{Maximize } Z=50X_1+45X_2 \quad [7.79]$$

Subject to:

$$20X_1+30X_2\leq 17{,}500 \quad [7.80]$$

$$15X_1+20X_2\leq 12{,}500 \quad [7.81]$$

$$X_1+X_2\geq 500 \quad [7.82]$$

$$X_1, X_2\geq 0 \quad [7.83]$$

Next, to convert the problem into a goal programming problem we need to define the goals and set some targets for each goal. Suppose that the future management sets $40,000 as the return goal from operation and 1,000 as the membership goal. Assuming both goals are equally important (i.e., identically ranked), the goal programming formulation then boils down to

$$\text{Minimize } Z=P_1d_1^{-}+P_2d_2^{-} \quad [7.84]$$

Subject to:

$$\text{Labor cost: } 20X_1+30X_2\leq 17{,}500 \quad [7.85]$$

$$\text{Maintenance cost: } 15X_1+20X_2\leq 12{,}500 \quad [7.86]$$

$$\text{Earnings goal: } 50X_1+45X_2+d_1^{-}d_1^{+}=40{,}000 \quad [7.87]$$

$$\text{Membership goal: } X_1+X_2+d_2^{-}d_2^{+}=1{,}000 \quad [7.88]$$

$$X_1, X_2, d_1^{-}, d_1^{+}, d_2^{-}, d_2^{+}\geq 0 \quad [7.89]$$

The problem can be solved using the simplex method, discussed earlier (not shown here). The solution of the problem would indicate that both goals have been underachieved: the earnings goal by $4,167 and the membership goal by 83. Additionally, the optimal solution for membership would produce 500 for the gymnasium (X_1) and 250 for the swimming pool (X_2).

As one would expect, the deviations from goals are quite common in goal programming and can be easily fixed if the decision makers are willing to make some changes in the parameters, as in sensitivity analysis. For example, assume that the management could restate the problem by reducing its goal levels by the amount of deviations to reduce the size of the feasible region which will make it possible for both goals to be achieved fully. In other words, if it were to set the earnings goal, say roughly, at $36,000 (instead of $40,000) and the membership goal at 900 (instead of 1,000), the management should have no difficulty in realizing its goals. In other words, it would achieve its targets.

In reality, the management could revise all the parameters of the model either individually or collectively to determine how sensitive the decision variables would be to these changes. Obviously, the more significant the impact of change, the more careful the management will have to be in order to determine how much change it wants in the decision variables. There is a clear tradeoff between changes that must take place in the parameters and the actual goal achievement. The ultimate choice, however, rests with the decision makers regarding how much tradeoff they are willing to accept without significantly affecting the goals or objectives they are trying to achieve.

Other Programming Models

The mathematical programming methods discussed above have two things in common. One, they are linear and, two, they are risk-free in that the objective functions and the constraint equations are linear and known with certainty. However, there are situations where linear models with certainty conditions may not produce the best solution to an optimization problem if the objective functions and the constraints are not linear and uncertain. The alternative is to use methods that are uniquely suitable for dealing with these types of problems. This section presents a brief reference to some of these methods; they are chance-constrained programming, quadratic programming, geometric programming, and convex programming. Of these, the last three methods belong to a category of programming commonly known as nonlinear programming (Kuhn & Tucker, 1951; Luenberger, 1973).

Chance-constrained programming is a classic case of mathematical programming under conditions of risk. It is based on a simple assumption that the system under consideration is random; that is, both the technical and the RHS coefficients are random rather than fixed or predetermined. Given this basic assumption, the objective is to optimize the expected value of a function subject to a set of constraints that are allowed to vary in relation to random variations in the system. Introduced by Charnes and Cooper (1959), the method has been widely used in engineering, finance, and economics, where uncertainties are quite common such as in recycling, exchange rates, price, demand, and supply.

As noted earlier, a fundamental characteristic of linear programming is the linearity of the objective function and the constraint equations. Although a great majority of the programming problems fall into this category, there are some that do not; they belong to the category of nonlinear programming. For instance, in quadratic programming, which belongs to this category, the objective is to optimize a quadratic objective function subject to a set of linear constraints. Another special case of nonlinear programming is geometric programming, which is particularly suitable for dealing with problems that have fractions and negative exponents in the objective function (Duffin, Peterson, & Zener, 1967). A further case of nonlinear programming is convex programming, in which the objective function is concave (curved inward on a two-dimensional plane) and all the constraint equations are convex (curved outwards). The convexity function has an important characteristic in mathematical programming in that it ensures that a local optimal solution is also the global optimal solution. What this means in operational terms is that one only needs to find a local optimum instead of an infinite number of local optima to reach the global optimum.

Chapter Summary

Both mathematical programming and classical optimization (discussed in the previous chapter) have one thing in common: they produce solutions that can help an organization carry out its activities efficiently (i.e., optimally). However, classical optimization works well in situations in which the number of constraints is limited but can become prohibitive as it increases in size. Also, it does not produce a "perfect" optimum in every situation. The problem can be significantly overcome with mathematical programming. What separates mathematical programming from classical optimization is the unique method of solution that produces optimal results. Five programming methods were discussed in the chapter: linear, integer, dynamic, heuristic programming, and goal programming. Of these, linear programming is by far the most widely used method. It is simple to understand, and relatively easy to use. However, linear programming is not without its limitations. For instance, the optimal solution produced by linear programming is useful insofar as the decision variables can be expressed in real numbers; as such, it is not suitable for problems that take non-negative integer values. Integer programming is uniquely suitable to dealing with this kind of problem. Operationally, it uses the same simplex algorithm as linear programming, but produces a much better solution for problems that require a discrete solution.

Dynamic programming, on the other hand, is considered more versatile than most programming methods in that it can be used in almost any situation and relationship· linear, nonlinear, deterministic, stochastic, continuous, and discontinuous. There are obvious advantages of using this type of method in that it allows a problem to be serially structured without any recycling or going back. In addition, there are multiple optimal points from which a decision maker can make a choice. This apparent flexibility has an advantage over other programming methods where the solution is restricted to a single optimal value. Heuristic programming, the fourth model discussed in the chapter, serves as a useful decision tool in situations in which it takes a prohibitive amount of time and cost to determine an optimal solution. By arriving at good-enough solutions using rules of thumb or approximation, heuristic programming can offer viable solutions to complex problems where conventional programming methods do not work.

Finally, the chapter provided a brief discussion of a programming model that deals with multiple goals, as opposed to a single goal, as in linear, integer, and other programming models, called goal programming. In goal programming, the goals of the decision makers are incorporated in the objective function, while the environmental variables, especially those outside the control of the management, are treated as constraints. Computationally, the model selects the solution set which satisfies the environmental constraints and produces a satisfactory solution for the goals, which are usually ranked in priority order. As for the solution methodology, like linear programming, it uses the simplex method for finding the optimum solution.[2]

Notes

1 When artificial variables are used for a maximization problem, which rarely happens, it is assigned an objective function value of –$*M* to force it out from the basis.
2 There is a large body of literature that has grown over the years that provides a thorough discussion of these methods. For an excellent discussion, see Hillier and Lieberman (1986).

References

Balas, E. (1965). An additive algorithm for linear programs with zero–one variables. *Operations Research, 13*, 517–546.
Bellman, R. E. (1957). *Dynamic programming.* Princeton, NJ: Princeton University Press.
Charnes, A., & Cooper, W. W. (1959). Chance-constrained programming. *Management Science, 6*(1), 73–79.
Charnes, A., & Cooper, W. W. (1961). *Management models and industrial applications of linear programming.* Englewood Cliffs, NJ: Wiley & Sons.
Dantzig, G. B. (1963). *Linear programming and extensions.* Princeton, NJ: Princeton University Press.
Duffin, R. J., Peterson, E., & Zener, C. (1967). *Geometric programming.* New York: Wiley & Sons.
Hillier, F. S., & Lieberman, G. J. (1986). *Introduction to operations research* (3rd ed.). San Francisco, CA: Holden Day.
Ignizio, J. P. (1976). *Goal programming and extensions*. Lexington, MA: Lexington Books.
Kuhn, H. W., & Tucker, A. W. (1951). Nonlinear programming. In J. Neyman (Ed.), *Proceedings of the second Berkeley symposium on mathematical statistics and probability* (pp. 481–492). Berkeley, CA: University of California Press.
Land, A. H., & Doig, A. H. (1960). An automatic method of solving discrete programming problems. *Econometrica, 28*, 497–520.
Lee, S. M. (1972). *Goal programming and decision analysis*. Philadelphia, PA: Auerbach Publishing.
Little, J. D., Murty, K. G., Sweeney, D., & Karel, C. (1963). An algorithm for traveling salesman problem. *Operations Research, 11*(6): 972–989
Luenberger, D. G. (1973). *Introduction to linear and nonlinear programming*. Reading, MA: Addison-Wesley.
Steuer, R. E. (1986). *Multiple criteria optimization: Theory, computation, and applications*. New York: Wiley & Sons.

8 Network Analysis

Planning, scheduling, and control are three of the most critical activities in an organization. The success of any project or program, public or private, depends on how effectively these activities are carried out and the kinds of operational tools one uses to undertake them. One such operational tool that is frequently used in this context is network analysis. For large and complex projects where time could be a major factor, network analysis plays a vital role by identifying the most efficient schedule for completing a project. Network analysis also provides an excellent tool for establishing the work sequence well in advance of the actual undertaking of a project, thereby allowing the decision makers the opportunity to monitor progress and correct potential problem areas. In addition to this, it helps in the allocation of resources by redefining the work relationship that can expedite the achievement of project goals and objectives.

A number of network problems have been developed to date. This chapter presents four such problems that are among the most frequently used in network analysis: the shortest-route problem, the minimal spanning tree problem, the maximal flow problem, and CPM and PERT for planning and control. However, before discussing these specific problems, the chapter briefly introduces several terminologies that serve as a precursor to network analysis and can be used as a guide to solving practical problems.

Basic Terminologies

Network analysis has its origins in the more general theory of graphs. A graph is an interconnected network of elements consisting of nodes called vertices and branches called edges or lines. A node is a point that is usually denoted by a circle and connected by one or several branches. Nodes generally represent locations such as communities, service stations, bus terminals, and so on, while branches represent flow of goods, services, messages, distance, time, and so forth. Branches in a graph may or may not have directions. If every branch in a graph has a direction, it is called a directed graph; if it does not, it is called an undirected graph. If, on the other hand, some of the branches are connected and some are not, the graph is called mixed.

Figure 8.1 presents a simple illustration of directed, undirected, and mixed graphs. As the figure shows, all three graphs have some directions; that is, we know their points of origin and destination. A graph with a clear direction is called a network.

When two nodes in a network are connected by a branch, they are said to be adjacent. The degree of a node is the number of branches leading into it, called indegree, as well as the branches leading away from it, called outdegree. For instance, in the graph in Figure 8.1b, the center node has an indegree of three and an outdegree of one because it has three branches leading into it and one branch away from it. A sequence of adjacent branches and nodes is called a path. The word path is often used interchangeably with network or graph. Occasionally, one may find a node connected to itself, in which case it is called a cycle.

The Shortest-Route Problem

The simplest among all network models, the shortest-route problem deals with finding the shortest route from an origin to a destination through a network of alternative routes. The shortest route is a common problem, for example, faced by a traveler traveling between two locations or a postman delivering mail to multiple points. The approach to solving the shortest-route problem involves three simple steps: (1) identifying the source node (i.e., the node from which the travel will originate); (2) determining the nodes closest to the source node such as the first closest, the second closest, the third closest, and so on; and (3) selecting the nodes with the shortest total distance from the origin.

To give an example of a shortest-route problem, consider a problem that most of us are familiar with. A group of students are planning to drive to a beach resort for Spring Break. There are several different routes the students can take. Obviously, their preference is not to waste any time and to take the route that will produce the minimum distance between the college campus and the resort. For convenience, assume that the shortest route will also be the route with the minimum cost generated by the savings from the shortest distance traveled.

Figure 8.2 shows a network of possible routes, along with the distance in miles between any two nodes. As can be seen from the figure, node A is the source node, F is the destination node, and the rest are intermediate nodes in

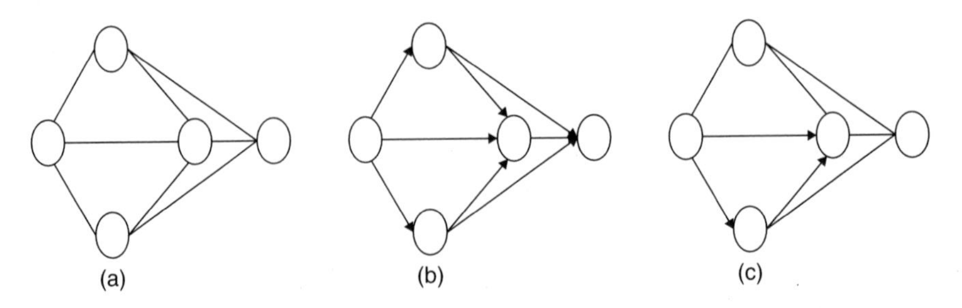

Figure 8.1 Three Simple Models of Graph or Network

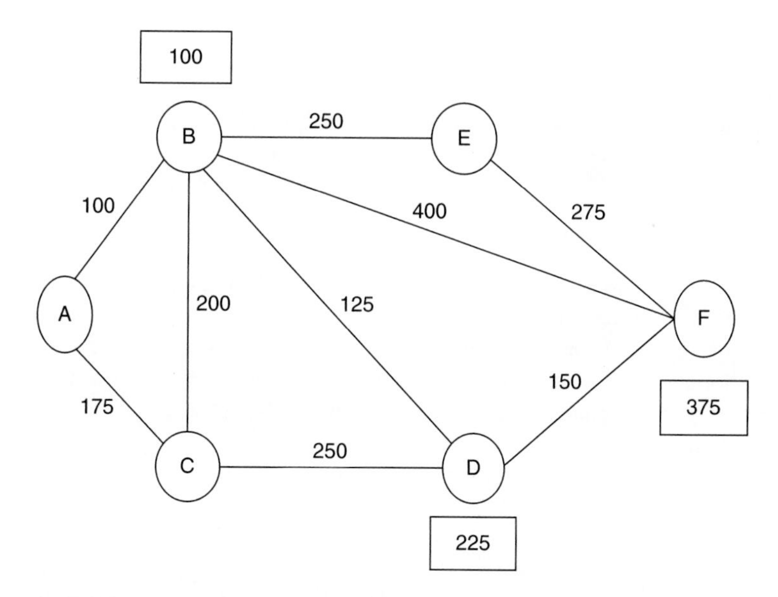

Figure 8.2 Network for the Shortest Route Problem

between the source and the destination node. As a matter of convention, the ordering of the notations used for a node in a network follows an alphabetical sequence beginning with the first and ending with the last, as one moves along any path to the completion of a project or activity.

To obtain the shortest route, we begin with any node, preferably the source node, find the nodes adjacent to it, and measure their distances. We then select the node with the shortest distance from the source node, which happens to be *B* in this case, and place the value of the distance in a rectangle. Next, we identify the nodes adjacent to *B*, take their distances from it, add them to the distance for *AB* obtained earlier, and select the pair with the shortest distance. The result produces the following distances, given in miles: $AB+BC=100+200=300$; $AB+BD=100+125=225$; $AB+BE=100+250=350$; and $AB+BF=100+400=500$. We select *BD*, since it has the shortest distance of all the adjacent nodes. Table 8.1 presents these distances in a pairwise fashion for all adjacent nodes.

Table 8.1 The Shortest-Route Network

A	B	C	D	E
$AB=100$ $AC=175$	$BC=200$ $BD=125$ $BE=250$ $BF=400$	$CD=250$	$DF=150$	$EF=275$

Note
Shortest route: *A–B–D–F*: $AB+BD+DF=100+125+150=375$ miles.

We repeat the process for node *D*. Since there is only one route from *D* to the final destination *F*, we add this to the shortest distances we obtained so far. The result is the shortest route from the campus to the beach resort the students are planning to drive. As can be seen both from the network diagram and the table, the shortest route is *A–BD–F*, with a combined distance of 375 miles; that is, $AB + BD + DF = 100 + 125 + 150 = 375$.

The Minimal Spanning Tree Problem

A special variation of the shortest-route problem is the minimal spanning tree problem. Like the shortest-route problem, it begins with a set of nodes and branches but, unlike the former, the nodes are not specified. As a result, instead of finding a shortest route through a completely specified network, the problem involves finding a set of branches that connects or spans the entire network. The term span or spanning means connecting a set of nodes in a network. It is also called a spanning tree because the solution to the problem yields a tree spanning or connecting these nodes in an optimal fashion (Elmaghraby, 1970).

Solving a minimal spanning tree problem for a network diagram is quite simple, and involves four basic steps: (1) start with any node, preferably the source node; (2) connect it to the adjacent nodes; (3) select the node closest to the nodes already connected; and (4) repeat the process until all the nodes have been connected. The result would produce the minimal spanning tree. Alternatively, one could use a table consisting of the same set of information, as in a network diagram, to determine the minimal spanning tree. For instance, construct a table with distances within the same nodes on the principal diagonal, which will be zero (since any distance from a node to itself must be zero), and the distance between different nodes on the off-diagonal, which will be a positive number. Some of the branches that do not produce the minimum total length for the network may not be fully connected, in which case one can use a symbol such as a letter to represent them.

Next, take each column in the table excluding the column containing the source node, identify the cell with the minimum distance, place it in a rectangle for easy identification, and connect the nodes corresponding to this cell. Then skip this column and start again with the remaining columns in the table. In the event that there is a tie between two or more cells, select any one arbitrarily. Repeat the process and connect all the nodes with minimum distances until the process is complete. The resulting spanning tree of connected nodes will produce the minimum total length for the network. This is shown in Table 8.2.

To illustrate the procedure, let us go back to the Spring Break problem, beginning with the network diagram first. For convenience, we start with the source node *A*, select the adjacent nodes *B* and *C*, and connect them with *A* (Figure 8.3). Since node *B* has the shortest distance from *A*, we select this node. Next, we connect it to two of its adjacent nodes, *D* and *E*, ignoring node *F*, because it is the farthest from *B* and can be more easily reached from *D* or *E*. Of the two nodes, *D* is much closer to *B* than *E*. Thus, we select *D* and connect it to its only

Table 8.2 The Minimal Spanning Distances between Nodes

From \ *To*	A	B	C	D	E	F
A	0	100	175	M	M	M
B		0	200	125	250	400
C			0	250	M	M
D				0	M	150
E					0	275
F						0

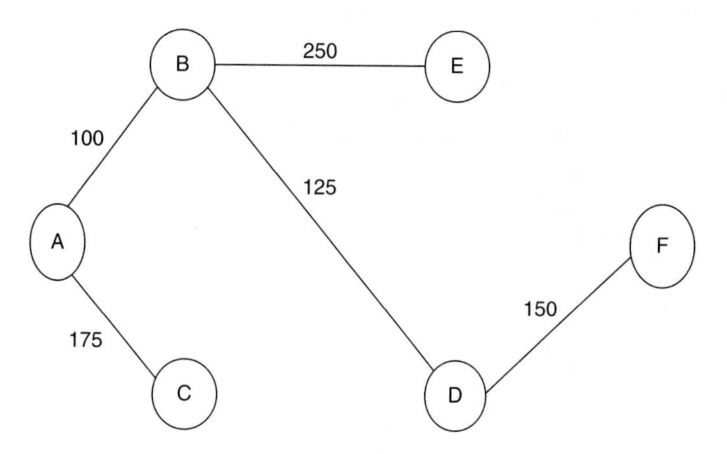

Figure 8.3 Network for Minimal Spanning Tree Problem

adjacent node F. We add the length of all the branches of the selected nodes, which produces a total of 800 miles; that is, $AB+AC+BD+BE+DF=100+175+125+250+150=800$. This is the minimum spanning tree or distance that is needed to connect all the nodes in the network at least once.

We could have obtained the same result from a direct analysis of the nodes in Table 8.2. As the table shows, the rows and columns represent the nodes, while the cells represent the distances between these nodes. Where the nodes are unconnected, they are usually represented by a letter, we will call it M (similar to big M used in the linear programming solution), which is an undefined mathematical quantity that does not have any realistic significance so far as finding the optimal distance is concerned. The assumption here is that the distance between any two unconnected nodes could be a large number, but we do not know how large. In fact, it could be so large that it would never end up being in the final solution; hence, the use of a letter to represent the arbitrary number.

An interesting feature of the table is that it is symmetric; that is, its lower and upper triangles are identical. Since one triangle is the exact opposite of the other, we include only one triangle (because including both in the table would not provide any new information that is not contained in the other triangle).

We begin with column *A*, which contains the source node, and exclude it from consideration since it does not have any nodes to the left of it. This leaves the table with five instead of six columns, *B* through *F*. We move to the next column to the right and select the *AB* cell, which has the smallest distance in that column (given that it is a triangular matrix). We place it inside the rectangle and exclude the column from further consideration, leaving the table with four columns instead of five, *C* through *F*. We repeat the process, each time identifying the cell in a column with the lowest distance, until all the remaining columns in the table are exhausted.

Now, if we add the cells with the lowest distance in each column, it should produce exactly the same length we obtained earlier. To see if this is true, we take the sum of the distances between the connected nodes, which will produce a total of 800 miles; that is, $100+175+125+250+150=800$, which corresponds to the total we obtained earlier. This is the minimum length of the network or spanning tree for the Spring Break problem. We can contrast this result with the solution for the shortest-route problem. For instance, the minimal spanning tree provides us with the minimum distances required to connect all the terminal locations (nodes) in a network. The shortest-route problem, on the other hand, shows the minimum (shortest) distance from an origin (node *A*, in our case) to a specific destination (node *F*).

The Maximal Flow Problem

The maximum flow problem deals with directing flows in a network in order to maximize the total flow from a source node to a destination point. An important characteristic of the maximal flow problem is the conservation of flow, which means that the flow entering a source node in a network must be equal to the flow leaving the network at the exit point. For instance, if 10,000 vehicles enter a network at an entry point, the conservation principle says that the same number must leave the network at the point of final exit, barring any unforeseen problems.

To solve the maximal flow problem, two things are necessary. First, find a path from the source to the destination point with information available on flow capacity for all the branches of that path, where a flow capacity is a positive number on a branch. Second, determine the branch with the smallest flow capacity and increase the flow on this path by reducing the current capacity of all the branches on this path by an amount equal to the branch with the lowest capacity. At the same time, increase the capacity in the reverse direction of all the branches on this path by an amount equal to the capacity of the lowest branch. When every path in a network from the source to the destination has at least one zero flow, the solution is considered optimal.

To give an example, consider a situation in which a state highway agency is planning to close down part of an interstate highway for repair. Let us say that the number of vehicles the stretch of the interstate being repaired can handle is 5,000 per hour. The agency, in collaboration with the nearest city transportation department, must determine whether the alternative routes through the city can handle the traffic flow for the duration of the repair. Assume that the city street route has a maximum capacity of 7,000 cars per hour. Figure 8.4 presents the possible routes for redirecting the traffic flow through the city. The number next to each node indicates the capacity of the branches (streets) in thousands of vehicles per hour. For instance, the number 4 on branch *AB* indicates that it has the capacity to carry 4,000 vehicles per hour to node *B*, and 3,000 from *B* to *A*. One of the branches, *AD*, shows an identical number for both nodes, which means that the capacity of the branch is identical in both directions.

To reroute the traffic flow according to the steps suggested above, we can start with any path. It does not matter which path we select since the same result will eventually be produced. Let us say that we start with the path *ABE*, which has two branches, *AB* and *BE*. The branch *AB*, with a flow capacity of 3,000, is the smaller of the two on this path. Thus, we decrease the capacity in the direction of flow by 3,000, and increase it in the reverse direction by 3,000. This will leave only 1,000 vehicles traveling from *A* to *B*, but increase the reverse flow by 6,000 from *B* to *A* (shown in parentheses).

We do the same for *BE*; that is, decrease the capacity by 3,000 in the direction of flow from *B* to *E*, and increase it in the reverse direction by an equal amount without affecting the total flow between *B* and *E*. We take the next path again with two branches, *AC* and *CE*, of which *AC* has the smaller flow of 2,000 vehicles per hour. As before, we decrease the capacity in the direction of flow by 2,000 and increase it in the reverse direction by 2,000. We now take the third and final path, *ADCE*, and repeat the process. Since every path has at least

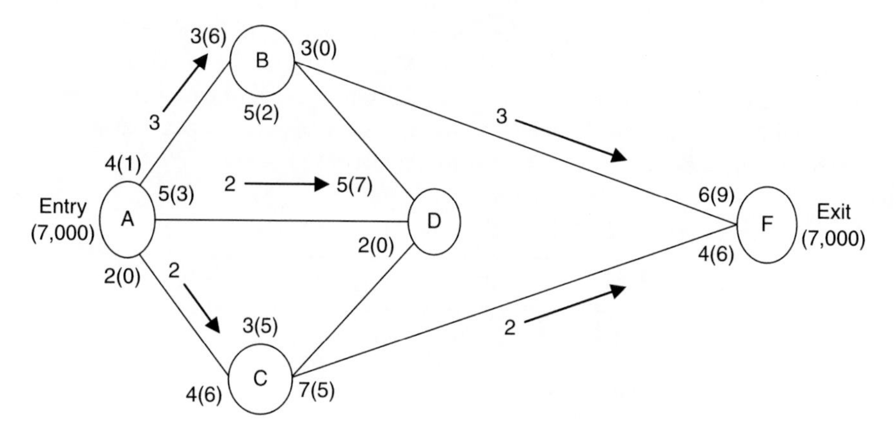

Figure 8.4 Network of Maximal Traffic Flow (in 1,000)

one 0 from the source to the destination, the solution is optimal; that is, we have obtained the network of maximal traffic flow for our problem.

Table 8.3 presents some of the same results shown in Figure 8.4, but is much less complicated to read. It shows the amount of flow that could be assigned by each path to reach the maximal flow. Note that the branch *BD* was not assigned any traffic at all, since it would not have made much difference in that it did not affect the network in producing the maximum traffic flow.

One final note about Figure 8.4: a careful look at it would indicate that the city streets (branches) have more capacity than is needed to handle the traffic flow. As noted earlier, they can easily handle up to 7,000 vehicles per hour, which is 2,000 vehicles more than the capacity of the stretch of the interstate being repaired. In other words, the city should have no difficulty in handling the traffic flow, if properly redirected through its streets.

The CPM and PERT Models

Two network models that have received more attention than any other model in the literature on network analysis are the critical path method (CPM) and the program evaluation and review technique (PERT) (Hillier & Lieberman, 1986). CPM was developed in 1956 by DuPont Company in collaboration with Sperry Rand Corporation as a tool for planning, scheduling, and control (Moder, Phillips, & Davis, 1983). PERT was independently developed in 1958 by Booz-Allen and Hamilton, a system consulting firm, working under the auspices of the Bureau of Ordnance, US Navy (Martino, 1970).

Interestingly, CPM and PERT are quite similar in that they were designed essentially for the same purpose and, as such, are treated together as one. However, there is one minor difference that separates the two: where CPM is used for activities whose completion times are known exactly or with precision, PERT is mostly used for activities whose completion times may or may not be known with precision. There is also a similarity between CPM/PERT and all other network models in that they all use the same type of network diagram, although they do not use exactly the same terminologies. For instance, instead of using nodes and branches for networks, CPM and PERT use terms such as activities and events. Activities are tasks an organization plans to complete within a specified time. Events are concrete milestones such as percentage of tasks completed at a point in time within an overall time frame.

Table 8.3 Assignment of Maximal Traffic Flow (in 1,000)

Path	AB	AC	AD	BE	CE	DC
ABE	3	0	0	3	0	0
ACE	0	2	0	0	2	0
ADCE	0	0	2	0	2	2
Total	3	2	2	3	4	2

Let us look at a simple example to illustrate how these two network models work. Suppose that a city is planning to build a convention center to generate additional revenue to supplement its current income. The city council feels it has the support of the majority of taxpayers for the project. However, the state law under which the city government operates requires that the decision to build a major facility must be formally approved by the voters in a city-wide referendum. Table 8.4 presents the principal activities the project will entail, beginning with the city council formally adopting the proposal to build the center, followed by voter approval, selection of project site, sale of bonds to finance the project (a standard government practice for large-scale projects), and so on.

According to the table, it will take the city somewhere between 26 and 50 months to complete the project. Obviously, some of these activities will take more time to complete than others, especially those related to physical construction and acquisition of materials, equipment, and so forth.

An important characteristic of CPM and PERT is that when two or more activities start at a point, say i, and end at a point, say j, an arrow must be added to prevent the two real activities from both originating and terminating on the same event. Such activities are called dummy activities. The dummy activities have a duration of zero, meaning they do not take up any additional time or resources but instead preserve the order of precedence. In other words, an activity cannot be started until the activities for the preceding events have been completed first. Figure 8.5 shows the dummy activity, t_{23}, given by the dashed

Table 8.4 Activities and Their Completion Time (Months)

From	*To*	*Activity Description*	a	m	b	e_t	s
1	2	Council decides to go ahead and call public referendum	1	2	3	2.0	0.333
2	3	Dummy activity (*d*)	0	0	0	0.0	0.000
2	4	Voters approve and the council decided to undertake the project	2	3	4	3.0	0.333
2	5	Council issues General Obligation bonds to finance the project	3	4	6	4.2	0.500
3	6	Site preparation and initial groundwork	3	4	6	4.2	0.500
4	6	Setting up utility lines for water, sewerage, and electricity	3	4	6	4.2	0.500
5	6	Sale of bonds	2	3	5	3.2	0.500
6	7	Physical construction of the project	8	10	12	10.0	0.667
7	8	Utility connection, acquisition of materials, equipment, etc., and installation	4	6	8	6.0	0.667
		Total	26	36	50	–	–

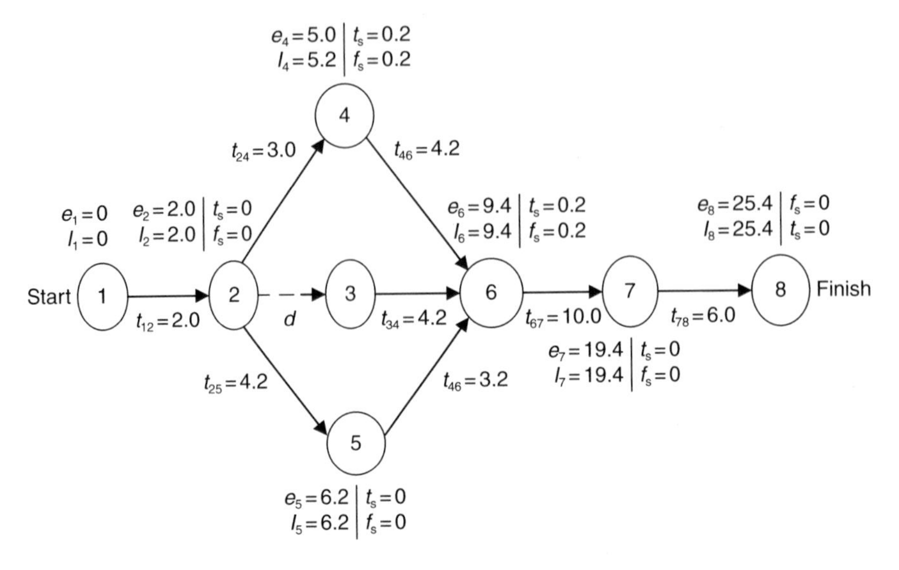

Figure 8.5 Network Diagram for CPM/PERT

lines, *d*, which can be undertaken only after the activity t_{12} has been completed; that is, only after the voters have approved the project.

The Expected Completion Time

Given the wide range of activities that are typically involved in a large project, it is quite likely that they will not always get completed on time. Decision makers are aware of this. As a result, they must know the alternatives that are available to them. Three types of time schedules are generally considered in planning a project activity: optimistic (the minimum time it takes to complete a task); pessimistic (the maximum time it can take to complete a task); and most likely (the time that lies somewhere in between). We will denote these three time schedules by the notations *a*, *b*, and *m*, respectively.

Let us say that the city is optimistic about the project and expects it to be completed at the earliest possible time. Therefore, according to the table, the city should be able to complete the project in 26 months. On the other hand, if the city had taken a pessimistic view, the completion time would be 50 months. By the same token, a most likely scenario would give the project 36 months to be fully completed.

In reality, however, very few projects get completed as planned or scheduled. A variety of factors – bad weather, early depletion of funds, non-availability of materials – contribute to frequent delays in project completion. Therefore, it is important to consider an average, called the expected time, for project completion. It is given by the expression:

$$e_t = \frac{1}{6}(a + 4m + b) \quad [8.1]$$

where e_t is the expected or average time at time t.

Since the expression used in Equation 8.1 is an average, it is quite likely that the actual time it will take to complete a project will vary (i.e., deviate from this average). We can present this deviation from the average, e_t, in terms of the following:

$$s^2 = \left[\frac{b-a}{6}\right]^2$$

$$\therefore s = \sqrt{\left[\frac{b-a}{6}\right]^2}$$

$$= \left[\frac{b-a}{6}\right] \quad [8.2]$$

where the terms s' and s stand for variance and standard deviation, respectively.

In general, the larger the variance or standard deviation, the greater the departure from the expected time of completion, and vice versa. For instance, the activity (1, 2) with $a=1$, $b=3$, and $m=2$ has an expected time of two months. That is:

$$e_t = \frac{1}{6}[1 + (4)(2) + 3] = 2 \text{ months}$$

The resulting standard deviation, s, is:

$$s = \frac{b-a}{6} = \frac{3-1}{6} = 0.333 \text{ months}$$

which means that the council should expect to receive the project approval in two months, with a standard deviation of 0.333 months.

Since the standard deviation is small, we can expect the council to be able to complete this task within this time. Similar averages for the remaining activities, along with their respective standard deviations, are also given in Table 8.4.

Optimal Completion Time

Of the three alternative schedules (optimistic, pessimistic, and most likely), the question facing the city is: what will be the optimal completion time for the project? An optimal completion time is the minimum time it will take for a project to be fully ready for operation, while making sure that each activity is carried out in the order in which they appear on a project schedule. To arrive at an optimal completion time, we use a simple procedure known as the earliest expected completion time. According to this procedure, the activity times are added for each possible path leading from the starting to the ending point, where the largest total becomes the earliest expected time. The procedure is simple, but can be time consuming if the network involves too many activities and events.

However, it is possible to simplify the procedure if we can define two parameters – an expected completion time t for activities i and j, t_{ij}, and an earliest expected completion time for an event j, e_j. Assume that we have a fixed value of j, say j', then the earliest expected completion time is

$$e_j' = \max\langle i \rangle [e_i + t_{ij}'] \qquad [8.3]$$

where e_i is the earliest expected completion time, with i ranging over all activities, from i to j'.

Applying this to all the activities in Table 8.4 for the convention center problem, we get the following expected times:

$$e_1 = 0$$

$$e_2 = \max\langle i=1 \rangle [e_1 + t_{12}] = 0 + 2.0 = 2.0$$

$$e_4 = \max\langle i=2 \rangle [e_2 + t_{24}] = 2.0 + 3.0 = 5.0$$

$$e_5 = \max\langle i=2 \rangle [e_2 + t_{25}] = 2.0 + 4.2 = 6.2$$

$$e_6 = \max\langle i=2, 4, 5 \rangle [(e_2 + t_{25}), (e_4 + t_{46}), (e_5 + t_{56})] = [(2.0 + 4.2), (5.0 + 4.2), (6.2 + 3.2)] = 9.4 \text{ (for } i=5)$$

$$e_7 = \max\langle i=6 \rangle [e_6 + t_{67}] = 9.4 + 10.0 = 19.4$$

$$e_8 = \max\langle i=7 \rangle [e_7 + t_{78}] = 19.4 + 6.0 = 25.4$$

The expected completion time, or e value, for each event indicates the earliest time in which an event could be accomplished, provided that each activity j on every path leading to the specific event is completed in exactly the e_j unit of time. The path followed to determine the expected value for the final event, called sink, is the critical path of the network.

Figure 8.5 also shows the values for the network problem for the convention center. In general, the sum of the activity times for the critical path must be greater than the sum of the activity times for any other path in the network. This holds good for our critical path 1–2–5–6–7–8, although by a very small margin; that is, 1–2–5–6–7–8 (25.4) > 1–2–4–5–6–7–8 (25.2) > 1–2–3–6–7–8 (22.2). What this means is that the city should not expect the center to be fully completed and ready for operation in less than 25.4 months.

A related measure frequently used in CPM and PERT is the latest allowable completion time. The latest allowable completion time is the opposite of the earliest expected completion time and serves as a check on the critical path obtained from the latter. It can be defined as the longest time it will take for an event to complete without delaying the completion date for the project, provided that all succeeding events are completed as scheduled.

To illustrate the method, assume that we have a fixed value for an event i, say i'; then the latest allowable completion time for the event can be written as

$$l_i' = \min\langle j \rangle [l_j - t_{ij}'] \quad [8.4]$$

where l_i' is the latest allowable completion time (with i ranging over all activities, from i' to j).

Interestingly, unlike the earliest expected completion time, the computation of the latest allowable completion time follows a backward or recursive order, starting with the latest event and moving to the first. Thus, the starting point for the latest allowable time is always the cumulative time for the critical path, obtained from the earliest expected completion time, which in this case happens to be 25.4 months.

To formally apply the method, we begin with this cumulative total and subtract the time value of each event, starting with the next to last and continuing backward all the way to the first. The computation is considered correct if the latest allowable time for the first event in the network turns out to be zero.

The following computations show the latest allowable completion time based on each of the eight events (Table 8.4):

$$l_8 = e_8 = 25.4$$

$$l_7 = \min\langle j=8 \rangle [l_8 - t_{78}] = 25.4 - 6.0 = 19.4$$

$$l_6 = \min\langle j=7 \rangle [l_7 - t_{67}] = 19.4 - 10.0 = 9.4$$

$$l_5 = \min\langle j=6 \rangle [l_6 - t_{56}] = 9.4 - 3.2 = 6.2$$

$$l_4 = \min\langle j=5 \rangle [l_6 - t_{46}] = 9.4 - 4.2 = 5.2$$

$$l_2 = \min\langle j=4, 5, 6 \rangle [(l_4 - t_{24}), (l_5 - t_{25}), (l_6 - t_{26})]$$

$$= [(5.2 - 3.0), (6.2 - 4.2), (9.4 - 4.2)] = 2.0 \text{ (for } j=4)$$

$$l_1 = \min\langle j=2 \rangle [l_2 - t_{12}] = 2.0 - 2.0 = 0$$

The result indicates the amount by which the city can delay an event without affecting the completion date for the center, as long as all the successive events are completed on time. For instance, the council can absorb a delay in its decision to contract out the project by 5.2 months, once the project has started and provided that no delay occurs at any of the remaining events.

Probability Estimates for the Critical Path

When a project does not get completed on time, it is usually due to uncertainty of events over which the decision makers do not have much control. Uncertainties can seriously damage a project schedule, costing an organization more in time and resources than the actual cost of the project. Therefore, once the critical path has been determined, it may be necessary to determine the effect uncertainties will have on project completion times. This can be achieved by calculating the probability estimates for the critical path.

To give an example, recall the critical path we obtained for the center: 1–2–5–6–7–8, with an expected project completion time of 25.4 months. The standard deviation corresponding to this is 2.667 months, obtained by simply adding the standard deviations for individual activities on this path (Table 8.4). That is:

$$S = s_{12} + s_{25} + s_{56} + s_{67} + s_{78} = 0.333 + 0.500 + 0.500 + 0.667 + 0.667 = 2.667$$

The information thus obtained on the earliest expected completion time and its standard deviation can now be utilized to determine the probability of completing the project within a specified time. Suppose the council wants to know what the probability will be for the project being completed within 30 months or less. This is a reasonable assumption on the part of the council since it must have some ideas regarding project completion to avoid any potential problem in the future.

Let us assume for the sake of argument that the project completion time has a standard normal distribution such that

$$z \approx N(0, 1) \qquad [8.5]$$

where z represents the standard normal distribution, 0 is its mean, and 1 is the corresponding standard deviation. Note that the rationale for using a z distribution is that it is much less restrictive than most other continuous distributions.

According to the distribution, the probability that the project will be completed within 30 months or less can be obtained first by transforming the target time into a z value, then finding the probability corresponding to it. To find the z value, we take the difference between the expected and the target completion time and divide the result by the observed standard deviation, as shown below:

$$\begin{aligned} p(t \leq 30) &= p[z \leq (x - e_t)/s] \\ &= p[z \leq (30.0 - 25.4)/2.667] \\ &= p(z \leq 1.7248) \\ &= 0.9562 \end{aligned} \qquad [8.6]$$

where x is the target date, as determined by the council, e_t is the expected date of project completion at time t, and s is the standard deviation.

The corresponding probability for the observed z value of 1.7248 (i.e., 1.73 after rounding off) is 0.9582, which can be obtained from any standard normal table. To answer the question, what this means for the council is that there is almost a 96 percent chance that the city will be able to complete the project within 30 months or less. Conversely, there is less than a 5 percent chance that it will not be able to complete it by the target date.

Slack Time

It is quite reasonable to assume that there will be occasional glitches that will cause delays in work schedules, making it difficult to complete a project by a target date. Most schedules recognize the potential for such delays and have provisions for what is known as slack time, but there is a limit to the amount of delay that can be absorbed without significantly affecting the project schedule or cost. The purpose of using slack times in network analysis is to determine the amount of time by which an activity can be delayed without affecting the schedule for project completion. Slack times are generally associated with noncritical activities; that is, those activities that are not in the critical path and whose effects will, therefore, be minimal on project completion. For critical activities, the slack time will always be zero, meaning that there is no room for delay.

Two types of slacks are commonly used in a network: total and free. Total slack for any two activities, i and j, is the difference between the maximum time available to schedule an activity and the estimated duration of that activity. The times within which activities are scheduled to be completed are obtained from the e_j' and l_i' values for each activity's beginning and end node. For instance, for activity (4, 6), the city must schedule to begin it no earlier than five months ($e_4=5.0$) and end it no later than 9.4 months ($l_6=9.4$). Therefore, the maximum time period available for completion of activity (4, 6) is $l_6-e_4=9.4-5.0=4.4$ months. Since it takes 4.4 months ($t_{46}=4.4$) to complete the activity (4, 6), the time by which it can be delayed is 0 months; that is, $l_6-e_4-t_{46}=9.4-5.0-4.4=0$. This the amount of total slack for activity (4, 6), meaning there is no slack.

We can now formally express the total slack as:

$$ts_{ij}=l_i-e_i-t_{ij} \qquad [8.7]$$

where ts_{ij} is the total slack for activities i and j, t_{ij} is the duration of the activities i and j, and l and e are the earliest expected and the latest allowable completion times, respectively.

Likewise, the total slack for activities (6, 7), based on the expression in Equation 8.7, is also 0. That is:

$$ts_{67}=l_7-e_6-t_{67}=19.4-9.4-10.0=0$$

The same procedure can now be applied to compute the slack for the remaining activities. Figure 8.5 (presented earlier) shows the slack for each activity in the network. According to the figure, most of the activities have zero slack, meaning that there is hardly any room for these activities to be delayed. Therefore, the city must complete the project on time.

Free slack is considerably different from total slack. It represents the amount of time by which an activity can be delayed without causing any delay in the immediate successor activities, and is given by the expression:

$$fs_{ij}=e_j-e_i-t_{ij} \qquad [8.8]$$

where fs_{ij} stands for free slack for activities i and j, e_j is the earliest time at j, e_i is the earliest time at i, and t_{ij} is the duration of activities i and j, as before.

For example, the free slack for activities (4, 6) and (6, 7), obtained from the expression in Equation 8.7, are 0.2 and 0, respectively. That is:

$$fs_{46}=e_6-e_4-t_{46}=9.4-5.0-4.2=0.2$$

$$fs_{67}=e_7-e_6-t_{67}=19.4-9.4-10.0=0$$

The remaining free slack can be obtained the same way.

Figure 8.5 also shows the free slack alongside each activity. It is perhaps worth noting that both free and total slack for the activities in the current example have identical values. When free slack equals the total slack, it means they do not share or have in common any slack with any of the successor activities. On the other hand, if the free slack is less than the total slack, it suggests that any delay in an activity would delay the successor activity beyond its starting date. The opposite is true when the free slack is greater than the total slack, meaning that it will not cause any delay in the successor activity.

Project Crashing and Activity Cost

A couple of additional factors must be taken into consideration when planning or scheduling a project: crashing and the cost associated with crashing. Crashing generally means reducing the project completion time ahead of its normal schedule by speeding up, or crashing one or more critical activities. This can be accomplished by adding more resources such as labor, materials, and equipment to those activities that are to be crashed. Obviously, the more resources a project employs, the greater is the cost associated with it. The purpose of crashing is to keep this cost to a minimum, while reducing the project completion time by as much as possible. Crashing is used in conjunction with CPM, not PERT, although it should not make much of a difference if it is carried out with the latter.

Two types of costs are generally associated with crashing: normal cost for normal completion time and crash cost for crash time. In general, there is an inverse relationship between crash time and cost, meaning that for each *unit* of reduction in time there is an equal increase in cost. This relationship between crash time and cost, called incremental cost, can be expressed in the following way:

$$I_C=(\Delta\text{Cost}/\Delta\text{Time})=(C_C-N_C)/(N_t-C_t) \quad [8.9]$$

where I_c is the incremental cost, C_C is the crash cost, N_C is the normal cost, N_t is the normal time, and C_t is the crash time.

To illustrate the cost, let us return to our convention center example. Assume that we have information on each of the terms in Equation 8.9 such that we can

Table 8.5 Normal and Crash Time and Cost

Activity	N_t *(Months)*	C_t *(Months)*	N_C *(1,000)*	C_C *(1,000)*	I_C *(1,000)*
1–2	2.0	1.0	100.00	250.00	150.00
2–3 (*d*)	0.0	0.0	0.00	0.00	0.00
2–4	3.0	2.0	50.00	150.00	100.00
2–5	4.2	3.0	75.00	150.00	62.50
3–6	4.2	3.0	5,000.00	6,350.00	1,125.00
4–6	4.2	3.0	500.00	750.00	208.33
5–6	3.2	2.0	500,028,500.00	175.00	104.17
6–7	10.0	8.0	3,500.00	31,250.00	1,375.00
7–8	6.0	4.0		5,000.00	750.00
Total	36.8	26.0	$37,775.00	$44,075.00	$3,875.00

calculate the incremental cost for each activity. Table 8.5 shows the normal and crash costs for the problem. For example, the normal time for activity (3, 6) is 4.2 months and the crash time is 3 months, while the cost of crashing is $6.35 million and the incremental cost is $1.125 million.

As the table shows, the total expected time (assuming all the activities associated with the project are normal) is 36.8 months, with a total cost of $37.775 million. Thus, if all the activities were crashed, the project time would be 26 months with a total cost of $44.075 million, which is obviously much higher than the normal cost. From the city's point of view, the objective then is to crash the project in such a way that it will be completed within 26 months, but its cost will not be much higher than the normal cost.

To achieve the city's objective we do the following: we start with the initial critical path (1–2–5–6–7–8) and the activity with the minimal incremental cost on this path. The activity in this case is (2, 5), with a minimum cost of $62,500. The amount by which this activity can be crashed is 1.2 months, obtained by taking the difference between normal and crash time. The crash cost associated with the new completion time is: $62,500 × 1.2 = $75,000. This is the amount by which the normal cost, N_C, will increase from $37.775 million to $37.775 + 0.075 = $37.850 million.

Next, we revise the network, adjusting for time and cost assigned to activity (2, 5), as shown in Figure 8.6. The new critical path is 1–2–4–6–7–8, since it has the lowest incremental cost, which can be obtained by following the steps in Equation 8.8. We repeat the process, in iterations, and keep repeating it until we get a value less than or equal to the crash time of 26 months.

The stopping rule for the process is to repeat it as many times as possible until all the activities in the critical path are completely crashed. Alternatively, repeat it until a point comes when the crash cost will be a minimum, or both. Since there was not much slack in our center problem to begin with, it was not possible to crash the entire critical path. Nevertheless, the process produced a minimum cost of $37.8175 million after 20 iterations, which is considerably less than the initial crash cost of $44.075 million.

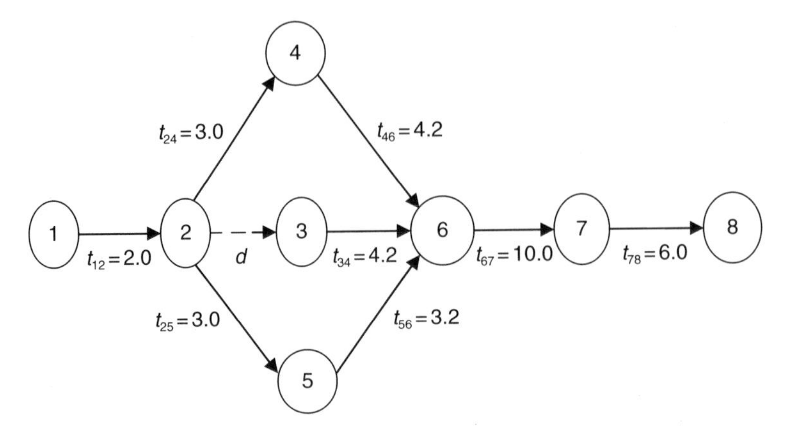

Figure 8.6 Network Diagram (after Iteration 1)

Table 8.6 presents the results of the final iteration. According to the table, we were able to crash only two out of eight activities by a combined total of 0.4 months, thereby reducing the normal completion time from 25.4 months to 25 months. Correspondingly, the normal completion cost also increased by a small margin of $42,500, from $37,775 million to $37,817.5 million. It is obvious that the more slack there is in a network the greater is the amount by which a project schedule can be crashed and the higher becomes the completion cost.

The final network involving crashed activities (1, 2) and (2, 5) is shown in Figure 8.7. As can be seen from the figure, the difference is negligible among all three network diagrams due in part to the fact that there was not enough initial slack time in our problem to begin with. It is perhaps worth noting that the cost

Table 8.6 Crash Time and Cost

Activity	*Crash Time (Months)*	*Crash Cost (1,000)*	*Revised Time,* N_t* *(Months)*	*Revised Cost,* N_t* *(1,000)*
1–2	0.2	30.00	1.8	130.00
2–3 (*d*)	0.0	0.00	0.0	0.00
2–4	0.0	0.00	3.0	50.00
2–5	0.2	12.50	4.0	87.50
3–6	0.0	0.00	4.2	5,000.00
4–6	0.0	0.00	4.2	500.00
5–6	0.0	0.00	3.2	50.00
6–7	0.0	0.00	10.0	28,500.00
7–8	0.0	0.00	6.0	3,500.00
Total	0.4	$42.50	36.4	$37,817.50

Note
*Revised.

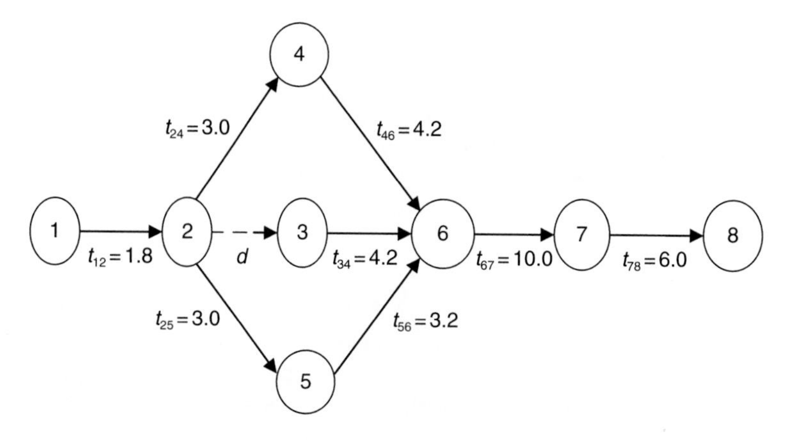

Figure 8.7 Network Diagram (after final Iteration)

data used in the current example include mostly direct costs, but a proper decision structure involving planning and scheduling must also include indirect costs. In other words, it must be based on direct as well as indirect costs. In general, direct costs are at their maximum when the crash time is at a maximum for a project. Indirect costs, on the other hand, increase as a function of time. This means that the longer a project takes to complete all its activities, the higher the indirect costs (i.e., costs on utility, interest, and so forth). Therefore, the optimal choice for a project manager should be to complete the project within a time frame at which these costs will be a minimum.

Networks with Uncertainties

This section briefly introduces a concept that has received some interest in network analysis, called the graphical evaluation and review technique, or GERT (Clayton & Moore, 1972). Like CPM and PERT, GERT consists of events and activities to develop a network, but its advantage lies in its ability to deal with uncertainties. In CPM and PERT, particularly in CPM, the branching from a node (activity) must always take place, while in GERT it is associated with a probability that it will be performed. As we all know, the probability of any activity taking place must range between 0 and 1. Thus, when it is 0, it indicates absolute certainty that the activity will not be performed. By the same token, when it is 1, it indicates absolute certainty that it will be performed, in which case it becomes deterministic (i.e., the same as CPM).

Another advantage of GERT is that it can deal with loops, which are not allowed in CPM and PERT. As noted earlier, looping in networks means that an event can be repeated (from a node back to itself) as many times as necessary. There are problems in the real world where it may be necessary to repeat an activity as part of solving a network problem. By allowing looping, GERT can deal with repetitions more effectively than any other network model. A further advantage of GERT is that, unlike CPM and PERT, its critical path does not have to be the path with the longest expected (average) elapsed time. In other words, it can deal with more than one critical path. In general, when more than one critical path exists in a network, GERT allows one to collect statistics on the relative frequency of occurrence for different (critical) paths and use it as a basis to calculate the relevant probability for each path. The network thus obtained then becomes a stochastic (probabilistic) network.

Chapter Summary

Network analysis plays an important role in planning and scheduling the activities of an organization in an optimal fashion. This chapter has briefly presented some of the well-known network models; in particular, the shortest-route problem, the minimal spanning tree problem, the maximal flow problem, and CPM and PERT. Of these, CPM and PERT are considered more sophisticated and are understandably more complex than any other network model, but all play a critical role in planning, scheduling, and controlling the activities of an organization. From a decision-making perspective, all network models have some built-in flexibilities that allow the decision makers an opportunity to incorporate any changes from the initial plan and study their impact on project completion. This, in turn, makes it possible to develop a planning and scheduling system that is sufficiently dynamic to respond instantaneously to changed conditions. It is this latter characteristic that makes these models particularly useful for decision making in an organization.

References

Clayton, E. R., & Moore, E. R. (1972). PERT versus GERT. *Journal of Management Science, 23*(2), 11–19.

Elmaghraby, S. E. (1970). *Some network models in management science.* New York: Springer-Verlag.

Hillier, F. S., & Lieberman, G. J. (1986). *Introduction to operations research.* San Francisco, CA: Holden-Day.

Martino, R. L. (1970). *Critical path networks.* New York: McGraw-Hill.

Moder, J., Phillips, C. R., & Davis, E. W. (1983). *Project management with CPM and PERT.* New York: Van Nostrand Reinhold Company.

Part III

Special Topics in Cost and Optimization

9 Games and Decisions

Decision makers do not always operate in a decision environment that is certain. While that is true for most organizations in the private sector, it can also apply to public organizations where decisions are not always made with perfect knowledge. Perfect knowledge is a condition in which all the information necessary to make a decision is known with certainty. Unfortunately, there is hardly a situation in the real world where one has all the information necessary or available to work with. As a result, decisions are often made with limited information or some uncertainties. Theoretically, one could think of information availability as a situation on a continuum between two extremes: complete certainty and complete uncertainty. Most decisions are made with information that lies somewhere in between.

With a few notable exceptions, much of our discussion in the previous chapters concentrated on methods that were deterministic in nature, meaning that all the information necessary to solve a problem was known with certainty. In this chapter we present several methods and techniques that are specifically designed to deal with conditions that are uncertain. Three types of uncertainty conditions are discussed in the chapter: risk or partial uncertainty, complete uncertainty, and conflict.

Decisions under Risk

Decision makers in government are not necessarily risk takers, but there are situations in which a decision has to be made under conditions of risk. A risk can be defined as a situation (event) in which one does not have complete information to make a decision, but where it is possible to estimate the outcome (i.e., the result of an event) with some degree of uncertainty or probability. Probability and uncertainty, therefore, go hand in hand. As long as there is an element of uncertainty in any decision process, the outcome will always be uncertain; it will be probabilistic. Probability thus becomes an integral part of all three conditions of uncertainty: risk, complete uncertainty, and conflict.

For most decision making under conditions of risk, the decision makers must know the courses of action that are available to them, including those that are not feasible. They must also be cognizant of the states of nature and their associated probabilities. States of nature are conditions or events over which a decision maker does not have control. Finally, decision makers must be able to determine the return, called payoff, from a particular course of action. In general, the greater the risk, the higher the payoff.

Expected Value

Common sense tells us that when an event or a state of nature is uncertain, the payoff will also be uncertain. The average payoff with uncertainties is called expected value (EV). EV plays a critical role in most decision problems involving risk. They determine the feasibility of a course of action. For instance, if there are n number of courses of action for a given problem, there will be n number of EVs. Each EV will then provide the decision maker with a choice in terms of the feasibility of that course of action. We can define EV as the sum of the payoffs for each course of action times the probabilities associated with each state of nature. It can be formally expressed as

$$EV(X_i) = \sum_{j=1}^{m} X_i P_j = X_1 P_1 + X_2 P_2 + \ldots + X_i P_m \quad [9.1]$$

where $EV(X_i)$ is the EV of the variable X for the ith course of action, and P is the probability corresponding to each state of nature j (for $i=1,2,3,\ldots, n$ and $j=1,2,3,\ldots,m$; and $i \neq j$).

Let us look at a simple example to illustrate this. Suppose that a state government is looking into two alternative measures to deal with its current revenue problem: a statewide lottery (X_1) and casino gambling (X_2). Since they are alternative measures, they cannot be pursued at the same time, meaning that only one alternative can be accepted at a time. Assume that both measures are expected to produce more or less the same amount of revenue, but their ability to generate the revenue consistently throughout the year will not be identical. Assume further that the taxpayers are indifferent as to which alternative is selected as long as the government does not raise taxes. Table 9.1 shows the payoff matrix for the two alternative measures.

Following Equation 9.1 and based on the information contained in Table 9.1, the EV for the two alternatives can be calculated as:

$$EV(X_1) = \$500(0.6) + \$500\ (0.4) = \$300 + \$200 = \$500 \text{ million}$$

$$EV(X_2) = \$300(0.6) + \$700\ (0.4) = \$180 + \$280 = \$460 \text{ million}$$

where $EV(X_1)$ is the EV of decision X_1 (statewide lottery) and $EV(X_2)$ is the EV of decision X_2 (casino gambling).

A cursory look at the expected value for each course of action would suggest that the government should go with the lottery option since it has a higher expected value or payoff: \$500 million as opposed to \$460 million for casino gambling. Let us assume that we do not know the probability distribution for the

Table 9.1 Payoff Matrix for Alternatives X_1 and X_2

Alternatives (Action X*)*	*State of Nature*	
	Low Demand (L_D) ($P_1 = 0.6$)	*High Demand* (H_D) ($P_2 = 0.4$)
Lottery (X_1)	\$500 million	\$500 million
Casino gambling (X_2)	\$300 million	\$700 million

states of nature in our problem. When the probability distribution is not known a priori, we can assign the probabilities arbitrarily (randomly) to the various states of nature, as long as the sum of these probabilities add up to 1. The logic of this argument is based on a fundamental law of probability that the sum of the probabilities associated with all possible states of nature for any given distribution must be equal to 1. Thus, if one assigns P_1 to one of our two states of nature, and $(1-P_1)$ to the other, then, by definition, $P_1+(1-P_1)$ must be equal to 1.

Let us say that we assign P_1 to the months when demand is low (i.e., when individuals are less willing to buy lottery tickets or spend money on gambling). We will call it the low demand state, and assign $(1-P_1)$ to the months when demand is high (i.e., when the opposite is true). We will call the latter the high demand state. The expected values for the two courses of action, based on the two states of nature, will then be:

$$EV(X_1)=500(P_1)+500(1-P_1)$$

$$EV(X_2)=300(P_1)+700(1-P_1)$$

Since we made an assumption that our taxpayers are indifferent to the course of action the government is likely to take, we can let the two equations be equal to each other, then solve for P_1 and $(1-P_1)$. That is:

$$EV(X_1)=EV(X_2)$$

$$500(P_1)+500(1-P_1)=300(P_1)+700(1-P_1)$$

$$500P_1+500-500P_1=300P_1+700-700P_1$$

$$400P_1=200$$

$$\therefore P_1=0.5$$

The result produces a probability of 0.5 for low demand months. The probability for high demand months will also be 0.5; that is, $1-P_1=1-0.5=0.5$. It is important to note that at these probabilities the expected value will be the same for both courses of action. However, if P_1 was greater than 0.5, this would have resulted in a higher expected value for X_1, while the converse would have been true if P_1 was less than 0.5.

Expected Opportunity Loss

Another concept frequently used in decision problems with risk is the expected opportunity loss, also known as regret. It is obtained by subtracting the payoff for a particular course of action from the payoff for the best course of action. Unlike an expected value, which is primarily used to maximize expected payoff for a given course of action, the expected opportunity loss is used to minimize the expected loss. There is a similarity between opportunity loss and a term we introduced earlier, opportunity cost. Both opportunity loss and opportunity cost essentially mean the same thing (i.e., the value an alternative course of action would have produced had it been pursued instead of the current course of action). It is the amount of opportunity loss (i.e., opportunity forgone).

To illustrate the concept of opportunity loss, let us look at the same problem, but add a third alternative for the government that it might issue savings bonds instead of licensing casino gambling or introducing a statewide lottery. We, therefore, have three alternatives instead of two. Our objective is to select the alternative that will produce the minimum expected loss to the government. Table 9.2 presents the opportunity losses for the three alternative courses of action.

According to the table, the opportunity loss for alternative X_2 low demand state is \$200 million; that is, \$500 million – \$300 million = \$200 million. Similarly, the opportunity loss for alternative X_3 for the same state is \$150 million; that is, \$500 million – \$350 million = \$150 million. It should be pointed out that the opportunity loss for the best course of action, alternative X_1 in our example is 0; that is, \$500 million – \$500 million = \$0.

The same procedure can now be used to calculate the expected opportunity loss for each alternative course of action for the high demand state. In principle, calculating the expected opportunity loss for a course of action boils down to taking its expected value, as can be seen from the following results:

$$EOL(X_1) = (0)(0.6) + 200(0.4) = 0 + 80 = \$80 \text{ million}$$

$$EOL(X_2) = 200(0.6) + (0)(0.4) = 120 + 0 = \$120$$

$$EOL(X_3) = 150(0.6) + 350(0.4) = 90 + 140 = \$230 \text{ million}$$

Table 9.2 Payoff Matrix for Alternatives X_1, X_2, and X_3

Alternatives (Action X)	*State of Nature*	
	Low Demand (L_D) ($P_1 = 0.6$)	*High Demand* (H_D) ($P_2 = 0.4$)
Lottery (X_1)	\$500 million	\$500 million
Casino gambling (X_2)	\$300 million	\$700 million
Savings bonds (X_3)	\$350 million	\$350 million
X_1	0	\$200 million
X_2	\$200 million	0
X_3	\$150 million	\$350 million

where the term *EOL* represents the expected opportunity loss for a given course of action.

The results clearly indicate that alternative X_1, the statewide lottery, is the best option since it produces the minimum expected loss to the government (i.e., $80 million, as opposed to $120 million for X_2, and $230 million for X_3). In other words, it has the lowest expected loss of the three.

Bayesian Analysis

In the examples presented above there was an implicit assumption that the information used in calculating the expected value or loss corresponding to a particular decision was limited. In other words, our decision makers did not have unlimited information. In reality, information available to a decision maker may not be limited and could be increased incrementally such that each additional piece of information will produce marginally improved decisions. Bayesian analysis, a method named after an eighteenth-century English clergyman by the name of Thomas Bayes, is frequently used to deal with decision problems of this nature (Raiffa, 1968).

Bayesian analysis is based on the principle that with additional information it is possible to revise (i.e., marginally improve) the probability of an event or outcome (defined as the result of an event, a state of nature, or a course of action). The revised probability is called the posterior probability. The probability associated with the initial state of nature or condition is called the prior probability. Therefore, the probability information one needs to move from prior to posterior, or from posterior to prior, depends on the condition that one or more events have already occurred. The latter is known as conditional probability and can be defined as the probability of an event, given the probability that one or more events have occurred.

In general, if one has information on both prior and conditional probability, one can easily calculate the posterior probability by using the Bayesian criterion called the Bayesian rule. The following expression is commonly used to obtain this probability:

$$P(E_j \mid O_1) = \frac{P(O_1 \mid E_j)P(E_j)}{\sum_{j=1}^{m} P(O \mid E_j)P(E_j)} = P(E_j|O) \quad [9.2]$$

where $P(E_j|O)$ is the posterior (revised) probability of an event j, given the occurrence of an outcome 0, $P(O|E_j)$ is the conditional probability of outcome 0, given the occurrence of the event j, and $P(E_j)$ is the marginal or prior probability of event j (for $j = 1, 2, 3, \ldots, m$). Note that the events $1, 2, 3, \ldots, m$ are mutually exclusive.

To illustrate how the Bayesian principle works, let us go back to the revenue problem again. Suppose that we know the conditional probability of each outcome, given the occurrence of each state of nature. We define these outcomes as O_1 for the low demand (L_D) state, and O_2 for the high demand (H_D) state. Let us say that conditional probabilities of the outcomes, O_1 and O_2, given that each state of nature has occurred, are as follows:

$$P(O_1|L_D)=0.7$$

$$P(O_2|L_D)=0.3$$

$$P(O1|H_D)=0.5$$

$$P(O2|H_D)=0.5$$

The probabilities presented above can be read in the following manner: if, for instance, the state of low demand exists in the future, the probability that the outcome O_1 can actually predict this state is 0.7. Similarly, if the state of high demand continues in the future, the probability that the outcome O_1 can actually predict the state is 0.5, and so forth. However, in order to find the posterior probability for our problem, we also need to have information on prior probabilities. Let us say for the sake of argument that we know these probabilities, which are $P(L_D)=0.6$ for the low demand state and $P(H_D)=0.4$ for the high demand state. We already know the conditional probabilities. Therefore, if the conditional probability of the state of low demand is known, given that it has occurred, $P(O_1|L_D)$, the posterior probability of this state occurring, given that it has occurred, can be obtained from the following expression:

$$P(L_D \mid O_1) = \frac{P(O_1 \mid L_D)P(L_D)}{P(O_1 \mid L_D)P(L_D) + P(O_1 \mid H_D)P(H_D)}$$

$$= \frac{(0.7)(0.6)}{(0.7)(0.6) + (0.5)(0.4)}$$

$$= \frac{0.42}{0.42 + 0.20}$$

$$= 0.677$$

The result indicates that the probability that the low demand state will occur in the future is 0.677, which reflects an improvement of 0.077 over the initial probability of $P(L_D)=0.6$; that is $0.677-0.6=0.077$. In other words, this is the amount by which the probability of the state of low demand will increase for an additional piece of information on its prior probability. We can repeat the process to calculate the remaining posterior probabilities. The result will be $P(L_D|O_2)=0.474$, $P(H_D|O_1)=0.323$, and $P(H_D|O_2)=0.526$.

Expected Utility

Decision makers do not always select alternatives that maximize the expected value of their decisions, although that is a common expectation. There are at least two explanations for this. First, it may not be in the best interest of an organization to accept a potential loss in the short run in order to realize a potential gain in the long run. Decision makers who follow this principle are called risk avoiders. Second, there are situations in which a decision maker may be willing to accept a loss today in order to realize a potential gain tomorrow. These types of decision makers are called risk takers.

In reality, however, most decision makers are not risk takers; instead, they select those decisions that will result in smaller expected payoffs with lower risk than decisions that will result in higher payoffs with greater risk. This common sense rationality in decision making can be better explained by a concept called utility (Von Neumann & Morgenstern, 1969). The term utility means the satisfaction one derives from a given course of action, but it has a different meaning when used in a decision-making context. In the latter case, it means the value or worth of an alternative, given the return it is expected to produce (expressed usually in dollar terms), as opposed to the expected returns from all other alternatives. Obviously, a rational decision maker will select the alternative that will produce the greatest expected return or utility.

Consider a problem in which a local government wants to determine whether it should or should not buy insurance to cover the risks for its firefighters. If the government decides to buy insurance, it will cost $25,000 per year, with a 0.5 percent [$0.005 \times 100 = 0.5$] probability that there will be any fire-related accidents. On the other hand, if it decides not to buy, then obviously it will not have to pay the premium. However, in the event that there is an accident leading to a serious injury or death, it will cost the government $1 million. What would be the rational course of action for the government in this case? There are two ways to go about it: either to use the expected cost (EC) or use the expected utility (EU) of each decision. Table 9.3 presents the data for both alternatives.

Thus, if the government decides to use the EC as a decision criterion, the choice will be to not insure since this will produce a higher EC: $25,000 as opposed to $5,000. The following shows the EC calculation for the two alternatives:

$$EC(X_1) = \$25,000$$

$$EC(X_2) = \$1,000,000\ (0.005) + \$0\ (0.995) = \$5,000$$

Table 9.3 Payoff Matrix for Alternatives X_1 and X_2

Alternatives (Action X)	*State of Nature*	
	Risk ($P_1 = 0.005$)	*No Risk* ($P_2 = 0.995$)
Insurance (X_1)	$25,000	$25,000
No insurance (X_2)	$1,000,000	0
X_1	–0.5	–0.5
X_2	0.00400	0

Alternatively, we can formulate the problem in terms of the EU it has to the government. Suppose that the government assigns a value of –0.5 to the premium it has to pay should it decide to buy insurance. Let us call it a util of –0.5, and –400 utils if it decides not to. The EU for each alternative would then be:

$$EU(X_1) = -0.5$$

$$EU(X_2) = -400(0.005) + 0(0.995) = -2$$

As the EU is higher for alternative 1 (i.e., –0.5>–2), the government should go with this alternative. In other words, it should buy insurance.

It is, however, worth pointing out that one must be careful in using utility as a decision criterion in determining an expected value since much depends on how one assigns these utils or utility values to a particular decision. The common sense notion, though, is that the assignment of these values depends on the risk-averse or risk-acceptance nature of a decision maker. In other words, the more risk a decision maker is willing to take, the lower the utility value he or she will assign to a given risk. This simple relationship between the nature of a decision maker and the utility value he or she assigns can be shown with the help of a utility curve, as in Figure 9.1.

The figure relates utility values to dollar amounts associated with each decision. According to the figure, the utility value will be different for different types

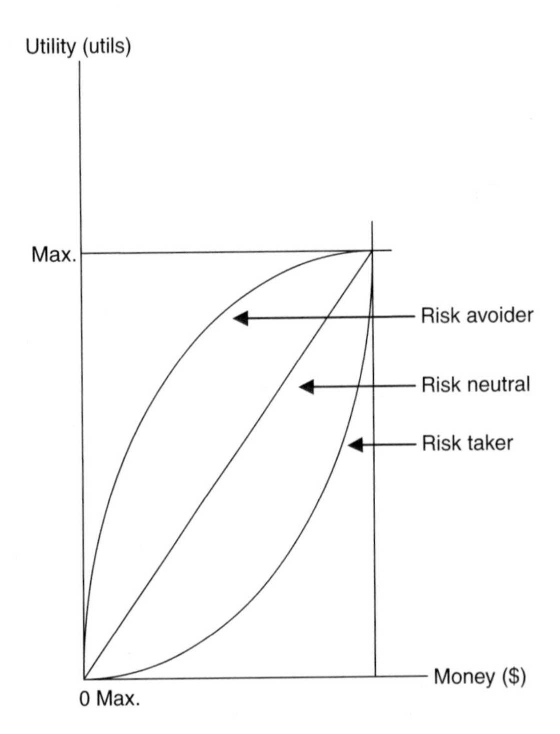

Figure 9.1 A Simple Utility Curve

of decision makers. For instance, the curve will be linear for a risk-neutral decision maker and curvilinear for others.

Decisions under Uncertainty

Decision making under uncertainty is a condition in which the probabilities of likely outcomes of a set of decisions are not known a priori. While the outcomes may be known, the decision maker may be ignorant of the probabilities of occurrence of those outcomes and, as such, cannot assign the probabilities to the states of nature. This makes decision making under uncertainty more complex than decision making under conditions of risk. However, several criteria have been developed over the years that can deal with this problem. This section discusses four of these criteria: Laplace, maximin, maximax, and minimax.

Laplace Criterion

The Laplace criterion is considered a subjectivist criterion in that it is based on a simple assumption that since the probabilities of future states of nature are unknown, we can assume them to be equally likely (Schlaifer, 1959). In other words, each state of nature is assigned the same probability of occurrence. The rationale behind this assumption can be found in a dictum called the Principle of Insufficient Reason (Keynes, 1921). The principle basically says that things in nature do not occur without a cause, even if the cause is a trivial one. Since we do not know the reasons for a state of nature to occur, we can assume that one is as likely to occur as another. So long as this assumption is considered valid, one can easily determine the expected payoff for each alternative (strategy) and select the alternative (strategy) that produces the largest expected payoff.

For a general case with n states of nature, we can formally express this as:

$$EV(X_i) = \frac{1}{n}\sum_{j=1}^{m} P_{ij} \qquad [9.3]$$

where $EV(X_i)$ denotes the expected value of the variable X for alternative i, and P is the probability associated with the alternative i and the state of nature j (for $i = 1, 2, 3, \ldots, m$).

We can use a simple example to illustrate the criterion. Suppose that a county government is planning to invest $1 million in one of three alternative investment instruments: money market funds (X_1), Treasury notes (X_2), and municipal bonds (X_3).[1] Assume that the government wants to invest all the funds in the alternative that will produce the highest expected return. Assume also that we do not know what the market conditions will be in the future, but we can assign the same probability to each of the conditions. Table 9.4 presents the payoff matrix for the decision alternatives for our government.

Table 9.4 Payoff Matrix for Investment Decisions

Alternatives (Action X)	*State of Nature*			
	Condition A	*Condition B*	*Condition C*	*Minimum Payoff*
Money market funds (X_1)	\$50,000	\$40,000	\$10,000	\$10,000
Treasury notes (X_2)	\$30,000	\$30,000	\$15,000	\$15,000
Municipal bonds (X_3)	\$20,000	\$20,000	\$20,000	\$20,000
Maximum payoff	\$50,000	\$40,000	\$20,000	

Based on the information provided in the table, the expected values for the three investment alternatives can be calculated in the following way:

$$EV(X_1) = (1/3)(50{,}000) + (1/3)(40{,}000) + (1/3)(10{,}000) = \$33{,}333.33$$

$$EV(X_2) = (1/3)(30{,}000) + (1/3)(30{,}000) + (1/3)(15{,}000) = \$25{,}000.00$$

$$EV(X_3) = (1/3)(20{,}000) + (1/3)(20{,}000) + (1/3)(20{,}000) = \$20{,}000.00$$

As the results indicate, both Treasury notes and municipal bonds produce a lower expected return than the money market funds, meaning that the government should consider investing in the latter.

Maximin Criterion

The maximin criterion was introduced by Abrabam Wald (Miller & Starr, 1960). Unlike the subjectivist Laplace criterion, it takes a pessimistic view of decision outcomes. It assumes that the nature would always be malevolent; that is, regardless of the alternative a decision maker takes, it will always produce the worst outcome. Therefore, the objective of the decision maker should be to select that alternative which will maximize the minimum payoff. That means the decision maker must first try to determine the worst outcome for each alternative, then select the one that would produce the maximum of the minimum payoffs.

Given this criterion of maximum of the minimum payoffs to the investment problem mentioned above, our decision maker would select the alternative of investing in municipal bonds because it would result in the maximum of the minimum values, which is \$20,000 (Table 9.4). It should be pointed out that as a decision criterion, maximin is ideal for a decision maker who is a risk avoider; that is, one who takes a conservative approach to the future.

Maximax Criterion

This criterion is the exact opposite of the maximin criterion in that it takes an optimistic view of decision outcomes. It is based on the notion that faced with different alternatives, a decision maker will select the alternative that will produce the maximum return. Thus, for our investment problem the maximum payoff that would result from an alternative would be Treasury bonds producing a maximum of the maximum payoff of \$50,000 (Table 9.4). Ideally, the criterion is suitable for decision makers who are risk takers; that is, those who are more optimistic about the future.

It is important to note that both maximin and maximax criteria take an extreme view of decision outcomes. In reality, decision makers are neither completely pessimistic nor completely optimistic. In fact, they are risk neutral and generally take a compromise position that lies somewhere between the two extremes. Leonid Hurwicz (1951) suggested a measure called coefficient of optimism that is frequently used in this context to measure the degree of risk associated with a particular decision. The coefficient, given by α, ranges between 0 and 1, where 0 represents complete pessimism and 1 represents complete optimism. Interestingly, if α is set equal to 0, the Hurwicz criterion becomes the same as the maximin criterion, and when it is set equal to 1, it becomes the same as the maximax criterion.

We can now apply the coefficient of optimism to the investment problem to determine which alternative would produce the best payoff. Assume that we have an α value of 0.75; our coefficient of optimism, the payoff for a risk-neutral decision maker, will then be:

$$X_1 = (50{,}000)(\alpha) + (10{,}000)(1-\alpha) = (50{,}000)(0.75) + (10{,}000)(0.25) = \$40{,}000$$

$$X_2 = (30{,}000)(\alpha) + (15{,}000)(1-\alpha) = (30{,}000)(0.75) + (15{,}000)(0.25) = \$26{,}250$$

$$X_3 = (20{,}000)(\alpha) + (20{,}000)(1-\alpha) = (20{,}000)(0.75) + (20{,}000)(0.25) = \$20{,}000$$

Looking at the results, one can see that since the money market funds (X_1) have the highest weighted value, this should be selected as the best alternative. However, there is a problem with the Hurwicz criterion, in that it is not clear how the α value is selected. There is no hard-and-fast rule for selecting this value. Therefore, the best one can do is to use a trial-and-error approach until a precise estimation of the degree of optimism is obtained.

Minimax Criterion

Originally developed by L. J. Savage (1951), the minimax criterion is based on the concept of opportunity loss, discussed earlier. The basic principle underlying the criterion is that once a state of nature has occurred, the decision maker may have regrets if the outcome (payoff) resulting from the alternative he or she has selected is less than the maximum payoff. Savage argues that under these conditions the decision maker should attempt to minimize the regret, which can be measured by the difference between the payoff actually received and the payoff that could have been received if the state of nature that was going to occur had been known.

Suppose that for our investment problem we assume that condition *A* has actually occurred. In other words, the government has invested in money market funds, meaning that it will have no regrets because it produces the maximum possible payoff ($50,000), but if it had invested in Treasury notes instead it would have lost $20,000; that is, $50,000–$30,000=$20,000. This is the measure of regret. Alternatively, if it had invested in municipal bonds, it would have experienced a regret of $30,000; that is, $50,000–$20,000=$30,000.

Similarly, if condition *B* had prevailed and the government had invested in money market funds, it would not have any regrets since it produces the largest possible payoff. Alternatively, if it had invested in Treasury notes, it would have experienced a regret of $10,000; that is, $40,000–$30,000=$10,000. In the same vein, if it had invested in municipal bonds, it would have had a regret of $20,000; that is, $40,000–$20,000=$20,000. The process could be continued for the remainder of the payoff matrix. Table 9.5 shows the regret matrix for all three strategies under conditions of *A*, *B*, and *C*.

Given the regrets in Table 9.5, which strategy should our government select? Savage maintains that to minimize the regrets the decision maker should select the minimum of the maximum regrets. Therefore, the government should invest in Treasury bonds since it produces the minimum of the maximum regrets, which is $10,000.

Decisions under Conflict

Our discussion up to this point focused on decision against nature. The tenet of the argument has been that the state of nature that occurs is independent of the selection strategy of the decision maker. When rational opponents are involved instead of the state of nature in a decision problem, we have decisions under conflict. The difference between the two is that a rational opponent will give careful thought to the strategy an opponent will select before selecting his or her own course of action. The decision theory that explains this kind of behavior is called game theory. The theory was first introduced by Von Neumann and Morgenstern in 1944 in a seminal work, entitled *The Theory of Games and Economic Behavior* (1944), and has since been applied to a wide variety of disciplines and problems.

Table 9.5 Regret Matrix for Investment Decisions

Alternatives (Action X*)*	*State of Nature*		
	Condition A	*Condition B*	*Condition C*
Money market funds (X_1)	0	0	$10,000
Treasury notes (X_2)	$20,000	$10,000	$5,000
Municipal bonds (X_3)	$30,000	$20,000	0
		Maximum regret	
X_1		$10,000	
X_2		$20,000	
X_3		$30,000	

Zero-Sum Game

In its bare essence, game theory analyzes the strategic behavior of individuals, called players, and how each player uses various strategies to his or her best advantage. Games are generally classified according to the number of players participating, which is usually two, but can involve three or more players. They can also be classified according to the payoff each player receives, called the value or outcome of the game. An outcome may be positive, negative, or zero. If it is positive, it is called a gain; if it is negative, it is considered a loss; and if it is zero, it means neither gain nor loss.

We start with a simple situation in which two communities, let us call them Player A and Player B, are competing for a federal grant of $25 million. Assume that the communities are comparable in size and economic background. Also, assume that each community has a set of strategies or game plans that it can use to produce the most favorable outcome for itself. Assume further that the total amount of grant could go to either community, or be divided between them in equal or varying proportion, depending on how successfully they are able to use their strategies.

Our problem is a classic case of a two-person, zero-sum game. It is called a zero-sum game because the gain by one player is considered a loss to another. Since the gain and the loss cancel each other out, it produces a sum of zero. It is also called a constant sum game because, regardless of the choices made or strategies used, the sum of the payoffs to both players is a constant, which in this case will be $25 million. Table 9.6 presents the payoff matrix for the two communities.

According to the table, each community has three choices: A_1, A_2, and A_3 for A; and B_1, B_2, B_3, for B. Let us say that A selects A_2 and receives $20 million as a result, which means that B loses $20 million that it could have received had A not selected this strategy. The sum of A's gain and B's loss produces a total of zero, that is, $\$20 \text{ million}_A + (-)\$20 \text{ million}_B = 0$. It is obvious that A will select A_2 because it produces the maximum gain for it. However, if A gains $20 million, then B will receive the remainder of the $25 million of the grant money, which is $5 million. The result is a constant sum of $25 million; that is, $\$20 \text{ million}_A + \$5 \text{ million}_B = \$25 \text{ million}$. Unfortunately, B does not emerge a clear winner from this simple two-person zero-sum game, but it can definitely try to improve on the choices it makes to have a more favorable outcome.

Table 9.6 Payoff Matrix for the Two-Person, Zero-Sum Game ($ Million)

	Community B's Strategies		
Community A's Strategies	B_1	B_2	B_3
A_1	8	0	13
A_2	20	15	18
A_3	17	0	5

Pure Strategies

When faced with a situation of the type discussed above, one should take a cautious approach, assume the worst, and act accordingly. This simple strategy is based on two basic principles: maximin and minimax. As noted earlier, the maximin principle is frequently associated with gains and the minimax principle with losses. According to the former, the objective of the decision maker is to maximize gains from among minimum possible outcomes, whereas according to the latter the objective is to minimize losses from among maximum possible outcomes.

Let us look at the problem again and try to identify the maxima and minima for the two communities. Table 9.7 shows these maxima and minima. As the table shows, the minimum gains for community A from the three strategies are: $A_1=0$, $A_2=\$15$ million, and $A_3=0$. Similarly, for community B, the maximum losses from the three strategies are: $B_1=\$20$ million, $B_2=\$15$ million, and $B_3=\$18$ million. Assume that the decision makers in both communities are cautious and to a degree pessimistic in that they will restrict their choices to these six scenarios to avoid any potential risk by selecting any other strategy. Therefore, barring any other alternative, A will choose $A_2=\$15$ million, which is the maximum of the minimum gains and *B* will choose $B_2=\$15$ million, which is the minimum of the maximum losses.

Since neither community is interested in changing its strategy to avoid any potential risk, it is called a pure strategy. For most game-theoretic problems, pure strategy exists when the solution has reached an equilibrium, known as a saddle point or steady state. An equilibrium or steady state is a solution from which no player has an intention to move.[2] In the current example, the equilibrium is reached when A selects A_2 and B selects B_2, producing a saddle point or equilibrium value of \$15 million, which means that A will gain \$15 million or B will lose \$15 million. Even though B will lose \$15 million, it will still gain \$10 million and will come out better off by \$5 million when compared with the earlier solution. In either case, the result will be a constant sum; that is, \$15 million_A+\$10 million_B=\$25 million.

In general, the solution produced by a pure strategy is always optimal (the best the players could do given the alternatives they have) as long as both players adhere to the principles of maximin and minimax. If one follows the principle and the other does not, the result will not be optimal.

Table 9.7 Payoff Matrix for the Two-Person, Zero-Sum Game: Pure Strategy (\$ Million)

	Community B's Strategies			
Community A's Strategies	B_1	B_2	B_3	Minimum of Row Gains
A_1	8	0	13	0
A_2	20	15	18	15
A_3	17	0	5	0
Maximum of Column Losses	20	15	18	

Mixed Strategies

It is quite likely that for many real-world problems involving pure strategies there are no saddle points, no equilibria, and no steady states. However, it is possible to obtain a steady state solution for these kinds of problems by using what is known as mixed strategies. In a mixed strategy, the players do not have any information or knowledge of each other's strategies. When the strategies are unknown, one can only guess what moves the other player will make. This, in turn, will increase the uncertainty of outcomes that will result from these decisions.

We can easily reorganize the data in the last example to see how a set of mixed strategies would produce a saddle point or equilibrium solution. As before, we begin with the assumption that both communities will play by the rules in which A will try to select those strategies that will produce the best result from the set of minimum gains, and B will select those strategies that will produce the best result from the set of maximum losses. Thus, community A would select strategy A_3 since it produces the maximum of the minimum gains for it, which is $8 million.

Community B, on the other hand, would select strategy B_3 since it produces the minimum of the maximum losses, which is $10 million. But if A is rational (which we assume it is), then it would select A_1, keeping in mind that B will most likely select B_3. A_1 is the second best choice for A since it has the second maximum of the minimum gains at $5 million. Likewise, B would select B_2 thinking that if it selects B_3, A would definitely select A_1. B_2 happens to be the second best choice for B since it has the second minimum of the maximum losses at $15 million. This, in turn, would force A to move back to the first choice of A_3 and the process will continue without reaching an equilibrium; thus, creating a decision loop. This is shown in Table 9.8.

The decision loop created by moving back and forth between the two sets of strategies by the two communities in our example produces a submatrix consisting

Table 9.8 Payoff Matrix for the Two-Person, Zero-Sum Game: Mixed Strategy ($ Million)

	Community B's Strategies			
Community A's Strategies	B_1	B_2	B_3	Minimum of Row Gains
A_1	20	→**15** →	**5**	5
A_2	18	13	0	0
A_3	12	**8** ←	**10** ←	8
Maximum of Column Losses	20	15	10	

	Community B's Strategies	
Community A's Strategies	B_2	B_3
A_1	15	5
A_3	8	10

of two rows and two columns. The bottom half of Table 9.8 shows this reduced matrix. Since in mixed strategies the players do not know of each other's strategies, they have no choice but to randomly select a set of strategies that will best serve their interests (i.e., their payoffs will be optimal). For instance, if A selects A_1 with a probability of P, and A_2, with a probability of $(1-P)$, then A's expected gains, assuming B selects B_2, will be

$$P(15)+(1-P)(8) \qquad [9.4]$$

Alternatively, if B selects B_3, then A's expected gains, by the same process, will be

$$P(5)+(1-P)(10) \qquad [9.5]$$

Assume that A does not know the strategies B will select. The expected gains of A for each of B's possible choices, therefore, must be equal; that is, in equilibrium, the expected gains of A for each possible strategy of B must be the same. We can, therefore, set the two equations equal to each other to obtain the probability of A's gains for each possible choice of B, as shown below:

$$P(15)+(1-P)(8)=P(5)+(1-P)(10)$$

$$15P+8-8P=5P+10-10P$$

$$15P-8P-5P+10P=10-8$$

$$12P=2$$

$$P=0.17$$

where all the values, except for P, are in million dollars.

As the result shows, the probability of A's gain for P is 0.17; that is, $2/12=0.17$, and for $(1-P)$ it is 0.83; that is, $1-0.17=0.83$, meaning that A would select strategy A_1 17 percent of the time. Similarly, the probability of B's losses for each possible choice of A_1 can be determined by setting the two equations for B's expected losses equal to each other, as given below:

$$P(15)+(1-P)(5)=P(8)+(1-P)(10)$$

$$15P+5-SP=8P+10-10P$$

$$15P-SP-8P+10P=10-5$$

$$12P=5$$

$$P=0.42$$

where all the values, except for P, are in million dollars. This results in $(1-P)=1-0.42=0.58$, which means that B would select strategy B_2 42 percent of the time, and B_3 58 percent of the time.

Using the basic information described above, we can now determine the equilibrium value in terms of the expected gains and losses for the two communities, i.e., the gains for A and the losses for B for respective strategies. That is:

[i] A's expected gains are:

If B selects B_2: (0.17)($15 million)+(0.83)
($8 million)=$9.19 million≈$9.20 million

If B selects B_3: (0.17)($5 million)+(0.83)
($10 million)=$9.15=$9.20 million

[ii] Similarly, B's expected losses are:

If A selects A_1: (0.42)($15 million)+(0.58)
($5 million)=$9.20 million

If A selects A_3: (0.42)($8 million)+(0.58)
($10 million)=$9.16 million=$9.20 million

The solution produces an equilibrium value of $9.20 million, which is the amount A is likely to receive and B likely to lose, based on the strategies used by the two communities. Like before, the sum of the two payoffs will be a constant, which is equal to the total amount of the grant; that is, $9.20 million_A +($25.00 million_B−$9.20 million_A)=$9.20 million_A+$15.80 million_B =$25 million. Fortunately for B, it comes out the bigger winner this time.

Role of Dominance

An interesting characteristic of the problem we have just discussed is that each community had exactly the same number of strategies but, in reality, this may not always be the case. A game in which one player has more choices than the other is denoted by a matrix of $(K \times M)$ or $(M \times K)$, where $M > K$. The notations M and K represent the number of strategies or choices available to each player in the game. If, for instance, community B had only two choices as opposed to three, then we would have an $M \times 2$ game with $M=3$ for A and $K=2$ for B. Likewise, if community A had two choices instead of three, we would have a $2 \times M$ game with $M=3$ for B and $K=2$ for A.

When all the elements in a column of a payoff matrix are greater than or equal to the corresponding elements in another column, then that column is said to be dominated. Similarly, when the elements in a row of a payoff matrix are less than or equal to the corresponding elements in another row, then that row is said to be dominated. When a particular strategy is completely dominated by another strategy, the dominated strategy can be eliminated from the payoff matrix since it is quite unlikely that a player would consider that strategy. The concept of dominance is useful for payoff matrices that are considerably large. By applying the rule of elimination, the matrix can be reduced to determine the optimal solution.

Consider the example in Table 9.7 again. Looking at the payoff matrix in the table, we can see that it is unlikely that A will select A_1 because if it does then B could select B_1 or B_2, in which case it will be worse off. Thus, we can say that row A_1 is dominated by A_2 and A_3; hence, it should be eliminated. In the same vein, it is unlikely that B will select B_1 because if it does and in response A selects A_2 or A_3, it will be worse off. That means B_1 is dominated by B_2 and B_3. By examining the reduced matrix further in Table 9.7, we can see that row A_3 is now completely dominated by row A_2. As such, we can further eliminate it from our reduced matrix, leaving it with one row, A_2, and two columns, B_2 and B_3. This is shown in Table 9.9.

As the table shows, community B has no choice but to select B_2 to minimize its loss. Therefore, the saddle point solution will be a pure strategy of A_2 for A, and B_2 for B, with an equilibrium value of $15 million. This is the amount A will gain and B will lose, producing a constant sum of $15 million$_A$ + $10 million$_B$ = $25 million for the game. This is close to what we obtained earlier in our solution to the problem under pure strategy.

A Note on Linear Programming Formulation of Game Theory

The procedures we have used so far to find the value of a game are useful as long as our players have a limited number of strategies, but when the number of strategies increases beyond a handful, the simple process of elimination will not be sufficient. One would need a more rigorous approach that is capable of dealing with a large number of strategies. Linear programming, discussed earlier, is an ideal alternative in this type of situation.

Linear programming deals with optimization; that is, maximizing or minimizing an objective function subject to a set of constraints, but when applied to game theory the objective function becomes one of finding a set of optimal strategies for each player in the game. To give an example, consider the federal grant allocation problem for the two communities again. Suppose that each community now has four strategies each, instead of three, as shown in Table 9.10. We can start by trying to find the probability associated with selecting a particular

Table 9.9 Reduced Payoff Matrix for the Two-Person, Zero-Sum Game Dominance ($ Million)

	Community B's Strategies	
Community A's Strategies	B_2	B_3
A_2	15	18
A_3	0	5
A_2	15	18

Table 9.10 Payoff Matrix for the Two-Person, Zero-Sum Game: Mixed Strategy ($ Million)

	Community B's Strategies			
Community A's Strategies	B_1	B_2	B_3	B_4
A_1	5	0	15	10
A_2	15	20	10	15
A_3	10	15	0	10
A_4	15	5	25	5

strategy by the two communities. Suppose further that the probability associated with each strategy for community A is given by the vector (P_1, P_2, P_3, and P_4) and for B by the vector (P_1, P_2, P_3, and P_4). Thus, if community B selects B_1, the expected mixed strategy payoff for community A is $5P_1+15P_2+10P_3+15P_4$. Similarly, if community A selects A_1, the expected mixed strategy payoff for community B will be $5P_1+0P_2+15P_3+10P_4$.

Assume now that our maximin and minimax principles hold good for both players; the objective of community A would then be to maximize the expected payoffs (gains) for different strategies of B. Similarly, the objective of community B would be to minimize the expected payoffs (losses) for various strategies of A. Assume further that the value of the game is V, then the expected payoff for A for different strategies of B can be expressed as:

$$5P_1+15P_2+10P_3+15P_4 \geq V \text{ for strategy } B_1 \quad [9.6]$$

$$0P_1+20P_2+15P_3+5P_4 \geq V \text{ for strategy } B_2 \quad [9.7]$$

$$15P_1+10P_2+0P_3+25P_4 \geq V \text{ for strategy } B_3 \quad [9.8]$$

$$10P_1+15P_2+10P_3+5P_4 \geq V \text{ for strategy } B_4 \quad [9.9]$$

where P is the probability of selecting a particular strategy by A and B.

Similarly, the expected payoffs for B for different strategies of A can be expressed as:

$$5P_1+0P_2+15P_3+10P_4 \leq V \text{ for strategy } A_1 \quad [9.10]$$

$$15P_1+20P_2+10P_3+15P_4 \leq V \text{ for strategy } A_2 \quad [9.11]$$

$$10P_1+15P_2+0P_3+10P_4 \leq V \text{ for strategy } A_3 \quad [9.12]$$

$$15P_1+5P_2+25P_3+5P_4 \leq V \text{ for strategy } A_4 \quad [9.13]$$

where P is the probability of selecting a particular strategy by A and B, as before.

Since the objective is to maximize the minimum of the gains and minimize the maximum of the losses, we can present the goals and strategies of the two communities, along with their corresponding probabilities, in a linear programming framework, as follows[3] for Community A:

$$\text{Minimize } V' = P'_1 + P'_2 + P'_3 + P'_4 \tag{9.14}$$

Subject to:

$$5P'_1 + 15P'_2 + 10P'_3 + 15P'_4 \geq 1 \tag{9.15}$$

$$0P'_1 + 20P'_2 + 15P'_3 + 5P'_4 \geq 1 \tag{9.16}$$

$$15P'_1 + 10P'_2 + 0P'_3 + 25P'_4 \geq 1 \tag{9.17}$$

$$10P'_1 + 15P'_2 + 10P'_3 + 5P'_4 \geq 1 \tag{9.18}$$

$$P'_1, P'_2, P'_3, P'_4 \geq 0$$

where V is a constant that is greater than 0, P is the probability of selecting a particular strategy by A and B, and P' is P_j / V (for $j = 1, 2, 3, 4$).

Similarly, for community B, it can be written as:

$$\text{Minimize } V' = P'_1 + P'_2 + P'_3 + P'_4 \tag{9.19}$$

Subject to:

$$5P'_1 + 0P'_2 + 15P'_3 + 10P'_4 \leq 1 \tag{9.20}$$

$$15P'_1 + 20P'_2 + 10P'_3 + 15P'_4 \leq 1 \tag{9.21}$$

$$10P'_1 + 15P'_2 + 0P'_3 + 10P'_4 \leq 1 \tag{9.22}$$

$$15P'_1 + 5P'_2 + 25P'_3 + 5P'_4 \leq 1 \tag{9.23}$$

$$P'_1, P'_2, P'_3, P'_4 \geq 0$$

where the terms of the equations are the same as before.

The simplex solution, when applied to the two models (not shown here), produces the following results:

A: $P'_1 = 0.0$, $P'_2 = 0.062$, $P'_3 = 0.0$, $P'_4 = 0.015$, and $V' = 0.077$

B: $P'_1 = 0.0$, $P'_2 = 0.0$, $P'_3 = 0.031$, $P'_4 = 0.046$, and $V' = 0.077$

Putting these values in terms of the original (variable) relationships $P'_1=P_1/V$, $P'_2=P_2/V$, $P'_3=P_3/V$, $P'_4=P_4/V$, and $V'=1/V$, we obtain the following:

A: $P_1=0.0$, $P_2=0.81$, $P_3=0.0$, $P_4=0.195$, and $V=\$12.98$ million

B: $P_1=0.0$, $P_2=0.0$, $P'_3=0.4$, $P'_4=0.6$, and $V=\$12.98$ million

As the results indicate, the expected payoff for A and B at equilibrium is \$12.98 million, which is the amount A is likely to gain and B likely to lose, given the probability that A will select strategies $A_1=0.0$, $A_2=0.81$, $A_3=0.0$, and $A_4=0.195$, and B will select strategies $B_1=0.0$, $B_2=0.0$, $B_3=0.4$, and $B_4=0.6$. What this means is that A will not select strategies A_1 and A_3 since the probabilities associated with them are equal to zero. Likewise, B will not select strategies B_1 and B_2 for the same reason. As before, the sum of the two values or payoffs is equal to the total amount of the grant available to the two communities, which is a constant at \$25 million; that is:

$$\$12.98\text{ million}_A+(\$25.0\text{ million}-\$12.98\text{ million})_B=\$12.98\text{ million}_A$$
$$+\$12.02\text{ million}_B=\$25\text{ million}$$

It is worth noting here that it is not necessary to run two separate linear programming solutions for our problem because the dual solution for B is also the solution for A. By the same token, the dual solution for A is the solution for B. That means one linear programming solution with its associated dual would have produced answers for both.

Finally, since every game is expected to produce a solution, we can likewise expect to find an optimal solution to the linear programming formulation for each game. As noted earlier, it is not necessary to solve the problems independently for each player. One only needs to solve either the primal or the dual to obtain the optimal strategy for both.

Dynamic Games

In our discussion of both pure and mixed strategies, we made an implicit assumption that the players do not follow any protocol in making their moves. In other words, they can move in any manner that seems appropriate. In reality, it is quite likely that they would follow some protocols in selecting their strategies, such as moving in sequence in response to each other's moves, as in a chess game. In other words, one player will make the first move, the second player will respond; the first player will then make a countermove, and so on until a final solution is obtained. When a game is played in response to each other's moves, it is called a dynamic game. Obviously, the result one would obtain from a dynamic game will be different from the one obtained from a normal game in which the players select their strategies simultaneously or in a random fashion.

To illustrate the difference between a normal and a dynamic game, let us revisit the two-communities game problem with dominance (Table 9.9). Recall that our analysis produced two dominant strategies, A_2 and B_2, and a solution corresponding to these strategies based on the assumption that both players made their moves simultaneously. Now suppose that A has the option to make the first move and B does not have any option other than to follow A, only after it has made the move. Figure 9.2 shows these choices for the two communities with the help of a tree diagram, known also as an extensive form diagram.

Given that A has the choice to make the first move, what should A select? For instance, if A selects A_2, the maximum payoff it can receive is \$18 million, provided that B selects B_3. On the other hand, if A selects A_3, the maximum payoff it can expect to receive is \$5 million. However, being a rational player and the fact that it had the option to make the first move, A would choose A_2. The question, then, is why should B select B_3? The answer is that it will not, because selecting this strategy will make it worse off by \$3 million. Instead, it will select B_2, which is a rational choice on its part, given that it has two options it must choose from once A has made the first move. The final solution will be A_2 for A and B_2 for B, with a corresponding value of \$15 million for A and \$10 million for B. This is the same solution we obtained earlier for the dominant strategies in Table 9.9.

Let us now reverse the situation, where B has the option to make the first move. The objectives of the two communities are also reversed, where B will now try to maximize its gains and A to minimize its losses, given the strategies they will have to choose from. Thus, if B selects B_2, the maximum payoff it can receive is \$15 million, provided that A selects A_2, but there is also the likelihood

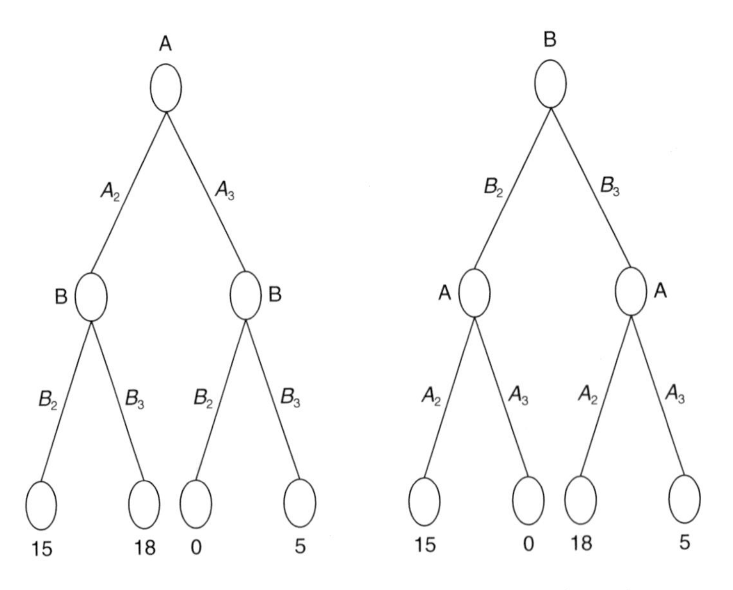

Figure 9.2 Dynamic Games with Mixed Strategies (Dominance)

that A might select A_3, in which case its payoff will be 0. On the other hand, if B selects B_3 the maximum payoff for it will be \$18 million, provided that A selects A_2 and there is no possibility of a \$0 payoff. Therefore, B will select B_3. What will A select? A will no doubt select A_3 because it will produce a payoff of \$5 million for B, which is the smaller of the two payoffs. The final solution of the game will be B_3 for B and A_2 for A. Unfortunately, B comes out much worse off in this case even though it had the first option to move.

Interestingly, however, if we keep the objectives of the two communities the same as before, where A will try to maximize its gains and B minimize its losses from the set of choices they make, the final solution, with B making the first move, will be B_2 for B and A_2 for A. The equilibrium values corresponding to this will be \$15 million for A and \$10 million for B. In other words, the outcome of the game will be the same as the one under dominant strategies, given the nature of the payoff matrix we have in this example.

Chapter Summary

In government, as in the private sector, decisions are often made with limited knowledge or information. When one makes decisions with insufficient information, it will invariably produce results that will be less than optimal. In most instances, the results will be probabilistic rather than deterministic. In other words, the results will reflect risk and uncertainty, although their degrees will vary from situation to situation. This chapter has provided a brief discussion of several commonly used methods for dealing with risk and uncertainty. They include, among others, expected value, Bayesian statistics, and game theory.

Of the methods discussed here, game theory deserves special attention because of its widespread use in a variety of disciplines and their problems. As a theory, it is both challenging and dynamic. For instance, it can deal with games that are zero-sum, as well as those that are non-zero-sum, although much of our discussion here focused on zero-sum. It can also deal with games that involve any number of players, from two to *n*, although *n*-person games are much more difficult to operationalize. Finally, it can deal with games in which the players have both partial and full information, along with their corresponding strategists, including those of their opponents. As indicated earlier, games with partial information are far more complex than games with complete information.

Notes

1 Money market funds, also called money market mutual funds, are short-term, low-risk securities that include certificates of deposits, commercial paper, repurchase agreements, and Treasury bills, among others. Since they are highly liquid, approximately one-third all US investments take place in money market funds.

2 There is a similarity here with an equilibrium condition called the Nash equilibrium, used in experimental economics, where the players are non-cooperating and know full well the equilibrium strategies of their opponents; as such, they will not have anything to gain or lose by changing their own position.

3 Since the four strategies available to each community exhaust the strategy set for the problem, the sum of their corresponding probabilities must be equal to 1; that is, $P_1+P_2+P_3+P_4=1$. Also, since every element of the payoff equations can be made positive by adding a suitable constant in all the coefficients on the left-hand side of these equations, we can assume $V>0$.

Let us, for convenience, divide each of the relationships in Equations 9.6–9.13 by V. The new relationship can then be written as

$$5P_1/V+15P_2/V+10P_3/V+15P_4/V\geq 1 \quad (1)$$

$$0P_1/V+20P_2/V+15P_3/V+5P_4/V\geq 1 \quad (2)$$

$$15P_1/V+10P_2/V+0P_3/V+25P_4/V\geq 1 \quad (3)$$

$$10P_1/V+15P_2/V+10P_3/V+5P_4/V\geq 1 \quad (4)$$

for community A, and:

$$5P_1/V+0P_2/V+15P_3/V+10P_4/V\leq 1 \quad (5)$$

$$15P_1/V+20P_2/V+10P_3/V+15P_4/V\leq 1 \quad (6)$$

$$10P_1/V+15P_2/V+0P_3/V+10P_4/V\leq 1 \quad (7)$$

$$15P_1/V+5P_2/V+25P_3/V+5P_4\leq \overline{V}\leq 1 \quad (8)$$

for community B.

We can simplify these relationships further by setting $P'_1=P_1/V$, $P'_2=P_2/V$, $P'_3=P_3/V$, and $P'_4=P_4/V$. The objective of community A will, therefore, be to maximize the value of the game by minimizing $1/V$. The minimization of $1/V$ comes from the fact that as V gets larger, $1/V$ gets smaller. Similarly, the objective of community B is to minimize the value of the game by maximizing $1/V$, as should be the case. Thus, we can rewrite Equations 1–8 in terms of their equivalent linear programming problem, as shown in Equations 9.14–9.23.

References

Hurwicz, L. (1951). Optimality criterion for decision-making ignorance. In *Cowler commission discussion series on statistics.* No 370. (Cited in R. D. Luce & H. Raitta (1958). *Games and decisions.* New York: Wiley & Sons.

Keynes, J. M. (1921). *A treatise in probabilities*. London: Macmillan. (Referred to by Keynes in the book, was developed originally by Bernoulli and subsequently expanded by Laplace and others).

Miller, D. W., & Starr, M. K. (1960). *Executive decisions and operations research.* Englewood Cliffs, NJ: Prentice-Hall, Inc. (For a discussion on Wald's criterion).

Raiffa, H. (1968). *Decision analysis.* Reading, MA: Addison-Wesley.

Savage, L. I. (1951). The theory of statistical decisions. *Journal of the American Statistical Association, 46*, 55–67.

Schlaifer, R. (1959). *Probability and statistics for business decisions.* New York: McGraw-Hill.

Von Neumann, J., & Morgenstern, O. (1944). *Theory of games and economic behavior.* New Haven, CT: Princeton University Press.

Von Neumann, J., & Morgenstern, O. (1969). *Analysis of decisions under uncertainty.* New York: McGraw-Hill.

10 Productivity Measurement

How to remain efficient in a world of rising costs is a concern for all organizations. Governments, in particular, have been looking for ways to bring their cost behavior in line with that of the private sector for a long time. Although the two do not function in the same manner or from the same behavioral perspective, the concern for efficiency, productivity, and improvement nonetheless remains at the center of all their activities. To a large measure, these concerns are guided by the same objective that motivates an organization to devise methods and suggest ways to better respond to the changing needs of those they serve. In recent years there has been a proliferation of these methods, from simple qualitative measures such as quality indices to complex analytical methods such as mathematical programming and multivariate production functions.

While it is not possible to cover every single development in a single chapter, this chapter focuses on those that are simple and relatively easy to use. The chapter discusses four such measures that lie at the heart of most discussions on productivity: productivity index, production function, learning curve, and time study. Although the notion of optimization may not be readily apparent in these techniques, they are implicit in the way in which they are applied and the results they produce in dealing with a specific problem.

Productivity Index

If there is a single pervasive term that can define the crux of the problem facing a government, it is *productivity*. Every decision a government makes has some elements in it that are directly or indirectly related to productivity, yet there is no general consensus as to how best to define or measure it; much depends on the specific needs of an organization. However, a common sense definition, one that has an intuitive appeal, is getting a job done in the most efficient way. Efficiency in this sense implies a quantitative relationship between input and output, usually expressed as a ratio. Based on this, we can define productivity as the level of efficiency with which the inputs needed to perform a task can be converted to outputs. In general, the higher the level of efficiency, the greater is the productivity, and vice versa. This section discusses two basic measures of productivity that have been extensively used over the years under varying conditions: productivity index and production function.

Single-Factor Productivity

An index is a measure of change. A productivity index measures the ratio of change in output to a change in input. In common sense terms, it means the quantity of goods produced or service provided with a given quantity of input. This relationship between input and output is the single most important factor in determining productivity; that is, how well an organization has been able to utilize the input resources to produce an output. For government, it is the conceptual analog of the number of automobiles manufactured or bushels of wheat produced in the private sector with a given level of input. Occasionally, the term efficiency is used interchangeably with productivity, although the two are not quite the same. Implicit in efficiency is the notion of a production possibility frontier (i.e., feasible levels of output, given the scale of operations, vis-à-vis input combinations). In general, the greater the level of output for a given level of input combinations, the more efficient the activity. By comparison, productivity is simply a ratio of output produced to input used.

There are different ways to measure productivity. The simplest way to measure productivity is to express it as an input–output ratio:

$$PI = O/I \qquad [10.1]$$

where PI is the productivity index, O is the output, and I is the input, and the expression O/I is the ratio of output to input.

The input–output ratio in Equation 10.1 is essentially an output index that can be defined as output per unit of input such as tons of garbage collected per week per worker, number of citations given by a patrol officer per month, number of major fires prevented by a fire district per year, and so forth. When a single input factor is used, such as labor, to produce a given quantity of output, it is called partial or single-factor productivity (SFP). It is worth noting that the expression in Equation 10.1 has an inherent weakness in that it does not take into consideration the base year. The base year serves as a reference point against which the output in any given year can be compared to see if there has been a change in productivity, given the change in input.

Equation 10.2 shows the productivity index with base year input and output:

$$PI = \left(\frac{O_i}{O_0}\right) / \left(\frac{I_i}{I_0}\right) \qquad [10.2]$$

where the subscript i indicates the reference year, 0 is the base year, and the rest of the terms are the same as before.

The numerator in Equation 10.2 represents an index of output and the denominator an index of input. The productivity index thus becomes a ratio of two indices: output and input. To give an example, suppose that a state welfare agency receives thousands of application each year for a service it provides. Last year, which is our base year, it processed 5,000 applications and used a total of 2,000 labor hours for the year, with an average of 2.5 applications per hour

$[O_0/I_0 = O_{t-1}/I_{t-1} = 5{,}000/2{,}000 = 2.5]$. This year, the agency processed 6,000 applications and used 2,500 labor hours with an average of 2.4 per hour $[O_t/I_t = 6{,}000/2{,}500 = 2.4]$. Assuming no changes in working conditions, the productivity index for the agency for processing applications would be 0.96, as shown below:

$$PI = \left(\frac{O_t}{O_{t-1}}\right) / \left(\frac{I_t}{I_{t-1}}\right)$$

$$= \left(\frac{6{,}000}{5{,}000}\right) / \left(\frac{2{,}500}{2{,}000}\right)$$

$$= (1.2)/(1.5)$$

$$= 0.96$$

The result indicates a decline in productivity of 4 percent $[(1-0.96) \times 100]$ from the previous year.

The decline in productivity, as observed in the above example, needs a little explanation. It indicates that if the agency had maintained the same level of productivity as in the base year, which is 2.5 applications per hour, it could have processed the applications this year with less labor, or processed more applications with the same amount of labor hours as in the base year, assuming everything else remained the same. In other words, instead of 2,500 labor hours it used to process 6,000 applications, it could have processed the same number with 2,400 labor hours $[6{,}000/(5{,}000/2{,}000) = 6{,}000/2.5 = 2{,}400]$, producing a saving of 100 labor hours $[2{,}500 - 2{,}400 = 100]$. Alternatively, it could have processed 7,500 applications; that is, 1,500 more instead of 6,000 applications $[(2{,}500 \times 2.5) - 6{,}000 = 7{,}500 - 6{,}000 = 1{,}500]$. Since any increase or decrease in productivity has a direct effect on cost, we can use this information to determine the actual savings it would have produced for the agency in cost terms. Let us say that it costs the agency \$20 per labor hour. Assume that the agency had used the same ratio as last year, that is, 2.5 applications per hour instead of 2.4 per hour this year, it would have saved the agency \$2,000 in processing costs $[((6{,}000/2.4) \times \$20) - ((6{,}000/2.5) \times \$20) = \$50{,}000 - \$48{,}000 = \$2{,}000]$.

In the example discussed above, we looked at a single service. In reality, most agencies provide multiple different services, although single service organizations are not uncommon. Assume now that the agency processes applications not for one, but for five different services. For convenience, we will call them *A*, *B*, *C*, *D*, and *E*. Table 10.1 shows the productivity indices for the five services: A=0.9778, B=1.028, C=1.04, D=0.9524, and E=1.0000. According to the table, there has been a productivity loss for Service A by 2.22 percent $[0.9778 - 1.0 = -0.0222 \times 100 = -2.22]$ and for Service C by 4.76 percent $[0.9524 - 1.0 = -0.0476 \times 100 = -4.76]$. On the other hand, there has been a productivity gain for Service B by 2.85 percent $[1.0285 - 1.0 = 0.0285 \times 100 = 2.85]$ and Service C by 4 percent $[1.04 - 1.0 = 0.04 \times 100 = 4.0]$. Apparently, there has been no change in productivity for Service E $[(1 \times 100) - 100 = 0]$. Overall, there

Table 10.1 Productivity Indices for Multiple Services

Service	*No. Applications Processed (O) (Last Year, t–1)**	*No. Applications Processed (O) (Current Year, t)*	*Ouput Index* (OI_t/OI_{t-1})	
A	125,000	110,000	0.8800	
B	75,000	90,000	1.2000	
C	50,000	65,000	1.3000	
D	40,000	40,000	1.0000	
E	25,000	20,000	0.8000	
Service	*No. Labor Hours Used (I) (Last Year, t–1)**	*No. Labor Hours Used (I) (Current Year, t)*	*Input Index* (II_t/II_{t-1})	*Productivity Index (OI/II)**
A	5,000	4,500	0.9000	0.9778
B	3,000	3,500	1.1667	1.0285
C	2,000	2,500	1.2500	1.0400
D	2,000	2,100	1.0500	0.9524
E	1,000	800	0.8000	1.0000

Notes
**t*–1 = base year; OI = output index; II = input index.

has been a marginal decrease in productivity for the agency by 0.026 percent [$(-0.0222+0.0285+0.04-0.0476+0)/5=-0.00026\times 100=-0.026$], obtained by taking the average of the productivity indices for the five services.

There is, however, one notable problem using simple average to measure overall productivity, in that it assumes a homogeneous input. In most cases, inputs are not homogeneous; they vary depending on the nature of the goods or services provided. If inputs are not homogeneous, then one needs to assign weights to the index to reflect variations in input characteristics. The result will be a weighted productivity index, as given below:

$$PI_W = \sum_{j=1}^{m} (w^j)\left(\frac{I_0^j}{I_t^J}\right)\left[\left(\frac{O_i}{O_0}\right)\right]^j \qquad [10.3]$$

where w represents the weights assigned to the input index for the jth program (for $j=1,2,3\ldots,m$). There is no precise guideline as to how to determine the weights, as long as they reflect the variations in the inputs used. By the same token, if output is not similar, one should use a weighted output index.

Multi-Factor Productivity

The example presented above has an interesting characteristic in that it deals with a single, homogeneous output (application processing) using a single input (labor hours). The assumption of a single homogeneous product or output with a single input allows one to take advantage of the measurement process where one only needs two sets of information to compute the productivity index. Consider another example, this time involving a county-run outpatient clinic that provides a wide range of non-emergency services (outputs) with multiple inputs such as physicians, nursing staff, technicians, equipment, indirect labor, indirect materials and supplies, etc. We will look at imaging service, in particular X-rays, say, chest X-rays with three inputs: direct labor (labor hours expended by technicians), capital utilization (percentage of time the equipment was used), and indirect labor (administrative and other services expressed as indirect labor hours). Table 10.2 shows the input and output data for the service over a five-year period.

Productivity ratios and indices for each of the three input variables are presented in Table 10.3. Let us elaborate on the results a little in Table 10.3. The productivity ratios were computed by dividing the output (the number of X-rays taken for each year) by the corresponding input variables – direct labor (X_1), capital utilization (X_2), and indirect labor (X_3). For convenience, we assumed that the machine costs $150,000, with no salvage value, has a lifespan of five years with a maximum of 15,000 operating hours, and can take a maximum of 15,000 X-rays over its lifespan, all of which can take place in a single year or multiple years. Productivity indices, on the other hand, were computed in two stages: (1) selecting a base year which, in this case, was the fourth most recent year ($t-4$) and setting it equal to 1.0, obtained by dividing the productivity ratio for the year by itself; that is, $0.1150/0.1150=1.0000$ for direct labor; and (2) dividing the productivity ratio for a given year by the base year ratio. For instance, the productivity index for direct labor for $t-3$ will be 1.0583; that is, $0.1261/0.1150=1.0965$, and so forth. As the table shows, there has been a consistent increase in productivity for the service with the exception of the current year, which declined by a small percentage from the previous year for all three input variables: by 9.21 percent for direct labor $[1.1074-1.0783)\times 100=9.21]$, 17.36 percent for capital utilization $[(1.5219-1.3483)\times 100=17.36]$, and 8.4 percent for indirect labor $[(1.0812-0.9972)\times 100=8.4]$.

It is worth noting that the productivity indices using multiple inputs (factors) are called partial factor productivity (PFP) because the indices are calculated individually for each input variable at a time, rather than for all the variables, taken together, as we did with the imaging service. While partial factory productivities

Table 10.2 Input and Output Data for the Outpatient Clinic

Year	*No. X-Rays for the Year (O)*	*Direct Labor (No. Hours) (X_1)*	*Capital Utilization (Capacity = 100%) (X_2)*	*Indirect Labor (No. Hours) (X_3)*
$t-4$	2,300	20,000	15,000	5,400
$t-3$	2,900	23,000	15,000	6,500
$t-2$	3,200	25,000	15,000	7,100
$t-1$	3,500	26,000	15,000	7,600
t	3,100	25,000	15,000	7,300

Table 10.3 Productivity Ratios (Part I) and Productivity Indices (Part II)

Part I

Year	*Productivity Ratio Direct Labor (O/X_1)*	*Productivity Ratio Capital Utilization (O/X_2)*	*Productivity Ratio Indirect Labor (O/X_3)*
$t-4$	0.1150	0.1533	0.4259
$t-3$	0.1261	0.1933	0.4462
$t-2$	0.1280	0.2133	0.4507
$t-1$	0.1346	0.2333	0.4605
t	0.1240	0.2067	0.4247

Part II

Year	*Productivity Index Direct Labor (PR_i/PR_0)**	*Productivity Index Capital Utilization (PR_i/PR_0)**	*Productivity Index Indirect Labor (PR_i/PR_0)**
$t-4$	1.0000	1.0000	1.0000
$t-3$	1.0965	1.2609	1.0477
$t-2$	1.1130	1.3914	1.0582
$t-1$	1.1704	1.5219	1.0812
t	1.0783	1.3483	0.9972

Notes
*ith year ($i=t, t-1, t-2$, and $t-3$); $t-4$=base year.

are conceptually simple and most common in productivity analysis, they do not provide an overall measure of productivity; for the latter, one needs to take into consideration all the input variables together. The simplest way to achieve that is to add the variables, but that would create a problem if the variables are different and expressed in different units. The problem can be avoided by expressing them in monetary terms, assuming one is able to find relevant price data for the variables. Where the price data are not readily available, one can use imputed price to provide a good approximation of the actual price. Table 10.4 shows the cost data for the input variables used in the example.

According to the table, the DLC was obtained by dividing the number of labor hours for a given year by the price of labor, which we assumed to be $30 per hour. On the other hand, the indirect labor cost was obtained by multiplying the total cost of X-rays for a given year by a fixed overhead percentage, which is a standard practice. We assumed the overhead cost to be 15 percent of the overall cost of operating the clinic. To obtain the total cost, we multiplied the number of X-rays for a given year by the cost of a chest X-ray, which we assumed to be $400 per X-ray. Finally, to obtain the cost of capitalization, we multiplied the total cost of the X-ray machine by the percentage of times it was used during the year; that is, ($150,000)(2,300/15,000)=$23,000, ($150,000)(2,900/15,000)=$29,000, and so forth. With the cost data thus obtained for the input variables, we now add them to produce the total cost of inputs. Next, we construct the productivity ratios and

Table 10.4 Cost Data (Part I) and Multi-Factor Productivity (Part II)

Part I

Year	*Direct Labor Cost (X_1)*	*Capital Utilization Cost (X_2)*	*Indirect Labor Cost* (X_3)*
$t-4$	20,000 × $30/hr = $600,000	$23,000	$920,000 × 0.15 = $138,000
$t-3$	23,000 × $30/hr = $690,000	$29,000	$1,160,000 × 0.15 = $174,000
$t-2$	25,000 × $30/hr = $750,000	$32,000	$1,280,000 × 0.15 = $192,000
$t-1$	26,000 × $30/hr = $780,000	$35,000	$1,400,000 × 0.15 = $210,000
t	25,000 × $30/hr = $750,000	$31,000	$1,240,000 × 0.15 = $186,000

Part II

Year	*Total Input Cost Z ($X_1+X_2+X_3$)*	*Productivity Ratio (O/Z_i)***	*Productivity Indices (PR_i/PR_0)*
$t-4$	$761,000	0.00302	1.0000
$t-3$	$893,000	0.00324	1.0728
$t-2$	$974,000	0.00329	1.0894
$t-1$	$1,025,000	0.00341	1.1291
t	$967,000	0.00321	1.0629

Notes
*First term represents the total cost of X-rays to the clinic; *i*th year ($i=t, t-1, t-2, t-3$); $t-4$ = base year.

corresponding indices for each of the five years using the same procedure we used earlier. Looking at the table, in particular the indices, it appears that the productivity for the imaging service has increased consistently, as before, except for the current year when it declined by 6.62 percent from the previous year.

In the example presented above, we used a single output (the imaging service), but in reality the clinic provides multiple different services. Thus, to get the total measure of productivity for the clinic as a whole we will need to include all the outputs and inputs. We can use the same procedure we used for multiple inputs, as shown below:

$$PI = \sum_{i=1}^{n} O_i / \sum_{j=1}^{m} I_j$$

$$= \frac{O_1 + O_2 + \ldots + O_n}{I_1 + I_2 + \ldots + I_m} \qquad [10.4]$$

where O_i is the *i*th output and I_j is the *j*th input (for $i = 1, 2, \ldots, n$ and $j = 1, 2, \ldots, m$). As before, to avoid the problem of adding apples to oranges where the outputs are different we will need to find a common denominator, such as the value of the outputs expressed in monetary terms.

Note that the term multi-factor productivity (MFP) is used interchangeably with total factor productivity (TFP). Although the two terms are used interchangeably, they are not essentially the same. TFP is the residue, similar to the error term in a regression, in that it reflects the portion of output not explained by the amount of input used in production. As such, it measures how efficiently the inputs have been used in the production of a good or service. Unfortunately, TFP cannot be measured directly and needs the use of a residual such as the Solow residual, which accounts for the effect on total output not explained by the inputs. Since the effect on output cannot be explained by the inputs, the growth in productivity is often attributed to growth in technology and other intangible factors that are difficult to measure. In general, the higher the unexplained portion of output growth, the greater the contribution of TFP. Interestingly, TFP is frequently considered as the real engine of growth in an economy, explaining in some cases as much as 60 percent of the overall growth (Easterly & Levine, 2001).

A Note on Effectiveness

A common concern in dealing with productivity indices is that they are narrowly defined in that they focus more on efficiency than on factors such as effectiveness or quality. Effectiveness simply means the extent to which an organization providing goods and services has achieved its goals and objective, or individuals consuming goods and services are satisfied with them. The definition applies to both public and private goods. However, measuring effectiveness is complicated, but it is much more so in the public sector because of the nature of the goods and services a government provides. Nevertheless, it is possible to measure effectiveness in government provided that one has sufficient data to work with. When data are not sufficient or readily available, one can use traditional measures such as questionnaire surveys to determine how satisfied the individuals are with the services they receive from their governments. Many state and local governments use citizen surveys to obtain citizen feedback on the goods and services they consume or to get their reaction on issues before making any decision that would affect their lives in the future (Hatry, 1981; Nayyar-Stone & Hatry, 2010). Although survey questionnaires do not necessarily produce an unbiased response, they can provide the government with some knowledge of effectiveness of the services it provides.

Although effectiveness is not always easy to measure, it is possible to incorporate measures that can show quality improvement in a program or service activity. A simple expression such as the one shown in Equation 10.5 is a good example:

$$O=(Q)(S)+[(Q)(F)](M) \quad [10.5]$$

where O is the output, Q is the total quantity of goods produced or service delivered, S is the percentage of success, F is the percentage of failure $(1-S)$, and M is the percentage of improvement.

To give an example, suppose that a state correctional facility releases on average 100 non-violent inmates per year under a reduced term of sentence plan.

It has a recidivism rate of 30 percent (F), and 70 percent rate of success (S) in that they do not return to the system. The personnel that run the facility observed that with a rehabilitation program the recidivism rate can be significantly reduced (M) by as much as 50 percent. Therefore, applying the expression in Equation 10.5, we can easily determine the number of individuals who could be successfully released with and without the program, as shown below:

$$
\begin{aligned}
O &= (Q)(S) + [(Q)(F)](M) \\
&= (100)(0.70) + [(100)(0.30)](0.5) \\
&= 70 + 15 \\
&= 85
\end{aligned}
$$

which indicates an improvement of 15 percent (i.e., 15 inmates who will not return to the system if they go through the rehabilitation program).

Suppose now that the department wants to increase the number of inmates to be successfully released by an additional 10 percent; from 70 to 80, which will increase the number by an additional five inmates to a total of 90. That is,

$$
\begin{aligned}
O &= (100)(0.80) + [(100)(0.20)](0.5) \\
&= 80 + 10 \\
&= 90
\end{aligned}
$$

Let us assume that we have some cost data on the facility. Let us say that it costs, on average, \$20,000 per year for an inmate with a total cost of \$2 million for 100 inmates for the year. The cost of the rehabilitation program is \$50,000 per year. Therefore, cost savings to the facility from the reduction in recidivism with the rehabilitation program will be \$250,000; that is, $TC_{W/O} - TC_W = (30 \times \$20{,}000) - [(15 \times \$20{,}000) + \$50{,}000] = \$600{,}000 - \$350{,}000 = \$250{,}000$, where $TC_{W/O}$ is the total cost of inmates who return to the system without the program, and TC_W that with the program.

Production Function

A formal and perhaps more effective way to measure productivity than using simple input–output ratio when more than one input factor is involved is to use a concept frequently used in microeconomic analysis, called production function. The term "production" means the transformation of inputs into outputs. The inputs could be anything (labor, capital, entrepreneurship, etc.) as long as they can be used in some meaningful combination to produce an output. The process that combines the input resources of an organization and transforms them into output is the technology. Technology is the knowledge of transforming input resources into outputs (goods and services). It is the state of the art, that is, the engine of progress that keeps a society moving forward. Thus, when one talks of technological change, it usually means an increase in knowledge based not necessarily on new ideas but on the improvement in the existing techniques of production.

A production function combines the notions of both technology and production into a precise mathematical relationship (Rasmussen, 2013). It specifies the maximum amount of output that can be produced with a given combination of inputs. Defined as such, a production function becomes synonymous with efficiency. It simply means that an output is efficient if it can be produced with no more resources than necessary. Alternatively, it must not be possible to produce the same quantity of output with less of any resources, other things being equal.

Linear Production Function

The simplest way to construct a production function is to assume a linear relationship between inputs and output. Although mathematically simple, a linear production function is not particularly suitable for measuring productivity. To illustrate the point, consider an output function that has a general expression of the form

$$Q=f(X_1,X_2,X_3,\ldots,X_n) \quad [10.6]$$

where Q is the quantity of output produced and the Xs represent the resources available as part of the technology used.

If, for instance, one could use K and L, instead of the Xs, as symbols representing capital and labor, respectively, the output function would become:

$$Q=f(K,L) \quad [10.7]$$

Assume now that there is a linear relationship between Q and the input variables K and L such that:

$$Q=\alpha K+\beta L \quad [10.8]$$

where α and β are the slope coefficients associated with K and L.

The constant terms indicate that for any change in K and L, output will change by a constant amount equal to the value of α and β. There is a problem with this expression in that it violates a fundamental condition of production relation between input and output, called the law of diminishing marginal returns. The law states that if technology and all inputs except one are held constant, then an increase in the variable input in equal increments will increase output up to a point, after which it will diminish. In conventional economic terms, it means that for the law to hold the marginal product of any input must be positive. In other words, α and β must be positive and they must be less than 1.

There is a mathematical corollary to this law which says that the first derivative of the function must be positive for α and β to be positive and that the second derivative must be negative to yield a diminishing marginal return. To show whether or not our linear production function satisfies these conditions, recall the equation:

$$Q=\alpha K+\beta L$$

Now, take the partial derivatives of Q with respect to K and L, and set them equal to 0:

$$\frac{\partial Q}{\partial K}=\alpha=0;\frac{\partial Q}{\partial L}=\beta=0$$

which produce two constants, α and β.

The constant terms indicate that the marginal products of K and L are fixed at all levels of production. What this means is that each additional unit of capital and labor would produce α and β units of output indicating a constant productivity, which is inconsistent with the law of diminishing marginal returns. To see if the function satisfies the second-order condition, where both a and β must be negative, we take the second derivatives of Q with respect to K and L:

$$\frac{\partial^2 Q}{\partial K^2}=0;\frac{\partial^2 Q}{\partial L^2}=0$$

which result in 0 in both instances, meaning that it is inconsistent with the law of diminishing marginal returns.

In addition to the mathematical inconsistencies, there is a further explanation of why a linear production function is not always useful. A linear production function implies that an output, Q, can be produced by a single input, K or L, and not both. For instance, if we assume $L=0$, then the output will be equal to αK; that is, $Q=\alpha K+\beta(0)=\alpha K$, which is positive only if K is positive. This means that an output can be produced without any use of labor. By the same token, if we assume $K=0$, then the output will be equal to βL; that is, $\alpha(O)+\beta L=\beta L$, which does not make much sense since output cannot be produced without the use of any capital. In other words, to produce an output one must have both labor and capital, not just one.

Nonlinear Production Function

The fundamental inconsistency with a linear production function makes it necessary to consider other, somewhat more complex, functional forms such as a nonlinear production function. One of the most widely used production functions is the Cobb–Douglas production function. Developed originally in the 1920s to measure productivity of firms (Cobb & Douglas, 1928), it is still considered the most popular among all production functions, and is given by the expression:

$$Q = AK^{\alpha}L^{\beta} \quad [10.9]$$

where A is a positive constant (intercept) and the rest of the terms are the same as before.

Like most mathematical relationships, the Cobb–Douglas production function is based on a number of assumptions: (1) it requires two inputs, capital and labor, expressed in terms of a multiplicative relationship; (2) both inputs must be positive for output to be positive (unlike the linear production function, where only one of the input variables needs to be positive); (3) its isoquants (i.e., locus of input combinations that produce the same level of output) are negatively sloped for positive values of K and L, as shown in Figure 10.1; (4) the marginal products are not constants, but vary with the level of input;[1] and (5) it is homogeneous of degree $\alpha+\beta$.[2]

Returns to Scale

All production functions, including the Cobb–Douglas production function, use a specific type of measure to determine the efficiency with which an input combination can be used to produce a given level of output, called returns to scale. Returns to scale express a relationship between input and output, assuming all

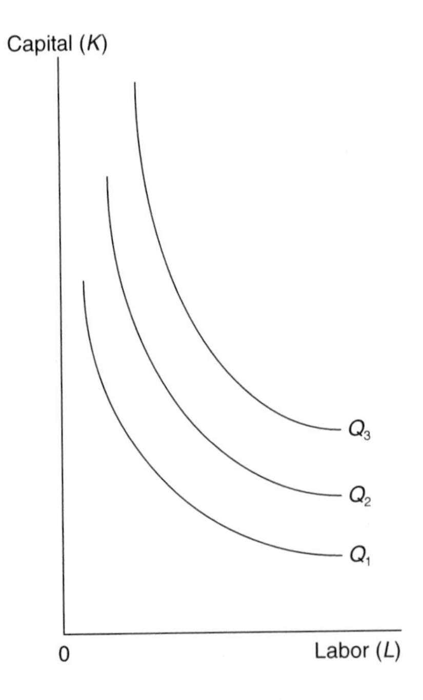

Figure 10.1 Returns to Scale

factors of production vary simultaneously in the same proportion. When the factors of production vary in the same proportion, one of three things can occur: (1) output can increase by a greater proportion than the increase in inputs, called increasing returns to scale; (2) output can decrease by a greater proportion than the increase in inputs, called the decreasing returns to scale; or (3) output can increase in the same proportion as the increase in inputs, called constant returns to scale.

To illustrate the point above, consider the production function with two inputs, capital and labor. Suppose that we increase the inputs by k percent, then output will increase by some proportion, q, yielding:

$$qQ=f(kK,kL) \qquad [10.10]$$

If $q=k$, the function has a constant return to scale; if $q>k$, it has an increasing return to scale; if $q<k$, it has a decreasing return to scale.

To determine the degree of homogeneity, we can write Equation 10.10 in terms of Equation 10.9:

$$qQ=A(kK)^{\alpha}(kL)^{\beta} \qquad [10.11]$$

Next, if we factor out the constant term, k, from Equation 10.11, it will give us:

$$qQ=Ak^{\alpha+\beta}K^{\alpha}L^{\beta} \qquad [10.12]$$

Note that the exponent of k, $(\alpha+\beta)$ in Equation 10.12, is the degree of homogeneity, which determines whether the function will have an increasing, decreasing, or constant return to scale. In general, if $(\alpha+\beta)>1$, it means an increasing return to scale; if $(\alpha+\beta)<1$, it means a decreasing return to scale; if $(\alpha+\beta)=1$, it means a constant return to scale. Figure 10.2 shows these returns to scale.

Let us look at a simple example to illustrate the returns to scale. Suppose that a local government is trying measure the efficiency of one of its services, say, solid waste collection, where efficiency is measured by the amount of solid waste collected by the department for a given amount of input. To keep the problem simple, assume that the government uses two inputs: labor (man hours) and capital (dollar worth of machines, tools, and equipment). Let us further assume that the government has experienced an increase in both labor and capital over time. To ensure that there are no significant qualitative differences in output over time, we assume the service to be a routine operation. This will allow us to avoid any potential measurement problems that are typical of data series with qualitative differences. Table 10.5 presents the data on input and output for the service.

We can now apply the production function ($Q=AK^{\alpha}L^{\beta}$) in Equation 10.9 to the data in Table 10.5. Since it is a nonlinear equation, we need to linearize it by taking logarithms on both sides so that the new equation becomes:

$$\ln Q=\ln A+\alpha\ln K+\beta\ln L \qquad [10.13]$$

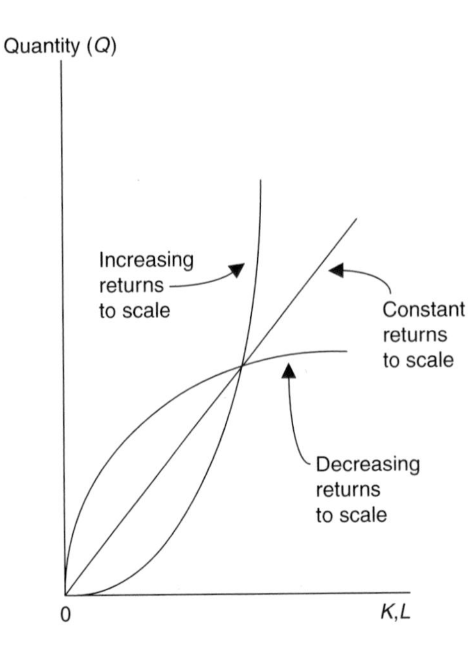

Figure 10.2 Typical Isoquants with Different Combinations of Capital and Labor

Table 10.5 Input–Output Data for Solid Waste Collection

Year	*Output (1,000 Tons)*	*Labor (Million Man-Hours)*	*Capital ($ Million)*
1 ($t-16$)	10.075	1.050	10.000
2 ($t-15$)	10.120	1.150	10.100
3 ($t-14$)	10.145	1.165	10.450
4 ($t-13$)	10.160	1.175	10.650
5 ($t-12$)	10.175	1.175	10.800
6 ($t-11$)	10.180	1.200	11.000
7 ($t-10$)	10.200	1.215	11.210
8 ($t-9$)	10.215	1.225	11.350
9 ($t-8$)	10.230	1.230	11.500
10 ($t-7$)	10.245	1.245	11.750
11 ($t-6$)	10.260	1.260	12.000
12 ($t-5$)	10.275	1.285	12.150
13 ($t-4$)	10.280	1.300	12.350
14 ($t-3$)	10.285	1.310	12.500
15 ($t-2$)	10.300	1.315	12.675
16 ($t-1$)	10.315	1.320	13.000
17 (t)	10.340	1.325	13.115

where ln represents the natural logarithm to the base 2.71828.

We apply the standard OLS method (discussed in an earlier chapter) to the expression in Equation 10.13. The results of the estimated equation, along with all the conventional statistics such as t values (in parentheses), R^2, and F ratio are presented below:

$$\ln Q = \ln A + \alpha \ln K + \beta \ln L = 2.1779 + 0.0568\ lnK + 0.0387\ lnL$$
$$(154.4610)\ (8.7385)\ (4.3483)$$

$$R^2 = 0.993\ F = 7{,}175.588\ D-W = 1.354$$

Looking at the results above, it appears that all three coefficients are significant at $p < 0.05$, given the observed t values (found in any standard statistics text). The high R^2 indicates a strong relationship between the output and input variables – explaining over 99 percent of the variation in solid waste collection. Overall, the model appears to be a good fit, as given by the high F ratio, although there appears to be a minor autocorrelation problem, which could be corrected with measures such as taking first differencing of the data, adding additional variables, expanding the number of data points, and so forth.[3]

Since the original equation was transformed using logarithms, we need to take the antilog of the transformed equation to express the estimated coefficients in terms of the original units. The estimated production function, with the constant term expressed in original units, can be written as:

$$Q = (8.8277) K^{(0.0568)} L^{(0.0387)}$$

To ensure that we can correctly interpret the results, we need to make certain that they conform to the properties of the production function. As noted earlier, the properties are ensured if $0 < (\alpha, \beta) < 1$, and by looking at the estimated coefficients it appears that they have been satisfied since both of them are less than 1 but greater than 0. That is:

$$0 < \alpha = 0.0568 < 1; \text{ and } 0 < \beta = 0.0387 < 1$$

Before continuing further with the results, let us digress a little on the elasticity concept we introduced earlier. Statistically, the estimated coefficients of a regression model can be treated as elasticities. Since the coefficients measure the change in the dependent variable (output) due to a unit change in the independent variables (inputs), they are essentially the same as elasticity. In other words, the estimates of α and β are the output elasticities of capital and labor. Therefore, according to our estimates, a 1 percent increase in physical capital will lead to a 5.7 percent increase in output. Similarly, for a 1 percent increase in labor (man hours) per week, output will increase by 3.9 percent, indicating an inelastic response (performance) in both cases. The inelasticity of the coefficients does not mean that the variables included in the model are irrelevant, especially if they are based on sound theoretical grounds; they can be the results of poor sampling, errors in data collection, and so forth.

Additional Tests of Significance

Once the model has been estimated and tested for significance, we need to determine the extent to which there has been a real change in productivity. To do so, we need to calculate the degree of homogeneity for the estimated production function. We do this by simply adding the estimated values of α and β, which comes out to be 0.0955 (0.0568+0.0387). By definition, then, it is homogeneous of degree 0.0955. Since it is less than 1 – that is, $(\hat{\alpha}+\hat{\beta})<1$ – we can safely suggest that there is a decreasing return to scale, meaning that the rate of increase in output has been less than the rate of increase in physical capital and labor input. Put simply, the solid waste collection has not been efficient in utilizing the input resources.

Before one is able to accept the above conclusion as reliable, it is necessary to further establish that the joint estimates of α and β are statistically significant (since we are looking at them together). To achieve this, we use an additional t test to determine the significance of the combined estimate of the parameters, as shown below:

$$
\begin{aligned}
t_{(\hat{\alpha}+\hat{\beta})} &= \frac{(\hat{\alpha}+\hat{\beta})-1}{SE_{(\hat{\alpha}+\hat{\beta})}} \\
&= \frac{(0.0568+0.0387)-1}{0.0065+0.0089} \\
&= \frac{-0.9045}{0.0154} \\
&= -58.7338 \qquad [10.14]
\end{aligned}
$$

where the denominator SE represents the combined standard errors of $\hat{\alpha}$ and $\hat{\beta}$ (errors in estimating the values of the two parameters).

The result of the additional test produces a t value of –58.7338. Since this value is much lower than the critical value of t at $p<0.01$ (found in any standard statistics text), we can say that $\alpha+\beta$ is significantly lower than 1. In other words, the decreasing returns to scale we observed in the data for the service is statistically significant. It is worth noting that although the example we used here looks at the efficiency of solid waste collection, the method can be applied to other public services such as public safety, public works, healthcare, recreation, education, and so forth. Another point that should be highlighted here is that even though we have used time-series data, the method can be equally applied to cross-sectional data, measuring the efficiency of multiple governments or agencies providing similar services.

Efficiency of Input Factors

Once a production function has been estimated and tested for the significance of the estimated parameters, it is possible to determine whether or not the input resources have been used efficiently. One way to achieve this is to set the ratio of marginal products of labor (i.e., output produced by the last unit of labor) and capital (i.e., output produced by the last unit of capital) equal to the ratio of price of labor and capital; that is:

$$\frac{MP_L}{MP_K} = \frac{w}{c} \qquad [10.15]$$

or

$$\frac{MP_L}{MP_K} - \frac{w}{c} = 0 \qquad [10.16]$$

where MP_L is the marginal product of labor and MP_K is the marginal product of capital, w is the wage or price of labor, and c is the price of capital.

In general, if $MP_L/MP_K < w/c$ (i.e., $MP_L/MP_K - w/c < 0$), there is an overutilization of labor relative to capital. By the same token, if $MP_L/MP_K > w/c$, (i.e., $MP_L/MP - w/c > 0$), there is an underutilization of capital relative to labor. In reality, it boils down to estimating a set of relationships between the rate at which one input is substituted for another along an isoquant called the marginal rate of technical substitution (MRTS) and the ratio of input prices for labor and capital. What this means is that for an input combination to be optimal for a given output, the MRTS must be equal to the ratio of input prices.

For the Cobb–Douglas production function, we can write Equation 10.16 as:[4]

$$\beta\frac{K}{L} - \alpha\frac{w}{c} = 0 \qquad [10.17]$$

which is the optimality condition for the Cobb–Douglas production function. From this, we can easily write:

$$\beta\frac{K}{L} - \alpha\frac{w}{c} < 0 \qquad [10.18]$$

for overutilization of labor. Similarly, we can write

$$\beta\frac{K}{L} - \alpha\frac{w}{c} > 0 \qquad [10.19]$$

for overutilization of capital.

What this means in common sense terms is that if the result produced by the efficiency criterion in Equation 10.17 is equal to 0, the department is optimally utilizing its labor and capital. If it is less than 0, it is overutilizing labor and if it is greater than 0 it is overutilizing capital. To see if there has been an overutilization of the input resources by the department, we need to find the value of each of the terms in Equation 11.17. Let us assume that we have information on the price per unit of labor and capita: $15 for labor and $0.10 for capital; that is, it costs the department $15 worth of labor and $0.10 worth of capital on average to produce (collect) one unit of output (i.e., one ton of solid waste). We already have the information on labor and capital (Table 10.5); we can take their averages from the table and use them as proxies for K and L. We thus have all the information we need for our purpose: $\hat{\alpha}=0.0568$, $\hat{\beta}=0.0387$, $\overline{K}=11.56$, $\overline{L}=1.16$, $\overline{w}=15$, and $\overline{c}=0.10$.

We can now apply the above information to the respective terms in Equation 10.17 to obtain the efficiency parameter by which to determine if there has been an overutilization of resources, as shown below:

$$\begin{aligned} \varphi' &= \beta\frac{K}{L} - \alpha\frac{w}{c} \\ &= 0.0387\left(\frac{11.56}{1.16}\right) - 0.0568\left(\frac{15}{0.10}\right) \\ &= 0.3857 - 8.52 = -8.1343 \end{aligned} \quad [10.20]$$

where φ' is the efficiency parameter.

Since φ' is considerably less than 0, we can say that there has been an overutilization of labor by the department. In fact, we can use this information to determine the optimum ratio of input resources, as well as optimal quantity of output that can be produced for the department.[5]

A word of caution may be in order here for how the average prices of labor (w) and capital (c) are calculated. Although for convenience we assumed the prices are given, in reality the process is a little more complicated, especially for the price of capital. In general, the average price of labor is calculated by dividing the total wages by the number of employees in an organization. However, the calculation of the average price of capital is not so straightforward. It requires that factors such as the cost and age of capital stock, the real rate of interest, and the rate of depreciation are taken into consideration in calculating this average. The convention is to use an approach that combines all three factors in a manner that best approximates the price of capital.

Other Forms of Production Function

The Cobb–Douglas production function is one of the many different types of productions functions that are used for measuring productivity. There is a range of such functions that can be used for this purpose. Two good examples would be fixed-proportion production function and constant elasticity of substitution (CES). The fixed-proportion production function is used when both capital and labor are used in fixed, constant proportions. As such, its isoquants are L-shaped instead of being negatively sloped and convex to the origin. What this means is that by increasing labor alone, one would not be able to increase output. Likewise, by increasing capital alone, one would not be able to increase output.

The CES, on the other hand, is among the most complex of production functions (Arrow, Chenery, Minhas, & Solow, 1961). As the name implies, it exhibits a CES between labor and capital, meaning that the production technology has a constant percentage change in factor proportions (the amount of labor and capital used) due to a percentage change in MRTS (the rate at which one factor is substituted for another along an isoquant to produce a given quantity of output). Operationally, CES incorporates several different production functions, including Cobb–Douglas, which explains part of its complexity. While most production functions deal with single output, there are variations of production functions that can deal with multiple outputs (Kumbhakar, 2010).

Limitations of Production Functions

Production functions are a powerful tool for measuring productivity, especially in predicting output that can be obtained from a mix of inputs. Besides productivity, they can also be used in long-term planning, but all production functions suffer from a common problem that can be directly attributed to data measures. For instance, the data frequently used in production functions are highly aggregated, which can overly simplify the indices used in measuring productivity. Also, when aggregate data are used, they may pose additional problems by restricting them to a narrow range of values. Therefore, data must be disaggregated whenever possible to have a more clear understanding of the differences in the characteristics of the variables being measured. This will not only improve the quality of indices, but will also provide a more effective measure of productivity change.

In addition to those mentioned above, there are other problems with the estimation of production functions, especially where the classical regression model using the OLS method is concerned. For instance, questions as to what causes an organization to produce less than the maximum output or how to account for variations in input combinations that produce the same level of output cannot be readily measured by OLS. An alternative is to use methods such as data envelopment analysis (DEA) that are much better suited to deal with this type of problem. Developed by Charnes, Cooper, and Rhodes (1978), DEA, also known as frontier analysis, is particularly suitable for measuring efficiency of organizations providing similar services. It can also be used to measure efficiency that is explicitly sensitive to both input and output mix and, as such, is more reliable than a production function. More importantly, DEA can be used to compare the units of an organization that are relatively inefficient (lie above or below the efficient frontier on a two-dimensional plane), as well as measure the magnitude of their inefficiency and suggest alternative courses of action that can minimize that inefficiency (Khan & Murova, 2015).

Besides DEA, there are other methods that can also be used for measuring efficiency, including some that are statistical in nature such as Tobin's two-step regression (especially in situations where outcomes of an interest are not fully observed in a sample), bootstraps (methods for assigning accuracy measures to sample estimates), as well as variations of frontier analysis that include both deterministic and stochastic models (approaches for measuring fixed and random shocks affecting the production process that is not directly attributable to the producer or the technology). Most of these methods, with the exception of deterministic and stochastic frontier analysis, are non-parametric methods, which have an advantage over conventional production functions in that they do not make any a-priori assumption about the distribution of the functions. As such, they are less restrictive than the conventional parametric methods. There is a wealth of literature that discusses these and other similar techniques for measuring efficiency and productivity.

The Learning Curve

It is basic human nature that a person engaged in a task will improve his or her performance by repeating it many times. This unique characteristic to search endlessly for improvement is the basis of progress in human societies. In most repetitive tasks improvement takes place smoothly and continuously, although there are exceptions to this rule. Progress resulting from a learning process is real and thus can be measured with a certain degree of precision. One particular approach that observes, measures, and analyzes this improvement in human performance over time is the well-known learning curve (Dejong, 1964). A learning curve shows the relationship between time and output from a repeated task; it is a measure of efficiency one gains in time and cost when a task is repeated n number of times.

The Learning Curve Function

The human learning process depicts certain observable patterns that can be expressed in terms of a mathematical relationship called the learning curve function. The learning curve function is an exponential function with negative slope and has the general form

$$Y = KX^{-\beta} \qquad [10.21]$$

where Y is the time expended to produce a unit of output, K is a constant required to produce the first unit, X is the amount of output, and β is the slope parameter that determines the learning rate. It is negative because the time per unit of output decreases with increasing effort.

Equation 10.21 has an important characteristic that constitutes the essence of the learning curve function. It states that each time a task X is undertaken (doubled, tripled, etc.), the time it takes to complete the task decreases by a fixed percentage, $\hat{a}$. If $\hat{a}$ is 10 percent, we call it a 90 percent learning curve; if it is 20 percent, we call it an 80 percent learning curve; and so on. It is quite common to express the learning curve in terms of a doubled production or doubled effort, which means that if it takes 10 minutes to produce the first unit, then for an 80 percent learning curve it will take 8 minutes [80 percent of 10, i.e., $10 \times 0.8 = 8$] to produce the second unit, and 6.4 minutes [80 percent of 8, i.e., $8 \times 0.8 = 6.4$] to produce the fourth unit, and so on. Figure 10.3 presents a simple illustration of an 80 percent learning curve. It shows a progressive improvement in productivity, but at a decreasing rate.

An interesting characteristic of the curve is that when plotted on double log paper, it becomes a straight line. In mathematical terms, it means that if one would take logarithms on both sides of Equation 10.21, it will produce a linear curve depicting a straight line. That is,

$$\log_{10} Y = \log_{10} K - \beta \log_{10} X$$

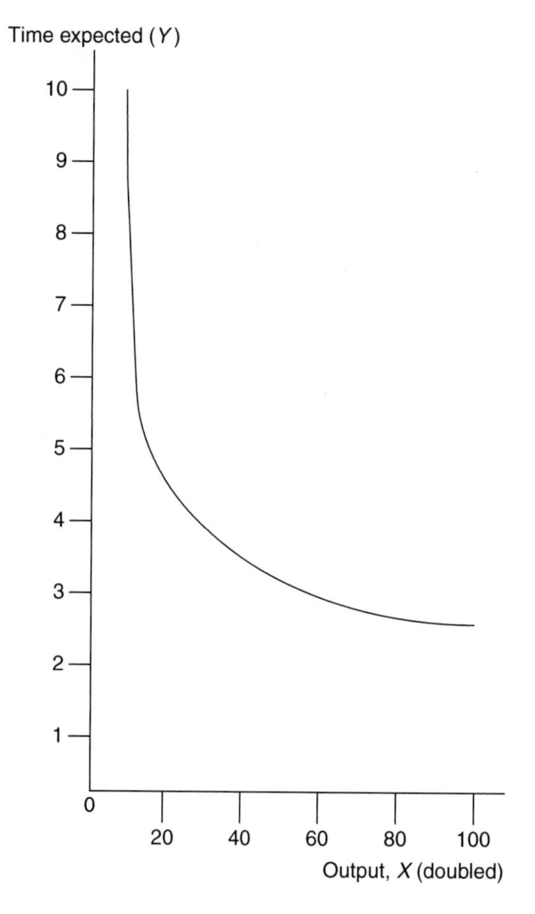

Figure 10.3 Eighty Percent Learning Curve (Approximate)

To give an example, let us go back to the application processing example we introduced earlier for a state welfare agency. Suppose that it takes an agent 25 minutes to process the first application and there are 250 applications the agent needs to process each month. How much time will it take, on average, to process all 250 applications and how much time will it take to process the last one, i.e., the 250th application, assuming a 90 percent learning curve? To ensure that the tasks are comparable, assume that the applications are from families with similar size, background, and income structure. The problem can now be solved in two simple steps. First, apply the equation to the problem to determine the learning rate, β, and then use this information to find the average of the learning curve function. Equation 10.22 presents a standard expression for finding the average of the learning curve function:

$$\text{Average} = \frac{K}{(1-\beta)}[(X+0.5)^{1-\beta} - (0.5)^{1-\beta}]/(X) \quad [10.22]$$

where all the terms of the equation have been defined earlier.[6]

To determine the learning rate, β, we take the ratio of time per unit of output (effort) to the rate of effort, both expressed in logarithms, without the constant term K. That is:

$$Y = X^{-\beta} \quad [10.23]$$

$$\log Y = -\beta \log X$$

$$(\log Y/\log X) = -\beta$$

Thus, for a 90 percent learning curve and assuming that the output doubles each time an effort is made, the rate will be:

$$-\beta = (\log 0.90)/(\log 2)$$

$$-\beta = -0.1520$$

Since it is negative, we multiply, for convenience, both sides of the expression by -1 to produce a positive value of β. That is:

$$\beta = 0.1520$$

Finally, to obtain the average, we substitute this value into Equation 10.22, as shown below:

$$\text{Average} = \frac{K}{1-\beta}[(X+0.5)^{1-\beta} - (0.5)^{1-\beta}]/(X)$$

$$= \frac{25}{(1-0.1520)}\left[(250+0.5)^{1-0.1520} - (0.5)^{1-0.1520}\right]/(X)$$

$$= (29.4811)[(108.31) - (0.5555)]/250$$

$$= 12.69 \text{ minutes}$$

which is what it will take on average to process 250 applications, which will cut the time by almost half, from 25 minutes to 12.69 minutes. From this, we can calculate the time it will take to process the last application. That is:

$$Y = KX^{-\beta}$$
$$= (25)(250)^{-0.1520}$$
$$= \frac{25}{(250)^{0.1520}}$$
$$= 10.801 \text{ minutes}$$

As expected, it will take the least amount of time to process the 250th application.

Production Progress Function

Our analysis of the learning curve function focused primarily on the learning effort of a single individual and the savings it would produce with an increasing number of efforts. How would this result differ from those produced for a group, or an organization as a whole? Organizational or group learning effort can be measured by a process called production progress function. This function, which measures and estimates the rate at which an organization learns to provide a good or service, follows the same negative power function that is used for individual learning. In general, the learning rate is much faster for organizations than it is for individuals. This is due in part to the fact that organizations make constant changes in the means of production either by adding capital to the production process or by making improvements in human learning skills, or both.

When organizational learning is considered, the emphasis is generally placed on reducing costs rather than on saving time, although the latter can be easily converted to determine the actual savings in costs. To illustrate the point, let us look at the auditing example again. Suppose it costs the agency \$35 to process the first application, how much will it cost to process, say, the 100th, 200th, and the 250th application? Assume a 90 percent learning curve, as before. The answer can be obtained by applying the learning rate to Equation 10.21 and then solving it for the respective units. That is,

$$Y(\text{cost of the 100th unit}) = KX^{-\beta} = (\$35)(100)^{-0.1520} = \$17.38$$
$$Y(\text{cost of the 200th unit}) = KX^{-\beta} = (\$35)(200)^{-0.1520} = \$15.64$$
$$Y(\text{cost of the 250th unit}) = KX^{-\beta} = (\$35)(250)^{-0.1520} = \$14.71$$

The results show a consistent decline in cost as the number of returns processed increases. The analysis can be expanded even further by asking what will be the average as well as the total cost of processing all 250 returns. To determine the average cost, let us look at Equation 10.22 again and substitute the appropriate values into the equation to obtain the average cost, as shown below:

$$\begin{aligned}\text{Average} &= \frac{K}{1-\beta}[(X+0.5)^{1-\beta} - (0.5)^{1-\beta}]/(X)\\ &= \frac{35}{(1-0.1520)}[(250+0.5)^{1-0.1520} - (0.5)^{1-0.1520}]/(250)\\ &= (41.2736)[(108.19)-(0.5556)]/250\\ &= \$17.77\end{aligned}$$

which produces a dollar value of \$17.77, meaning that it would cost the agency on average \$17.77 to process an application instead of \$35, when processed individually.

Finally, with information on average cost, we can easily find the total cost of this function by multiplying the average cost, *AC*, by the quantity of output, *Q*:

$$\begin{aligned}TC &= (AC)(Q)\\ &= (\$17.77)(250)\\ &= \$4{,}442.50\end{aligned}$$

The result produces a total of \$4,442.50 for the operation; this is how much it will cost to process all 250 applications.

From the above, we can also calculate the net savings the learning experience will generate for the agency by taking the difference between the two total costs: one, when a return is paid at the rate of \$35 per application, and the other when it is paid at the rate of \$17.77; that is, (250)(\$35)−(250)(\$17.77)=\$8,750.00−\$4,442.50=\$4,307.50, which is the net savings to the agency from the learning curve, based on a total of 250 applications processed. The savings would be even greater if the learning rate could be assumed to be more than 90 percent. In other words, with a higher learning rate, the average cost of production will go down, which, in turn, will result in greater savings.

Time Study

Time is a valuable measure of efficiency and performance. The task an individual performs or an organization undertakes is frequently measured in terms of the time needed to produce it. In a real sense, it is the content of the task and the time spent to accomplish it that determines the true worth of a task. Developed originally by Frederick Taylor in 1912, time study establishes a method of determining the time standard for a given task based on its work content. Its objective is to determine a set of reliable and consistent standards for tasks performed by an organization. With consistent measures, an organization can determine its output capacity, develop a work schedule to maximize resource use, and balance the workforce with available work in advance of production or service delivery.

This section presents four of the most frequently discussed elements of time study: time element, work sampling, standard time, and performance index.

Time Element

Time study begins by dividing an activity into groups, called elements. Elements are segments of an activity with readily measurable cutoff points. They are mutually exclusive, meaning that no two elements contain the same tasks. Theoretically an activity could include any number of elements but, in practice, only those elements that are of sufficient duration are included in a time study. With longer duration, more information is available on a given element, thus making it possible to measure the elements objectively. Measurement is generally performed with a stopwatch, a well-designed study form containing all the relevant information on each and every single element, and occasionally with the help of an electronic calculator to work-up the study. It can also be accomplished by trained observers walking around or riding through a work area. In special cases, time-lapse photography or security cameras can be used.

To get an accurate measure of performance, one must be able to evaluate the time spent on the elements of an activity against a time standard. Time standards determine the minimum number of time cycles needed to complete an activity, while allowing for possible delay due to fatigue, unavoidable circumstances, machine interference, and so on. Cycles can vary in length from less than a minute to over an hour. When longer cycles are involved, it may be necessary to break down the activities further into their component elements to facilitate data collection, analysis, and measurement.

Work Sampling

Time study is based on a sampling procedure known as work sampling. Work sampling requires that adequate data are collected from (statistical) samples to have a reasonably accurate standard. The term *sample* here means the number of times an activity or its elements have to be observed in order to determine how efficiently the activity is being carried out. The number of times the elements are to be observed is called the sample size.

Three things are necessary to determine a sample size: the cost one incurs in collecting data, the margin of error one is willing to tolerate for a given sample, and the amount of confidence one wants to establish on the estimates of the population characteristics, such as mean, variance, standard deviation, etc. In general, there is an inverse relationship between sample size and the margin of error one is willing to tolerate; that is, the larger the sample the smaller the margin of error, and vice versa. However, with larger samples the cost of data collection will also be higher.

Work sampling practitioners use markedly different approaches to determine a sample size. Some specify an error margin beforehand to determine the sample size, while others determine the sample size that is feasible based on cost constraints and then check to see if the resulting error will justify the study. We use the first option (which is more widely used) to determine the sample size by specifying a margin of error in the following manner:

$$e = \frac{(z_{\alpha/2})(\sigma)}{\mu\sqrt{n}} \qquad [10.24]$$

where e *is* the error margin, μ is the mean (cycle) time for the population from which data are collected, σ is the population standard deviation, α is the probability of error (probability of obtaining an observed estimate by chance), and z is the standard normal value corresponding to $1-\alpha$, called the confidence level (for a two-tailed distribution).

Since the upper limit of probability is always equal to 1, $1-a$ represents the confidence level (i.e., the probability of not obtaining the estimate by chance). If, for instance, a is 2 percent, then $1-\alpha$ will be 98 percent, which is the amount of confidence one should have in an observed estimate. In general, an a value higher than 10 percent is not considered acceptable or statistically significant.

From Equation 10.24, we can obtain a new equation for determining the sample size by rewriting it as

$$e = \frac{(z_{\alpha/2})(\sigma)}{\mu\sqrt{n}}$$

$$e\mu\sqrt{n} = (z_{\alpha/2})(\alpha_{n})$$

$$\sqrt{n} = \frac{(z_{\alpha/2})(\sigma_n)}{e\mu}$$

$$n = \left[\frac{(z_{\alpha/2})(\sigma_n)}{e\mu}\right]^2 \qquad [10.25]$$

where n is the sample size and the rest of the terms are the same as before.

Let us look at a simple example to illustrate this. Consider a case in which a county hospital wants to evaluate the performance of one of its departments in terms of the time it takes to complete a medical procedure. Let us say that the mean time for this procedure, nationally, is 25 minutes, with a standard deviation of five minutes. The hospital administration wants to know the sample size, i.e., the number of procedures it must observe to determine the efficiency of the department, in particular the individuals doing the procedure. Assuming a confidence level of 95 percent with an error margin of 2.5 percent (a reasonable statistical assumption), the desired sample size for the problem would be

$$n = \left[\frac{(z_{\alpha/2})(\sigma_n)}{e\mu}\right]^2$$

$$= \left[\frac{(1.96)(5)}{(0.025)(25)}\right]^2$$

$$= 245.86 \approx 246$$

This means that a sample of 246 observations would be needed to determine the standard time for the department against which individual performance can be measured and evaluated. Note that the z value of 1.96, as shown here, can be directly obtained from a standard normal curve, or from any standard normal (z) table for a two-tailed distribution (assuming a confidence level of 95 percent).

Equation 10.25 assumes that we already know the population mean and standard deviation, but what if we do not know them? In that case, the rule of thumb is to have some ballpark estimates of these parameters either by analogy from a similar study completed elsewhere, or based on the findings from a previous study of the same problem. A better alternative yet would be to take a small sample (about 10 percent or so of the actual sample), calculate the sample statistics (mean and standard deviation), and use them as proxies for the population parameters, μ and σ. This is known as pre-testing.

Standard Time

Once the sample size has been determined, it becomes relatively easy to calculate the standard time for an activity. In general, the standard time is calculated as the product of three factors: (1) an average time factor (*ATF*), based on the time spent by individuals performing the same task; (2) a performance factor (*PF*) that depends on whether or not an organization is working at 100 percent level of its capacity; and (3) a time allowance factor (*TAF*) to account for potential loss of time due to causes over which one may not have any control. It can be formally expressed as

$$ST_i = (ATF)(PF)(TAF) \qquad [10.26]$$

where ST_i is the standard time for the ith activity (for $i = 1, 2, 3 \ldots, n$) and the rest of the terms are the same as before.

To determine the standard time for the problem in our example, assume that the mean time for the procedure, based on a sample of 246 cases (i.e., observation of procedures), is 23 minutes with a performance factor of 100 percent and a time allowance of 15 percent above the normal time. The standard time for the procedure for the department would then be

$$\begin{aligned} ST_i &= (ATF)(PF)(TAF) \\ &= (23)(1.00)(1.15) \\ &= 26.45 \end{aligned}$$

or approximately 26 minutes. In other words, it is the standard we can use to measure how well the department is doing against the national standard, as well as to evaluate the performance of the individuals doing the procedure. It is worth noting that although we did not quite calculate the actual time for the procedures, we can say hypothetically that the result appears to be close to the national average (the value we used to determine the sample size), meaning that on average the department is not doing a bad job. However, the result may be different for individual cases.

Performance Index

An important characteristic of all time studies is the rating an individual receives for job performance in an organization. The reliability of a time standard cannot be ascertained unless a proper performance rating has been completed during the course of a study. Generally speaking, there are no hard-and-fast rules for rating the performance of an individual. However, the information one obtains for an organization can serve as a basis for measuring individual performance.

Let us go back to the problem again and see how we can measure individual performance based on the information we have for the department. Our objective, vis-à-vis that of the hospital, is to measure the performance of the individuals carrying out the procedure. Accordingly, we must compare the individual time with all the allowances against the standard time for the department. Equation 10.22 presents a simple expression that can be used to compare the individual time with the standard time, as shown below:

$$IP_{ji} = \left(1 - \frac{T_{ji}}{ST_i}\right) \times 100 \qquad [10.27]$$

where IP_{ji} is an index of performance that measures individual performance (i.e., the efficiency level for the *j*th individual *j* for *i*th activity), ST_i is the standard time for the *i*th activity (obtained from Equation 10.26), and T_{ji} is the time expended by the *j*th individual on activity *i* (for $i = 1, 2, 3 \ldots, n$, and $j = 1, 2, 3, \ldots, m$).

Assume now that it takes 32.25 minutes for the *j*th individual to complete the procedure, as opposed to the standard 26.45 minutes obtained for the department from our observation of 246 procedures. Therefore, by substituting the above information into Equation 10.27, we can obtain the individual's efficiency level. The result will be 78.07 percent.

$$\begin{aligned} IP_{ji} &= \left(1 - \frac{T_{ji}}{ST_i}\right) \times 100 \\ &= \left(1 - \frac{32.25}{26.45}\right) \times 100 \\ &= (1 - 1.2193) \times 100 \\ &= -21.93 \approx -22 \end{aligned}$$

which is about 22 percent below the standard [$100 - 122 = -22$]. The rule of thumb is that as long as the index does not fall below 10 percent, it is generally considered acceptable, but in this case it clearly reflects a gross inefficiency on the part of the individual.

We can extend the example further to determine the efficiency learning for the *j*th individual, which may not be the same as what the individual actually earns. Suppose now that the individual in question works full-time (eight hours per day, 40 hours per week) doing nothing but the procedure. Let us say that the going rate for the procedure is $125 per hour. Also, assume that the individual is paid according to the time he or she spends on the procedure. Therefore, the earning for the individual based on this time will be $1,000 per day [$\$125 \times 8 \times 1.00 = \$1,000$], or $5,000 per week [$\$1,000 \times 5 = \$5,000$]. In theory, however, the individual should receive $780 per day, or $3,900 per week [$(\$1,000 - \$1,000 \times 0.22)) \times 5 = \$780 \times 5 = \$3,900$] if he or she is to be paid according to the current level of his or her efficiency (which is about 22 percent below the standard).

As noted before, to be able to effectively rate one's performance, one must be able to compare the tasks an individual is expected to perform (the standard) and the actual level of performance. However, it should be emphasized that the rules and expectations must be different for different types of tasks individuals perform even within the same organization. For instance, the time it takes to deliver a certain type of healthcare cannot be compared with the time it takes to repair a bridge or provide fire protection because there are no commonalities among these services, but it is possible to compare the performance of different individuals provided that the individuals being compared are engaged in exactly the same type of task or activity (as in our example of the medical procedure).

Other Measures of Performance

Although simple in approach, the method suggested here can also be used for time-series data. In other words, it can be used to measure both individual and organizational performance based on the time spent on a given activity over time, in which case it becomes a learning curve problem. However, there are other more sophisticated measures one can use to evaluate both individual as well as organizational performance. A good example will be a measure called objective rating, which is often used to compare the efficiency level of individuals for various activities and their elements within an organization (Mundel, 1978). According to this measure, a work standard is developed based on a set of agreed upon performance standards, which serve as a base. All individual tasks are compared against this standard and the deviations from the standard then provide a measure of efficiency.

Two other measures occasionally used in this context are synthetic rating (Barnes, 1980) and the Westinghouse performance rating (Westinghouse, 1985), both of which require human judgment applied to a set of criteria or attributes, such as dexterity, certainty, effectiveness, physical application, and so on, which function as a benchmark for comparison.

A Note on Motion Study

It is difficult to have any discussion on time study without some references to motion study since the two terms, time and motion, are used interchangeably. The motion study was introduced independently by Frank Gilbreth in 1911, about the same time Frederick Taylor introduced the time study, to analyze the labor process, in particular the work motions using scientific laws to improve efficiency without distracting physical and mental strength. His theory was based on a simple concept that there is one best way to perform a task and, therefore, one best way to improve efficiency. The task is based on a number of basic activities, determined through a categorization process called "therbligs." The therbligs, which number around 15, include actions such as search, find, select, grasp, position, and assemble. To determine the motion that is efficient, the therbligs are plotted on a Sumo Chart (Simultaneous Motion Chart) along with the time each motion takes. Examining the chart, it is possible to determine which therblig is inefficient (i.e., takes too long to complete) and, as such, should be eliminated by rearranging the work. The principle underlying the motion study is that by analyzing a job in its smallest possible elements, it is possible to synthesize a method of performance for the job that will be efficient.

From an operational perspective, the manner in which motion studies are carried out is similar to the method used in time studies: (1) select the best method to analyze a job; (2) break down the job into its smallest elements; (3) observe the time it takes to complete each element; (4) rate the worker's performance; (5) calculate the average and standard deviation; and (6) calculate the standard time. However, like time studies, the method is not without its weaknesses, some of which would also apply to time studies. For instance, observers of job performance may not be proficient, individual action may not sufficiently reflect group action, job pressure may contribute to worker mistakes, workers may not cooperate, and so forth. In spite of their weaknesses, both time and motion studies are useful as operational tools for increasing efficiency, reducing waste, and improving working conditions, among others.

Chapter Summary

Productivity is vital to efficient operation of an organization. It is the principal instrument by which one determines the extent to which an organization is utilizing its resources efficiently. This chapter has presented several measures of productivity. They are: productivity index, production function, the learning curve, and time study. Productivity indices, in particular production functions, are useful in determining the level of efficiency in an organization, especially where multiple inputs are concerned. Additionally, because of their analytical sophistication and underlying statistical assumptions, production functions can be more easily subjected to statistical tests for reliability than those produced by conventional productivity indices.

From a slightly different perspective, the learning curve provides a useful alternative to measuring efficiency and productivity. It can be used in a variety of situations and organizational settings, but there are limits to how much it can accomplish. For instance, one may unrealistically assume that the learning rate can be infinitely increased, regardless of the nature of the job being performed. For the vast majority of jobs, the human ability to learn is a function of what one calls psychomotor conditions such as a person's age, his or her nervous system, and how much the person has been required to learn in the past, which vary from individual to individual.

Finally, time study is a method of establishing an allowable time standard for performing a task based on the measurement of work contents. The purpose of time study is to determine a set of reliable standards by which an organization can measure the efficiency of its operations, usually by investigating the difference between actual and standard times. Reliable time standards are also useful as a tool by which an organization can increase its output capacity by making the best use of its labor and equipment, thereby reducing waste in the process.

Notes

1 The marginal products for a Cobb–Douglas production function are

$$\frac{\partial Q}{\partial K} = \alpha A K^{\alpha-1} L^{\beta} = \alpha \frac{A K^{\alpha} L^{\beta}}{K} = \alpha \frac{Q}{K} \tag{1}$$

$$\frac{\partial Q}{\partial L} = \beta A K^{\alpha} L^{\beta-1} = \beta \frac{A K^{\alpha} L^{\beta}}{L} = \beta \frac{Q}{L} \tag{2}$$

both of which are positive. The second derivative of Q with respect to K and L are

$$\frac{\partial^2 Q}{\partial K^2} = \alpha(\alpha-1) A K^{\alpha-2} L^{\beta} \tag{3}$$

$$\frac{\partial^2 Q}{\partial L^2} = \beta(\beta-1) A K^{\alpha} L^{\beta-2} \tag{4}$$

meaning that α and β are less than 1; that is, $0<\alpha, \beta<1$. Therefore, the result satisfies the requirements of the law of diminishing marginal returns.

2 Degree in a homogeneous function is a measure of returns to scale: if the degree is equal to 1 (i.e., if $\alpha+\beta=1$), it indicates a constant return to scale; if it is greater than 1 (i.e., if $\alpha+\beta>1$), it indicates an increasing return to scale; if it is less than 1 (i.e., if $\alpha+\beta<1$), it indicates a decreasing return to scale.

3 Considered a part of the overall statistical tests, especially where time-series data are involved, the objective of $D-W$ statistic is to test the null hypothesis that there is no autocorrelation problem in a model; if there is autocorrelation, it violates one of the conditions of causal modeling. The *DW* statistic extends between 0 and 4 ($0<d<4$), where d stands for *DW* statistic. The rule of thumb for interpretation is that there is no autocorrelation if *DW* statistic is around 2; it is positive if it is between 0 and 2; and it is negative if it is between 2 and 4. The presence of autocorrelation means that the predictions based on OLS estimates will be inefficient, although the estimated coefficients may be unbiased. The simplest way to correct the problem is to take first differencing

of the data, add additional variables as well as data, if available, to make sure that the model was not misspecified, among others.

4 To understand how this relates to the Cobb–Douglas production function, we can write, from Equation (1), the marginal product of labor and capital as:

$$MP_L = \beta\left(\frac{Q}{L}\right) \tag{5}$$

$$MP_K = \alpha\left(\frac{Q}{K}\right) \tag{6}$$

Next, we rearrange the terms in Equations (5) and (6) to obtain the MRTS, i.e., the rates at which the inputs can be substituted for one another, as:

$$\frac{MP_L}{MP_K} = \left(\frac{\beta}{\alpha}\right)\left(\frac{K}{L}\right) \tag{7}$$

where MP_L / MP_K is the ratio of marginal products of labor and capital.

Since the ratio of marginal products of labor and capital is equal to the ratio of their corresponding input prices (Equation 10.15), we can further write Equation 10.15 in terms of Equation (7), so that the optimality condition will now become:

$$\left(\frac{\beta}{\alpha}\right)\left(\frac{K}{L}\right) = \frac{w}{c} \tag{8}$$

If we further multiply both sides of Equation (8) by α and rearrange the resultant expression, the efficiency criterion, with some algebraic manipulations, becomes:

$$\beta\left(\frac{K}{L}\right) - \alpha\left(\frac{w}{c}\right) = 0 \tag{9}$$

Equation (9) is the optimality condition for the Cobb–Douglas production function and is equivalent to the expression $MP_L / MP_K = w / c$ in Equation 10.15. From this, we can easily write:

$$\beta\left(\frac{K}{L}\right) - \alpha\left(\frac{w}{c}\right) < 0 \tag{10}$$

for overutilization of labor, and:

$$\beta\left(\frac{K}{L}\right) - \alpha\left(\frac{w}{c}\right) > 0 \tag{11}$$

for overutilization of capital.

5 To determine the optimum combination of inputs as well as the optimal quantity of output, we do the following: (1) take the partial derivatives of the estimated Cobb–Douglas production function, $Q=(8.8227)K^{0.0568}L^{0.0387}$, with respect to K and L, and set them equal to zero, which will give us the marginal products of labor (MP_L) and capital (MP_K); (2) take the ratio of the two marginal products; and (3) substitute these ratios into Equation 10.15 and solve for capital (K) and labor (L), which will give us the optimum quantity of labor and capital. Next, we plug the values of K and L into the estimated equation $Q=(8.8227)K^{0.0568}L^{0.0387}$, and then solve for Q. Ideally, one should apply these optimality conditions to see if there has been an over- or underproduction output but, in this case, it was not considered necessary since the department needs to collect the solid waste, regardless of the quantity produced.

6 Equation 10.22 is derived from the equation of the learning curve (10.21) if it can be

assumed that it is a continuous, instead of a discrete, function (because of the cumulative effect of learning) and integrated with an appropriate change of limits:

$$\text{Average} = [\int_{0.5}^{X+0.5} KX^{-\beta} dX]/X$$

$$= [K\int_{0.5}^{X+0.5} X^{-\beta} dX]/X$$

$$= K\left[\frac{X^{1-\beta}}{1-\beta}\right]_{0.5}^{X+0.5} /X$$

$$= \frac{K}{1-\beta}[(X+0.5)^{1-\beta} - (0.5)^{1-\beta}]/X$$

References

Arrow, K. J., Chenery, H. B., Minhas, B. S., & Solow, R. M. (1961). Capital labor substitution and economic efficiency. *Review of Economics and Statistics, August*, 225–250.

Barnes, R. M. (1980). *Motion and time study: Design and measurement of work.* New York: Wiley & Sons.

Charnes, A., Cooper, W. W., & Rhodes, E. (1978). Measuring the efficiency of decision making units. *European Journal of Operations Research, 2*(6): 429–444.

Cobb, C. W., & Douglas, P. H. (1928). A theory of production. *American Economic Review, March* (Supplement), 139–165.

Dejong, J. R. (1964). Increasing skill and reduction of work time. *Time and Motion Study, October*, 28–41.

Easterly, W., & Levine, R. (2001). It's not factor accumulation: Stylized facts and growth models. *The World Bank Economic Review, 15*(1), 177–219.

Hatry, H. P. (1981). *Practical program evaluation for state and local governments.* Washington, DC: The Urban Institute.

Khan, A., & Murova, O. I. (2015). Productive efficiency of public expenditures: A cross-state study. *State and Local Government Review, 47*(3), 170–180.

Khumbakar, S. C. (2010). Estimation of multiple output production functions. Paper presented at the North American Productivity Workshop, Rice University, Houston, Texas, June 2–5.

Mundel, M. E. (1978). *Motion and time study: Principles and practices.* Englewood Cliffs, NJ: Prentice-Hall.

Nayyar-Stone, R., & H. P. Hatry (2010). *Using survey information to provide evaluative citizen feedback for public service decisions.* Washington, DC: The Urban Institute.

Rasmussen, S. (2013). *Production economics.* New York: Springer.

Westinghouse, G. E. (1985). *Work Measurement IIE Microsoftware.* Atlanta, GA: Industrial Engineering and Management Press.

11 Quality Control

Quality is critical to effective management of an organization. In recent years, there has been a significant interest in quality control in government resulting from two interrelated developments: (1) an increase in public demand for lean and efficient government; and (2) an increase in public expectation that the goods and services they receive from government in return for their tax dollars are of the highest quality. Quality control deals with various methods, tools, and approaches for diagnosing the lack of quality, defined commonly as nonconformity of standards in materials, machines, and processes. In government, as in the private sector, quality control remains a major concern. Quality control can assure both – in particular the latter – by maintaining quality specifications throughout the production or delivery process (i.e., from the acquisition of materials, supplies, and equipment to the final delivery of the goods and services).

This chapter focuses on an aspect of quality control that lies at the heart of all control activities, called process control (Ott & Schilling, 1990). In particular, it looks at the causes of variation in a process and discusses the methods commonly used to explain those variations. For the most part, the methods used in quality control are based on simple graphs, charts, and diagrams; as such, they require very little analytical skill to construct, although the concepts underlying them are fairly rigorous.

Defining Quality

Defining quality is not as simple as it may seem. Quality is fundamentally relational; there is no universal definition of quality, but intuitively, as well as in practice, we know what the term means when we purchase a good or use a service. However, based on the purpose for which the term is used, there are different ways to define quality. For instance, from an economic point of view, we can define quality as the value for the price we pay for a good or service, and the utility we derive from its consumption. Similarly, from a production or manufacturing point of view, quality can be defined as suitability of a product for use without any defect, in that it will not easily breakdown. Likewise, from a measurement point of view, quality can be defined as conformance to standards, similar to what we discussed under cost control, which simply means how well a good or service meets the target and its tolerance limits set by individuals who have designed or provided it. Since deviations from standards are relatively easy to measure, quality control studies frequently use conformance of standards as the principle guiding tool for quality measurement. However, quality is not always related to a good or service; it can also apply to an organization and its performance, such as decrease in crime rates for a community or increase in graduation rates for a school district.

Regardless of how quality is defined, the primary objective of any organization is to ensure that the goods and services it provides are of the highest quality and that they conform to specifications (standards). To make sure that quality is maintained throughout the process, one needs to understand the factors that affect quality and take appropriate measures to control them. This is generally known as process control. For any organization, process is the most important element that determines how successful an organization is in carrying out its activities. Everything revolves around process. If the process is inoperative or not managed efficiently, it will produce ineffective results. A process is simply a series of activities that transform inputs into outputs. Inputs generally include factors such as materials (e.g., data, raw materials, information), machines (e.g., instruments, tools, equipment), labor (e.g., experience, skill, training), and management (e.g., organizational structure, style, relationship).

Variations in Quality

A necessary condition for quality control is that the factors described above must be of the highest quality to ensure quality results. For example, when quality materials are used in a production or delivery process they will reduce waste, increase performance, and lengthen the life of a good or service. Similarly, quality machines will reduce frequent breakdowns, minimize wear and tear, and smooth out the production or delivery process. Likewise, quality labor will ensure the skills an organization needs to increase its performance and maintain quality throughout the process. Finally, quality management will ensure better working conditions for its labor and guarantee a work environment that will be conducive to productivity.

As long as there are no variations in these conditions during a production or delivery process, organizations should not have any difficulty in maintaining quality control. However, variations do occur and in some cases they are inevitable, even for services that are repetitive. Two types of factors frequently contribute to these variations: chance factors and assignable factors. The presence of chance factors in a process means that certain variations are bound to occur in an organization due to causes that behave in an unpredictable fashion, such as sudden changes in weather. They are often beyond human control and, as such, cannot be prevented in most instances. Consequently, one must allow for these variations when providing goods and services. Variations of this nature are called allowable variation.

The assignable variation, on the other hand, is caused by nonrandom factors such as defective materials, poor handling of machines, incorrect work procedure, and so on. As such, they are manageable, meaning they can be identified and corrected before they can cause major damage to the process. Variations due to these factors are called preventable variation. Since one does not have much control over the random causes of variation, the most one can do in quality control is identify the assignable causes of variation and make every effort to keep this variation under control. Most of the tools used in quality control, including the ones discussed here, are essentially geared toward that.

Use of Control Charts

Perhaps the single most important tool used in quality control, in particular for analyzing variations in a process, is a control chart. A control chart is an effective means for discovering and correcting the assignable causes of variation outside the pattern of random causes. Developed originally by W. A. Shewart in 1924, and subsequently popularized by Deming (Deming, 1950; Kilian, 1992) when he applied them to management, they are essentially pictorial devices used for detecting unnatural patterns of variation (defined as lack of control in data resulting from repetitive processes).

Since control charts are mostly pictures, they are simple to construct and easy to interpret. Furthermore, they can tell us whether a process is in control by visually comparing the location of quality characteristics on a chart. This section discusses the essential properties of a control chart and the steps commonly used in constructing these charts.

Properties of a Control Chart

All control charts have one characteristic in common: they consist of three horizontal lines; a central line to indicate the desired level of the process; a line above the central line called the upper limit or boundary of control; and a line below the central line, called the lower limit or boundary of control. Together, the upper and lower boundaries determine the control limits within which one can expect to find most sample points (observations) in a dataset. In general, when the points lie inside these limits the process is said to be in control. When they lie outside the limits, it is said to be lacking control. Figure 11.1 shows a typical control chart.

For most control charts, the central line represents the average or mean of the distribution of data points. Therefore, any point that lies outside the control line represents a deviation from the mean. In a typical control chart, the deviations are given by $\pm 3\sigma$, where $+3a$ represents the upper limit of control and $-3a$ represents the lower limit.

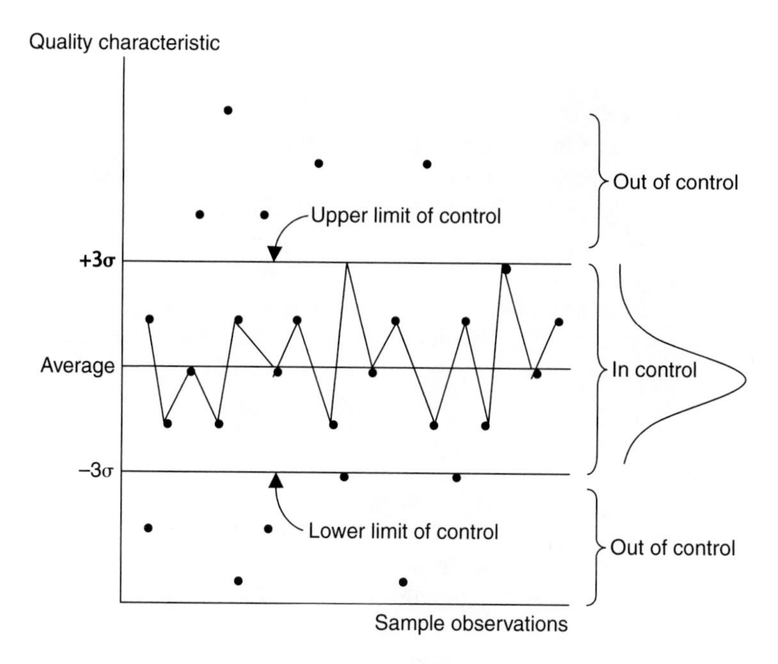

Figure 11.1 A Typical Control Chart

Steps in a Control Chart

Constructing a control chart is quite simple. However, it requires that a number of conditions are satisfied before the charts can be constructed. These conditions typically are: (1) checking for errors in the instruments used; (2) selecting a set of subsamples from a population from which data are collected for analysis; (3) calculating the relevant statistics for each subsample; (4) setting the control limits; and (5) constructing the control chart. Each of these steps is briefly discussed below.

Checking for Errors

The construction of a control chart begins with measurement; in particular, measurement in handling machines and tools, in designing instruments, in reading data, etc. For any production or delivery problem, errors can always enter into the measurement process due to faulty use of machines and tools, a lack of clear knowledge in dealing with these tools, or a lack of clear understanding of quality characteristics. Since the conclusions drawn from control charts depend on the variability of the quality being measured, it is important that the mistakes in reading data or handling tools are kept to a minimum so that one is able to draw valid and reliable conclusions from these charts.

Selection of Subsamples

The rationale for using subsamples in a control chart is to make comparisons of performance statistics possible. A subsample is a fraction of a sample drawn from a given population. There are no hard-and-fast rules for determining how many subsamples one must select in a quality control study and how many units each subsample must contain. However, the rule of thumb is to have four or five units in each subsample, with the number of subsamples not exceeding 25. The idea behind restricting the number of subsamples to 25 is that it should not be too few such that there is not enough information in a dataset, and it should not be too many such that it will include the entire dataset. The latter could be costly, as well as time consuming. In reality, the number can vary depending on the process being analyzed, but the number of units to be included in a subsample always remains fixed.

Calculation of Relevant Statistics

Once the sample size and the number of subsamples to be included in a study have been determined, the calculation of actual statistics for various charts can begin. This is not a difficult exercise since the statistics used in quality control are relatively simple. What is important, however, is the need to understand the distribution of the population from which these subsamples have been drawn. There are two simple explanations for this: first, the knowledge of population distribution which makes it possible to know a priori what statistics to use in a study. Second, especially where control charts are involved, it can provide clear guidance regarding how to use those statistics to set the limits for a chart. The latter is particularly important in quality control because without these limits it will be difficult to determine the extent to which a process is in control.

Setting the Control Limits

Control limits lie at the heart of quality control. As noted earlier, these limits set the boundaries against which quality measures are compared to see whether a process is out of control. Therefore, to understand what determines these limits, it is important to go over some basic (statistical) assumptions that underlie them since the factors that determine them are fundamentally statistical in nature. (Shewart was the first to recognize the statistical nature of these limits.)

We start with a simple assumption that we have a population parameter, which we will call Φ, and a set of sample observations, $x_1, x_2, x_3, \ldots, x_n$. Assume now that for our observations $x_1, x_2, x_3, \ldots, x_n$ we can define a statistic that corresponds to them in that it is a function of these observations. Let us say that the statistic is t. Since it is a function of the observations $x_1, x_2, x_3, \ldots, x_n$, we can write it as $t = t(x_1, x_2, x_3, \ldots, x_n)$.

Let us say that our t statistic has a distribution, given by:

$$E(t) = \mu_t, \text{ and Variance}(t) = \sigma_t^2 \qquad [11.1]$$

where $E(t)$ is the expected value of t, μ is the mean, and σ^2 is the variance, respectively.

Assume further that t is approximately normally distributed (a common assumption in statistical analysis when one is dealing with a small sample, usually 30 or less). Then

$$P[|t-E(t)|<3\sigma_t]=0.9973 \quad [11.2]$$

$$P[\mu_t-3\sigma_t<t<\mu_t+3\sigma_t]=0.9973 \quad [11.3]$$

where P is the probability of control limits, μ is the mean, and σ is the standard deviation.

Note that 0.9973 represents the area under a normal curve that lies within $\pm 3\sigma$. From a quality control point of view, this means that the probability that a random value of t will lie outside the limits $u_t \pm 3\sigma_t$ is 0.0027 $[1-0.9973=0027]$, which is quite small. Hence, if it is normally distributed, the limits of variation would be $\mu+3\sigma t$ and $\mu-3\sigma t$; that is, between the upper and the lower limits of control. If, for the ith sample, the observed t lies between these limits, we can say that the process is in control. Therefore, any variation that may exist among the sample observations can be explained by the presence of chance factors. On the other hand, when the observed t for the ith sample lies outside the limits the process is considered out of control, meaning that some assignable factors have created the problem.

Interestingly, if the assumption of normality of the t statistic does not hold, the argument presented above in support of 3σ control limits may not be strictly valid. In reality, quality characteristics are seldom normal. Consequently, when one is dealing with a set of observations that comes from a population that is not normal, one must rely on alternative measures such as Chevychev's inequality[1] to determine the nature of the distribution. Fortunately, however, even for a population that is not normal, the 3σ limits are almost universally accepted as a reliable guide (Farnum, 1994; Grant & Leavenworth, 1996).

Constructing the Charts

Finally, as shown in Figure 11.1, the control charts are plotted on a rectangular coordinate axis, with the vertical scale representing the statistical measures or quality characteristics such as X-bar and R, and the horizontal scale representing the sample observations (numbers). As noted earlier, for a process to work successfully, points on a given chart should always lie between the control limits. When the points do not lie inside these limits, it means the process is out of control. From a decision-making point of view, it then becomes important for the decision makers to know why the process is out of control; in particular, what has caused it.

In general, when a point resides outside the control limits it is an indication that assignable causes, rather than random factors, have contributed to the problem. Although in principle that is the case, in reality one needs to look at the distribution of all the data points, even those that are within the limits, to make certain that there is no unusual pattern that could tell us otherwise.

Types of Control Charts

There is a wide range of control charts one can use in quality control. However, based on the characteristics they measure, they fall into two broad categories: variable control charts and attribute control charts. Variable control charts deal with variables whose values are continuous, such as income, height, weight, distance, etc. Attribute control charts, on the other hand, deal with variables whose values are discrete, such as success–failure, conformity–nonconformity, pass–fail, taste, color, etc. Since the variables used in variable control charts are continuous, the methods commonly used for these charts are measures of central tendency and dispersion such as mean, standard deviation, and range. In contrast, since the variables used in attribute control charts are discrete and not continuous, the measures of central tendency and dispersion used for these charts are based mostly on proportions and frequencies (i.e., number of occurrences of a given attribute of a variable in a dataset).

This section discusses several different control charts that include both continuous and attributive variables; they are *X*-bar, *R*, *s*, *p*, *d*, *c*, and *u* charts.

X-*Bar and* R *Charts*

As noted earlier, the purpose of quality control is to detect the assignable causes of variation in a process. The control limits on the *X*-bar and *R* charts are placed in such a way that they reveal the presence or absence of these causes of variation based on several statistical measures such as the mean ($\bar{X}$), range (R), and standard deviation (s). In general, the *X*-bar chart is used to measure the variability in the location (i.e., the mean of the process) and the *R* chart is used to measure the range of control in the process.

Calculation of X-Bar and R *Statistics*

The calculation of *X*-bar and *R* charts begins with the calculation of a set of means, ranges, and standard deviations for each sample or subsample. The procedures for determining these statistics are quite simple. For instance, let X_{ij} (for $i=1,2,3\ldots,n$) be the measurements on the *j*th sample (for $j=1,2,3,\ldots,m$), then the mean $\bar{X}$, the range R, and the standard deviation s for the *j*th sample can be expressed as:

$$\bar{X}_j = \sum_{i=1}^{n} X_{ij} / n \qquad [11.4]$$

$$s_j = \sqrt{\sum_{i=1}^{n} \left(X_{ij} - \bar{X}\right)^2 / (n-1)} \qquad [11.5]$$

$$R_j = \max<i>X_{ij} - \min<i>X_{ij} \qquad [11.6]$$

for $j = 1, 2, 3, \ldots, m$ and $i = 1, 2, 3, \ldots, n$.

From the above, one can obtain a set of new means for $\overline{X}$, R, and s, called the aggregate means. Aggregate means serve a useful purpose in quality control by providing a base against which individual statistics (mean, range, standard deviation, and so on) can be compared to determine if a process is out of control.

Equations 11.7–11.9 explain how these new means are constructed for $\overline{X}$, R, and s:

$$\overline{\overline{X}} = \sum_{j=1}^{m} \overline{X}_j / m$$

$$= (\overline{X}_1 + \overline{X}_2 + \overline{X}_3 + \ldots\ldots.. + \overline{X}_m) / m \qquad [11.7]$$

$$\overline{R} = \sum_{j=1}^{m} R_j / m$$

$$= (R_1 + R_2 + R_3 + \ldots\ldots\ldots + R_m) / m \qquad [11.8]$$

$$\overline{s} = \sum_{j=1}^{m} s_j / m$$

$$= (s_1 + s_2 + s_3 + \ldots\ldots. + s_m) / m \qquad [11.9]$$

where $\overline{\overline{X}}$, $\overline{R}$, and $\overline{s}$ represent the new means; that is, the mean of means, the mean of ranges, and the mean of standard deviations, respectively.

Additional Measures

Setting the control charts for $\overline{X}$ and R charts depends on two additional measures: standard error of $\overline{X}$ and the sampling distribution of R. A standard error is simply the standard deviation of a sampling statistic such as the sampling mean (i.e., the mean of the means of a sampling distribution, based on repeated samples). For instance, if σ is the process standard deviation (i.e., the standard deviation of the population), the standard error of the sampling mean is given by $\sigma/\sqrt{n}$. That is,

$$SE(\overline{X}_j) = \sigma / \sqrt{n} \qquad [11.10]$$

where SE is the standard error of the mean $\overline{X}$ for the jth sample, σ is the corresponding standard deviation, and n is the sample size.

The sampling distribution of the range R, on the other hand, is given by the expected value, which can be written as a function of a constant that changes with the sample size. That is,

$$E(R)=k\sigma \tag{11.11}$$

where $E(R)$ is the expected value of R, k is a constant that changes with sample size n, and σ is the same as before.

Setting the Control Limits for X-Bar Chart

Setting the control limits for any chart, including the X-bar and R charts, also requires that we state in advance whether or not we know the population parameters, μ and σ. If we know the population parameters, it is understandable that we set the limits based on the known parameters, meaning that we do not have to estimate them; we can simply plug them in. If we do not know them a priori, then we must first obtain the estimates of the parameters from the observed data before setting the limits.

We start with the assumption that we know the population parameter u (i.e., the population mean) and its corresponding standard deviation, σ. The 3σ control limits for X-bar charts, based on our earlier discussion of setting the control limits, can be written as:

$$UCL_{\bar{X}} = \mu' + A\sigma' \tag{11.12}$$

$$LCL_{\bar{X}} = \mu' - A\sigma' \tag{11.13}$$

where UCL and LCL are the upper and lower limits of control, u' is the known population mean, σ' is the known population standard deviation, and A is a constant, similar to k, representing $3/\sqrt{n}$ that changes with sample size n, as noted earlier.[2] In reality, information on the constant terms can be obtained directly from published sources such as a quality control table, which can be easily plugged into the equations to determine the limits.[3]

If we do not know the population parameters, we can use the sample range as an estimate of the variability of the process to construct the control limits. We know from our knowledge of measures of dispersion that range (R) is simply the difference between the highest and the lowest value in a sample (Equation 11.6). The advantage of using range is that it can provide useful information about the variability of the data. In general, the higher the range, the greater is the variability in the data. Accordingly, the control limits for the X-bar chart can be written as

$$UCL_{\bar{X}} = \bar{\bar{X}} + A_2\bar{R} \tag{11.14}$$

$$LCL_{\bar{X}} = \bar{\bar{X}} - A_2\bar{R} \tag{11.15}$$

where $\bar{R}$ is the mean of ranges, A_2 is a constant representing three standard deviations of range that changes with sample size n, similar to k, and the rest of the terms are the same as before.[4] Interestingly, the values of the constant terms such as A_2 for different samples can be obtained directly from a quality control table, as one would find in a standard t or z table.

Setting the Control Limits for R *Chart*

The R chart, like the X-bar chart, plays an important role in quality control; it measures the variability of a process. While the X-bar chart measures the central tendency, the R chart measures the dispersion of the process, but the methods for developing and using the R chart are essentially the same as the X-bar chart. The central line represents the mean range, $\overline{R}$, and $\pm 3\sigma$ represents the upper and lower limits of control.

Like the control limits for the X-bar chart, the 3σ control limits for the R chart can be presented in the following way:

$$UCL_R = D_4\overline{R} \qquad [11.16]$$

$$LCL_R = D_3\overline{R} \qquad [11.17]$$

where D_3 and D_4 are constants that change with the sample size n.[5]

As before, information on D_3 and D_4 can be obtained from any standard table for control charts, such as the one produced by the American Society for Testing Materials (ASTM, 1951), similar to statistical tables one would find in any standard statistics textbook. They can also be tabulated by hand for different values of n, but the process will be cumbersome and laborious. Table 11.1 presents the information commonly used for constructing a control chart for a small number of observations.

Using X-Bar and R *Charts Together*

Although the X-bar and R charts use different variables in quality control, X-bar and R, both variables play an important role in explaining the variation in quality; as such, it is important to take both measures into consideration when constructing a control chart. Let us look at a simple example to illustrate this. Suppose that the Human Resource Department of a county government wants to investigate the absentee problem that has become a major concern for several of

Table 11.1 A Sample of Factors Used in a Control Chart and Corresponding Values

n	A	A_2	D_3	D_4	B_1	B_2	B_3	B_4
2	2.121	1.880	0	3.267	0	1.843	0	3.267
3	1.732	1.023	0	2.574	0	2.574	0	2.568
4	1.500	0.729	0	2.282	0	1.808	0	2.266
5	1.342	0.577	0	2.114	0	1.756	0	2.089
6	1.225	0.483	0	2.004	0	1.711	0	1.970

the departments in recent months. Let us say that the government allows its employees two weeks of paid vacation and two weeks of unpaid leave of absence during the year. The absentee problem occurs when more leave is taken than this entitlement grants. Table 11.2 presents the data for a sample of 10 departments for each of the five working days for an entire year. The table also presents information on the corresponding mean, range, and standard deviation for each department.

To determine if the process is in control, we can now apply Equations 11.14 and 11.15 for *X*-bar and Equations 11.16 and 11.17 for *R* charts, as shown below:

$$UCL_{\bar{X}} = \bar{\bar{X}} + A_2\bar{R} = 42.7 + (0.557)(19.7) = 53.673$$

$$LCL_{\bar{X}} = \bar{\bar{X}} - A_2\bar{R} = 42.7 - (0.557)(19.7) = 31.727$$

for the *X*-bar chart, and

$$UCL_R = D_4\bar{R} = (2.114)(19.7) = 41.646$$

$$LCL_R = D_3\bar{R} = (0.00)(19.7) = 0$$

for the *R* chart.

As the results of the control limits show, the process is out of control for three of the ten departments (1, 6, and 7) for the *X*-bar chart since they lie outside the control limits. However, the same conclusion cannot be drawn for the distribution of the range for the *R* chart since all the data points lie within the control limits. Note that although we have not constructed the actual charts, we can still identify the departments (i.e., the sample units) that are out of control by visually

Table 11.2 Absentee Data and Related Statistics for Quality Control (Hours)

Dept.	*M*	*T*	*W*	*R*	*F*	*Total*	*Sample Mean*	*Sample Range*	*Standard Deviation*
1	60	55	57	50	72	294	58.8	22	8.23
2	58	47	40	45	53	243	48.6	18	7.02
3	32	30	30	28	45	165	33.0	15	6.86
4	45	44	40	37	60	226	45.2	23	8.87
5	47	45	38	32	48	210	42.0	16	6.82
6	27	25	25	28	35	140	28.0	10	4.12
7	34	30	22	20	48	154	30.8	28	11.19
8	43	38	32	34	55	202	40.4	23	9.18
9	56	47	45	42	68	258	51.6	26	10.55
10	53	48	48	39	55	243	48.6	16	6.19

Notes
$n=10$; $\bar{\bar{X}}=42.7$; $\bar{R}=19.7$; $A2=0.557$; $D3=0$; $D4=2.114$; $\bar{s}=7.90$; $B3=0$; $B4=2.089$.

comparing the statistics we have obtained for our sample with the upper and lower limits of control for a given chart.

Additionally, the results produced by the two charts create an interesting dilemma for the decision makers since no definitive conclusions emerge from the charts. In reality, when one is faced with a situation in which a clear-cut picture does not emerge from the charts, the convention is to construct the R chart first. If the chart indicates that the process is out of control, there is no need to construct the X-bar chart until the process is brought under control. The X-bar chart then serves as a checking device to ensure the process is in control.

It should be worth noting that it is not necessary to use both standard deviations and range at the same time because for small samples the standard deviation, s, and its corresponding range, R, behave identically. In statistical terms, this means that they move in the same direction; that is, if s is small, R is likely to be small, and vice versa. However, for large samples an extreme outlier in either direction can have a significantly large effect on range, while its effect on the standard deviation will be much less. What this means is that for analyzing variability involving small samples, the range can be used as a substitute for standard deviation with little loss of efficiency. Since range is almost as efficient as standard deviation in small samples, it is preferred to standard deviation in quality control.

s *Chart*

Although both R and s charts are not necessary in the same problem, especially when the sample size is small, the s chart becomes meaningful as the sample size increases. In general, one must have a sample size of at least 10 to justify the use of the chart. Since our sample size is 10, we can at a minimum try to see if it will produce a different result from that we obtained for the R chart. Like the X-bar and R charts, we can construct the control limits for the s chart under two conditions: one, when the population standard deviation, σ, is known and, two, when it is not known. As we mentioned before, when it is known, we can use it directly to set the limits and when it is not known, we can always estimate it from the observed data before setting the limits.

The control limits for the s chart when the population standard deviation is known can be written as:

$$UCL_S = E(s) = 3SE(s) = B_2\sigma \quad [11.18]$$

$$LCL_S = E(s) = 3SE(s) = B_1\sigma \quad [11.19]$$

where UCL and LCL are respectively the upper and lower limits of control for the s chart, $E(s)$ is the expected value of s, and B_1 and B_2 are constants that change with the sample size n, especially B_2, and whose values can be obtained from any standard table for control charts.

Similarly, the control limits for the s chart, when the population standard deviation is not known, can be written as:

$$UCL_S = E(s) = 3SE(s) = B_4\sigma \qquad [11.20]$$

$$LCL_S = E(s) = 3SE(s) = B_3\sigma \qquad [11.21]$$

where B_3 and B_4 are constants, similar to B_1 and B_2, and the rest of the terms are the same as before.

We can now apply the 3σ control limits (Equations 11.20 and 11.21) to the absentee problem to see which of the departments are out of control, as shown below:

$$UCL_S = B_4\sigma = (2.089)(7.90) = 16.503$$

$$LCL_S = B_3\sigma = (0)(7.90) = 0$$

A cursory glance at the results would indicate that none of the departments is out of control since their standard deviations fall within the control limits. In other words, the process is in control. The result seems to be consistent with what we observed earlier for the *R* chart, as the case should be for small samples.

p *and* d *Charts*

The quality control charts discussed above deal with variables whose distributions are continuous. As such, they belong to the category we defined earlier as variable control charts, but when one has to deal with variables whose values are discrete and not continuous, the appropriate charts to use would be attribute control charts. The two attribute charts commonly used in quality control are *p* and *d* charts. As the term implies, the *p* chart is used primarily for determining the proportion (fraction) of nonconformities in a process, whereas the *d* chart is used for determining the actual number of nonconformities.

To determine if a process is in control, especially when one is dealing with attributive variables, one needs to look at the distribution of the data in a sample to see if they conform to population characteristics; in other words, to see if the sample or samples under consideration have the same population proportion. In general, when all the samples or subsamples can be determined to have the same population proportion, *P*, the process is said to be in control. To put it differently, if d is the number of nonconformities in a sample of size n, the sample proportion that is nonconforming is $p = d/n$, where d is a binomial variate with parameters n and P. The term binomial means that there are only two outcomes that can result from a trial, such as conforming and nonconforming, where the trials are identically and independently distributed. In other words, all the trials are exactly the same and the outcome of a trial does not depend on the outcome of any other trial.[6]

Setting the Control Limits for the p *Chart*

As before, to set the control limits it is important for us to know whether we know the population parameter (i.e., population proportion). If we know the population proportion, P, the 3σ control limits for p can be written as:

$$UCL_p = P + 3\sqrt{PQ/n} \quad [11.22]$$

$$LCL_p = P - 3\sqrt{PQ/n} \quad [11.23]$$

where P' is the population parameter P, and UCL and LCL are the upper and the lower limits of p, respectively.[7]

If we do not know the population proportion, P, the 3σ control limits for p can be written as:

$$UCL_p = \overline{p} + 3\sqrt{\overline{p}(1-\overline{p})/n} \quad [11.24]$$

$$LCL_p = \overline{p} - 3\sqrt{\overline{p}(1-\overline{p})/n} \quad [11.25]$$

where $\overline{p}$ is the mean of the proportion of nonconformities and the rest of the terms are the same as before.[8]

Setting the Control Limits for d *Chart*

If, instead of using p (the fraction of nonconformities in a sample), we use d (the number of nonconformities), the control limits for the d chart will be:

$$E(d) \pm 3SE(d) = np \pm 3\sqrt{nP(1-P)} \quad [11.26]$$

From the above, and assuming that we know the population fraction, P, we can write the 3σ control limits for the d chart as:

$$UCL_d = nP' + 3\sqrt{nP'(1-P')} \quad [11.27]$$

$$LCL_d = nP' - 3\sqrt{nP'(1-P')} \quad [11.28]$$

where P represents the known value of P and the rest of the terms are the same as before.

It is worth noting that since p cannot be negative, the LCL should not be negative. In the event that it comes out to be negative, one should treat it as zero. Also, for the d chart, as the construction of the limits depends on nP, the d chart is often called the np chart. It can also be construed as an adaptation of the p chart in that it focuses on the number of nonconformities (nP) rather than on the fraction (p) of nonconformities.

Let us look at a simple example to illustrate this. Suppose that a state correctional agency has recently made surprise visits to several of its correctional facilities in the state and found a good number of them to be in violation of the state-mandated guidelines, especially on matters related to health and sanitation. Our objective is to determine if this violation is any indication that the facilities are out of control. Table 11.3 presents the data on violation for 15 of the facilities observed during the visits by the state officials.

To determine which of the facilities are out of control, we can apply Equations 11.27 and 11.28 to the problem and obtain the respective values for the upper and lower limits of control for the *p* and *d* charts.

As before, we start with the *p* chart, as shown below:

$$UCL_p = \bar{p} + 3\frac{\sqrt{\bar{p}(1-\bar{p})}}{n} = 0.347 + 3\frac{\sqrt{0.347(1-0.347)}}{20} = 0.418$$

$$LCL_p = \bar{p} - 3\frac{\sqrt{\bar{p}(1-\bar{p})}}{n} = 0.347 - 3\frac{\sqrt{0.347(1-0.347)}}{20} = 0.276$$

where *p* was obtained by solving the expression $\Sigma d_j / \Sigma n_j = 104/300 = 0.347$.

We do the same for the *d* chart. That is,

$$UCL_d = n\bar{p} + 3\sqrt{n\bar{p}(1-\bar{p})} = (20)(0.347) + 3\sqrt{(20)(0.347)(1-0.347)} = 13.327$$

$$LCL_d = n\bar{p} - 3\sqrt{n\bar{p}(1-\bar{p})} = (20)(0.347) - 3\sqrt{(20)(0.347)(1-0.347)} = 0.553$$

We can interpret the above results the same way one would interpret the results for *X*-bar and *R* charts; that is, if all the data points fall within the control limits,

Table 11.3 Violations by State Correctional Facilities

Facility Number	*Items Inspected*	*Number of Violations,* d	*Fraction of Violations,* p
1	20	3	0.15
2	20	7	0.35
3	20	12	0.60
4	20	8	0.40
5	20	3	0.15
6	20	10	0.50
7	20	6	0.30
8	20	4	0.20
9	20	13	0.65
10	20	11	0.55
11	20	2	0.10
12	20	9	0.45
13	20	4	0.20
14	20	7	0.35
15	20	3	0.15

Note
$n=20$; $n\bar{p}=6.940$; $\bar{p}=0.347$.

the process is in control. If not, it is out of control. Therefore, when all the data points in a sample lie within the control limits, any variation that may exist in a process should be attributed to the chance factors, although one should not completely rule out the possibility that some preventable factors can always find their ways into the process; statistically, nothing is impossible.

Returning to the control limits for the *p* chart in our example, we can see that facilities 3, 6, 9, 10, and 12 are clearly out of control, meaning they must be immediately reported for correction. There are at least three other facilities in the current example (2, 4, and 14) that also show some indications of deterioration, which is evident from the fact that they lie between the mean (i.e., the central line, not shown here) and the upper limit of control; that is, between 0.347 and 0.418.

For the *d* chart, none of the facilities seems to be out of control, although the facilities identified in the *p* chart as out of control come close to being out of control here. Furthermore, an analysis of the mean for the *d* chart should reveal that it is quite high (6.94 to be exact), meaning that further investigation is needed for the facilities that lie between this average and the upper limit of control (i.e., between 6.940 and 13.327). In general, when the results of the two charts do not coincide, one must use the average (the central lines) to see how far they are located from this average and make a decision on the basis of their location.

Before leaving the discussion on *p* and *d* charts, let us introduce two new terms that are useful in a control chart – one called high spot and the other low spot. A high spot is a situation in which the sample observations lie outside the upper control limit. A low spot, on the other hand, is a situation in which the observations lie below the lower control limit. In general, the presence of high spots indicates a deterioration of quality, while the presence of low spots indicates an improvement. However, it should be emphasized that the presence of low spots does not necessarily mean that quality has actually improved. One must carefully analyze the improvement to make certain that it is not due to any slackness in inspection.

c *and* u *Charts*

Another type of chart that is frequently used in quality control is the *c* chart. Unlike the *p* or *d* chart, the *c* chart, also called the count chart, is an attribute control chart that is used to determine the number of nonconformities or defects in a unit, where a unit is defined as the sample that provides the data for inspection or investigation. An important requirement of the *c* chart is that the inspection unit must be the same (i.e., constant) for each sample. The term "the same" or "constant" in this case means that the inspection unit must always represent an identical area of opportunity for the occurrence of nonconformities, which can be a group of units or a single unit on which the number of nonconformities or defects are counted. The logic behind a constant area of opportunity is that opportunities for nonconformities in any sample are as good as any other sample.

In other words, each and every sample in a study has the same probability of containing a given number of nonconformities, provided that the samples under consideration have more or less the same characteristics.

While the area of opportunity is fixed, the sample size can vary. For many real-world problems, the sample size is usually large, meaning that the opportunities for nonconformities to occur (i.e., of finding defects in a sample) are more numerous in large samples than they are in small samples. What this means in statistical terms is that the probability of nonconformity being present in any spot is small. This seems logical since with a large n, the probability of finding any nonconformity, given by $1/n$, should be small as n increases in size. The probability distribution that best approximates this behavior in a sample is the Poisson distribution.

Setting the Control Limits for c *Chart*

Unlike most statistical distributions, the Poisson distribution has a unique characteristic in that its mean and variance are identical. Thus, if we can assume that c is a Poisson variate with parameter λ, then

$$E(c)=\lambda \text{ and } Var(c)=\lambda \qquad [11.29]$$

where $E(c)$ is the expected value of c.[9]

Assume now that we know the value of the population parameter λ; the 3σ control limits for the c chart, with known λ, can be written as:

$$UCL_c = \lambda' + 3\sqrt{\lambda'}$$

$$LCL_c = \lambda' - 3\sqrt{\lambda'} \qquad [11.30]$$

where λ' represents the known value of λ.

If, on the other hand, we do not know the value of the population parameter λ, we can always estimate it by taking the average of the nonconformities for a unit. Assume that c_j is the nonconformity for the jth unit, the estimate of λ, based on c, can then be written as:

$$\overline{c} = \sum_{j=1}^{m} c_j / n$$

where c is the estimate of λ.

From the above, in particular when the value of the parameter λ is not known, we can write the control limits for the c chart as

$$UCL_c = \overline{c} + 3\sqrt{\overline{c}} \qquad [11.31]$$

$$LCL_c = \overline{c} - 3\sqrt{\overline{c}} \qquad [11.32]$$

where UCL_c and LCL_c are, respectively, the upper and lower limits of control for c. Note that since c cannot be negative, the LCL cannot be negative. Therefore, when it is negative, it should be regarded as zero.

Setting the Control Limits for the u *Chart*

Earlier, we made an assumption that the area of opportunity for the c chart is fixed, which may not always be the case. When the opportunity area is not fixed (i.e., when it varies from sample to sample), the alternative is to create a standardized statistic, given by u (the population mean). The u in this case is obtained by dividing c (the number of nonconformities per area of opportunity) by n (the sample size or the number of inspection units per area of opportunity). The chart that represents this is called the u chart.

The control limits for the u chart, based on the above assumption, can thus be written as:

$$UCL_u = \bar{u} + 3\sqrt{\bar{u}/n} \qquad [11.33]$$

$$LCL_u = \bar{u} - 3\sqrt{\bar{u}/n} \qquad [11.34]$$

where the UCL and LCL are the upper and lower limits of control for u, respectively, and the rest of the terms are the same as before.

To illustrate these limits, let us look at an example. Suppose that the Environmental Protection Agency which is responsible for monitoring the environmental standards in the country, has recently come under attack from Congress because of a report that accuses the agency of being slack in enforcing federal regulations for a number of service industries. In response, the agency has decided to investigate several of these industries to determine how critical the situation is and whether it needs to be corrected. Table 11.4 presents the data on violations by these industries for the range of services they provide.

To determine the extent of violation by these industries, in particular to determine if these violations are really out of control, as the report claims, we can apply Equations 11.31, 11.32, 11.33, and 11.34, corresponding to the upper and lower limits of control for both charts, as shown below:

$$UCL_c = \bar{c} + 3\sqrt{\bar{c}} = 23 + 3\sqrt{23} = 23.0 + 14.4 = 37.4$$

$$LCL_c = \bar{c} - 3\sqrt{\bar{c}} = 23 - 3\sqrt{23} = 23.0 - 14.4 = 8.6$$

for c chart, and

$$UCL_u = \bar{u} + 3\sqrt{\bar{u}/n} = 6.14 + 3\sqrt{6.14/10} = 6.14 + 2.34 = 8.48$$

$$LCL_u = \bar{u} - 3\sqrt{\bar{u}/n} = 6.14 - 3\sqrt{6.14/10} = 6.14 - 2.34 = 3.80$$

for u chart.

Table 11.4 Industrial Violations of EPA Standards

Sample Industry	*Number of Services*	*Number of Violations (Nonconformities, c_j)*	*Nonconformities Per Unit, u_j*
A	5	21	4.20
B	4	38	9.50
C	6	15	2.50
D	3	46	15.33
E	7	11	1.57
F	4	14	3.50
G	5	24	4.80
H	2	31	15.50
I	6	18	3.00
J	8	12	1.50

Note
$C=23$; $u=6.14$.

The results of the control limits for the *c* chart indicate that only two, B and D, out of a total of ten industries investigated by the agency, are out of control. Of the two, D in particular appears to be more out of control than B, whose *c* value is marginally above the upper control limit. As for the *u* chart, only three, including the ones identified for *c*, seem to be out of control. Two additional industries, C and J, in fact, lie below the lower limit of control, meaning that these industries have improved their standards. However, no attempt was made here to ascertain to what degree this improvement may have been due to an oversight or slackness in investigation by the agency.

Errors in Control Charts

Once we have estimated the values of the population parameters (i.e., population characteristics), based on a sample drawn from the same population, we ask ourselves if the estimated parameters were obtained by error (chance) and conduct specific tests to ensure that the estimates are reliable, so that we can draw the conclusion with a certain degree of confidence that they were not obtained by chance. Similarly, we can ask ourselves if the statistics we observed for various control charts such as mean, range, standard deviation, proportion, and so forth were obtained by error, and conduct appropriate tests (i.e., hypothesis tests) to determine with a certain degree of confidence that they were not obtained by error.

From a statistical point of view, if we draw the wrong conclusion we will be in error. Two types of errors are commonly associated with statistical tests that also apply to control charts: Type I and Type II. The Type I error occurs when a process thought to be out of control is, in fact, not out of control. Likewise, the Type II error occurs when a process thought not be in control is, in fact, out of control (Table 11.5).

To give an example, suppose that we are conducting a hypothesis test for an *X*-bar chart for process control, with a population mean (μ) of 50. The null hypothesis for

Table 11.5 Errors in Quality Control

	Process	*In Control*	*Out of Control*
Hypothesis (null)	In control	No error	Error – Type II
	Out of control	Error – Type I	No error

the test is that the process is in control if μ is equal to 50 and out of control if it is not equal to 50, which is a standard two-tailed test (because the rejection region lies on both sides of the tail in a normal distribution). We know from our elementary knowledge of inferential statistics that a null hypothesis is a statistical statement that we reject on the empirical ground that it is not true, so we will not be in error. In other words, we will not commit a Type I error by incorrectly rejecting the hypothesis when it is true. Put differently, when we correctly reject the hypothesis, by default, we accept the alternative hypothesis, which is what we are trying to establish.

To test the hypothesis that $\mu \neq 50$, meaning that it could be higher or lower than 50, we follow a number of steps called hypothesis testing procedures: (1) set a significance level called the α level (i.e., probability of rejecting a null hypothesis when it is true, which is usually set at 0.05 or 0.01); (2) collect data from a sample and calculate the relevant statistics, which in this case will be mean (i.e., sample mean, $\overline{X}$) and its corresponding standard error, $s/\sqrt{n-1}$), where $n-1$ is the degrees of freedom; (3) find the observed t (z) value, depending on the size of the sample (i.e., t value, if the sample size is ≤ 30; z value, if otherwise); (4) compare the observed t (z) against its critical (theoretical) value from a standard t (z) table; and (5) draw the relevant statistical conclusion; that is, reject the null hypothesis if the observed t (z) is greater than or less than the critical value for a two-tailed test, greater than for a right-tailed test, and less than for a left-tailed test.

Let us say that we have collected data on 10 observations and have calculated their mean and standard deviation, which came out to be $\overline{X}=56$ and $s=5$. We set the α level at 0.05 and follow the test procedures in sequence, starting with the null and alternative hypotheses:

[1] H_0: $\mu = 50$

H_1: $\mu \neq 50$

[2] $\alpha = 0.05$

[3] $$t = \frac{\overline{X} - \mu}{s/\sqrt{n-1}}$$

$$= \frac{56 - 50}{5/\sqrt{10-1}}$$

$$= \frac{6}{1.67}$$

$$= 3.5928 \approx 3.60$$

Having calculated the t statistic, we now compare it against its critical value that can be found in any standard t table. Since the observed $t=3.60$ is higher than the critical $t^*=2.262$ with nine degrees of freedom $[n-1=10-1=9]$ for a two-tailed test for $\alpha=0.05$ (i.e., $\alpha/2=0.05/2=0.025$), we can reject the null hypothesis that the process mean is 50 with a 95 percent degree of confidence $[1-0.05=0.95]$ and accept, by default, the alternative hypothesis that it is not 50.

While being able to reject the null hypothesis at a given level of α, it does not necessarily guarantee that our observed t was not due to error. For the latter, we need to find the p value – the probability associated with the critical t (i.e., the probability that the observed t did not occur by chance). With nine degrees of freedom and $t=3.60$, the probability that the observed t did not occur by chance is between 0.01 and 0.005 (i.e., $0.005<p(t)<0.01$) that can be obtained from any standard t table. For a one-tailed test; that is, if we were to conduct a right- or left-tailed test, it would be between 0.005 and 0.0025, and in both cases the results would be statistically significant.

Process Capability

An important aspect of quality control is to be able to evaluate if the process is in control. The various control charts we have discussed so far help us determine that, but they do not tell us if the process has met the specifications we set a priori. Most organizations, public as well as private, have certain specifications that they want the process to meet when providing a good or service. Specifications are tolerances or limits of quality characteristics that we are willing to accept. When a process is able to meet the specifications, or even exceed them, it is called process capability. For instance, say, a local school district sets a target graduation rate for high school students at 85 percent with a standard deviation of ± 5, which means that the acceptable range or limits of graduation for the district will be between 80 and 90 percent (i.e., 85 ± 5). For most organizations, the acceptable limits are based on past experience, quality of personnel, machine capability, resource base, and so forth. To ensure that the service provision has met the acceptable limits, process variation must not exceed process specification, as set by the decision makers.

While the control charts provide an effective means for monitoring a process, unfortunately they do not ensure process capability. For an organization to be able to provide a good or service that is acceptable in that it falls within the acceptable limits, the process must be capable and in control before its provision can take place. To determine whether the process has met the expectations, one needs to use measures other than the control charts that can provide a precise knowledge of process capability. Two measures are commonly used for this purpose: process capability ratio, C_p, and process capability index, C_{pk}. The former is called six-sigma (6σ) and the latter three-sigma (3σ), because the sigma (σ) provides the width or range of acceptable limits for measuring process capability. The rationale for using six-sigma, as opposed to three-sigma, which is commonly used in quality control, is that most measurement processes lie within $\pm 3\sigma$ with a total of six standard deviations that cover over 99 percent of the area under a normal curve, which is almost the entire curve.

Process Capability Ratio, C_p

Process capability ratio shows how well a good or service meets the range specified by the design limits. In general, if the design limits are larger, in that they exceed expectations, than the limits specified in the process, the average of the process can be allowed to be flexible. In other words, it can move in either direction from the center. The capability ratio is obtained by taking the ratio of specification range to process range. The specification range shows the difference between the upper specification limit (USL) and the lower specification limit (LSL), while the process range shows the limits given by the sigma, which in this case is six:

$$C_p = \frac{USL - LSL}{6\sigma} \qquad [11.35]$$

where C_p is the capability ratio, *USL* and *LSL* are the upper and lower specification limits, and 6σ is the process limit.

Three possible ranges of values are commonly used in the capability ratio that can help us interpret the results: (1) $C_p = 1$: the process has met the specifications; (2) $C_p \geq 1$: the process has exceeded the specifications; and (3) $C_p \leq 1$: the process is not capable of meeting the specifications. The first indicates that the process is minimally capable; the second that it exceeds the minimal capability; and the third that it is below the minimal capability.

Process Capability Index, C_{pk}

Like the capability ratio, the capability index also tells us how well a good or service meets the range specified by the design limits. While the capability ratio is useful in measuring process capability, it has an inherent weakness in that it assumes that the variability in a process lies entirely on the specification, which may not always be the case. This is where the capability index is considered a better measure in that it provides two options and selects the one that produces the minimum value of the two to explain the variability.

The capability index (C_{pk}) is given by the expression, as shown below:

$$C_{pk} = \min\left[\frac{USL - \mu}{3\sigma}; \frac{\mu - LSL}{3\sigma}\right] \qquad [11.36]$$

where μ and σ are, respectively, the mean and standard deviation of the process and the rest of the terms are the same as before.

Let us look at a simple example to illustrate the two indices. Suppose that the average time it takes to complete the blood work for a county-run hospital laboratory is 28 minutes ($\mu = 30$), with a standard deviation of two minutes ($\sigma = 2$).

The standard time for the work is 27 minutes, with an upper specification limit of 30 minutes ($USL = 30$) and a lower specification limit of 24 minutes ($LSL = 24$). The administration wants to know if the laboratory is capable of meeting the performance standard.

We can now apply the information to both measures, starting with the capability ratio:

$$C_p = \frac{USL - LSL}{6\sigma}$$

$$= \frac{30 - 24}{(6)(2)}$$

$$= 0.50 \text{ minutes}$$

which is less than 1. Since the ratio is less than 1, it does not meet the target specification. In other words, the process is below the minimum capability.

Next, we apply the information to the capability index:

$$C_{pk} = \min\left[\frac{USL - \mu}{3\sigma}; \frac{\mu - LSL}{3\sigma}\right]$$

$$= \min\left[\frac{30 - 28}{(3)(2)}; \frac{28 - 24}{(3)(2)}\right]$$

$$= \min[0.333; 0.67]$$

$$= 0.333 \text{ minutes}$$

which is the lower of the two. Since there are two choices, the rule of thumb is to consider the lower (minimum) value. Unfortunately, both choices came out to be less than 1, indicating that it does not meet the target specification. In other words, it is below the minimum capability. Interestingly, both measures, C_p and C_{pk}, indicate that the process fails to meet the minimum expectations of quality.

Other Measures of Quality Control

In addition to the control charts discussed above, there are other measures that one can also use to ensure quality control in an organization. The number of these measures has been steadily increasing over the years; in particular with recent interest in quality control as a means for increasing productivity. To go over the entire breadth of the measures developed to date is beyond the scope of this chapter. However, a number of these measures, although less analytical in nature than those discussed here, are worth mentioning because of their increasing popularity with practitioners. Important among them are check sheets, histograms, flow charts, Pareto charts, cause-and-effect diagrams, and scattergraphs, which, together with the control charts, constitute what some would call the basic seven tools of quality control (Montgomery, 2005; Tague, 2004).

Check Sheets. Check sheets are simply charts used for gathering both quantitative and qualitative data. The idea is to start with an exhaustive set of data and gradually eliminate through a descriptive process the ones that are not relevant to the problem. With easy availability of computers, check sheets can be easily maintained electronically and updated periodically to ensure thoroughness of the process. Since the data are checked against sources, manner of collection, their relevance to the problem, and so forth, they are often called tally sheets.

Histograms. Histograms provide a visual picture of the characteristics of a population based on a sample drawn from the same population. They are essentially frequency distributions of the data, presented in graphical form. Besides providing a visual picture of the characteristics of the variables being measured, they can also tell us if the distribution of the data approximates a bell (normal) curve, if it is positively or negatively skewed, the nature of the outliers, and so forth.

Flow Charts. Flow charts break down a problem into its constituent components; for instance, a process into a set of subprocesses in a sequential manner. Breaking down a process into a sequence of subprocesses makes it possible to analyze the subprocesses in greater detail, as well as helps us understand the causes of the problem and to find appropriate solutions for a subprocess that causes the problem. Construction of flow charts involves using symbols such as an oval or circle (to indicate a beginning or end), an arrow (to indicate the direction of flow), a square or rectangle (to indicate a process), a diamond (to indicate a decision), and so forth.

Pareto Charts. Named after the famous nineteenth-century Italian economist and sociologist Vilfredo Pareto, these charts are constructed using both bars and line graphs. The bars represent frequencies of data such as cost or income, while the lines represent the cumulative total. The bars are presented in descending order, with the tallest on the left and the shortest on the right, so that one is able to make a visual comparison of the factors that are more important than others as one moves from the left to the right. The construction of the charts is based on a principle called the Pareto principle, which says that one only needs to look at a small percentage of sources to determine the cause of the problem. For instance, we can look at 20 percent of the data or sources to determine that they cause 80 percent of the problem.

Cause-and-Effect Diagram. Also called the fishbone chart, this was developed by Kuoru Ishikawa in 1943 to determine the cause-and-effect relationship in a problem. It is called a fishbone diagram because it resembles a fish skeleton with a head, a spine, and bones (branches) connected to the spine. The method for using the diagram is simple: define the goal or problem to be solved at the head, then fill in the branches with factors that cause the problem through a process of brainstorming until the best possible solution is found to address the problem. The factors typically include labor, materials, equipment, and procedures.

Scattergraphs. These are primarily used to show the cause-and-effect relationship between two variables, an independent variable (cause) and a dependent variable (effect), on a two-dimensional plane. The relationship is presented as a set of coordinates on the plane, with the dependent variable on the vertical axis and the independent variable on the horizontal axis. While the scattergraphs are extremely useful in providing an initial impression of the cause-and-effect relationship, they do not provide a precise measure of the relationship; for the latter, one needs to use methods such as correlation, regression, and other statistical measures.

In addition to the above, there are other measures that are also used in quality control, such as stratification (breaking down a whole category into smaller, related subgroups to identify possible causes of a problem), benchmarking (adopting the best practice in order to excel in a given job or activity), reengineering (redesigning an existing process to improve the service costs and quality attributes), and gap analysis (developing an understanding of goods and services provided from multiple perspectives), to name just a few; the list continues to grow. Interestingly, Kuoru Ishikawa in 1968 formally introduced many of these charts and diagrams to solve problems involving large-scale industries, referred to them as part of the basic tools of quality control.

Costs in Quality Control

The primary objective of quality control in any organization, public or private, is to reduce the costs of its operation. If control efforts do not lead to any savings in costs or cost reduction then, in principle, there is no need for quality control. However, for most control activities, cost savings are real and they remain a principal objective of quality control for most organizations.

Three types of costs are generally associated with quality control: assessment costs, prevention costs, and nonconformance costs. Assessment costs are costs an organization incurs in measuring quality characteristics to ensure that they conform to quality standards. These costs typically include the costs of inspection, labor as well as materials, costs of approval or certification when organizations meet quality standards, and so on. Prevention costs involve costs when organizations take measures to prevent poor quality of products or performance. Examples of prevention costs include the costs associated with quality planning, design, and development of quality measurement instruments, quality training, and so on. Since the assessment and prevention costs are incurred by an organization to assure or conform to quality, they are often called conformance costs. Finally, nonconformance costs, also called failure costs, occur when an organization fails to meet quality standards. This may be due to poor quality of labor, materials, and overhead (i.e., expenses accumulated for a service up to the point when nonconformities are detected).

Although exact statistics are not available, nonconformance costs constitute the lion's share of quality costs for an organization. A rule of thumb breakdown would put it at about 50 percent, with the rest divided between assessment and prevention costs. In general, there is an inverse relationship between prevention costs and the costs of assessment and failure: when prevention costs increase, they lead to better quality conformance which, in turn, reduces assessment costs. With lower assessment costs, the costs of nonconformance are also expected to decrease. Figure 11.2 shows a diagrammatic presentation of how improvements in quality can reduce the overall cost and that reduction is optimal at the point where the conformance and nonconformance costs intersect.

Interestingly, when both conformance and nonconformance costs are present in a good or service provision, the optimal or minimum cost will be at the lowest point on the total quality cost curve, obtained by vertically adding the nonconformance and conformance cost curves.

Measures of Quality Costs

Decision makers are often interested in comparing quality costs of one time period with that of another to assess the overall quality of a good or service. Since the services provided vary both in costs and quantity over time, it is not always possible to compare quality costs in absolute dollar terms. Consequently, one needs to relate the cost data for a given period to a base that measures the degree of activity for that period similar to the indirect cost allocation method, discussed earlier.

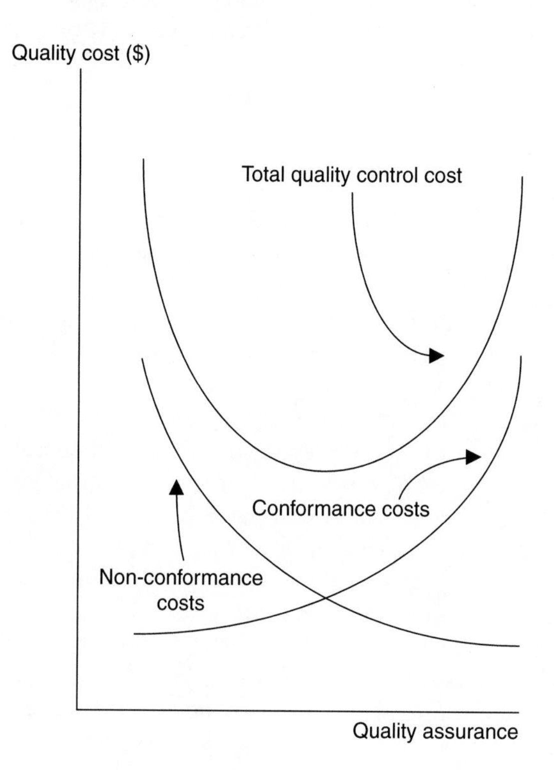

Figure 11.2 Optimal Quality Cost

Use of Direct Labor as a Base

To illustrate the point above, think of a situation in which the cost of direct labor is proportional to the quantity of goods produced or service provided. Let us say that 25 units of direct labor are used to produce 75 units of a good, say, X, 40 units of direct labor are used to produce 120 units of a good, say, Y, and so on. Thus, using direct labor as a base, quality costs can be expressed as a percentage of direct labor. The advantage of using direct labor as a base is that by treating quality costs as a percentage of the base, it can avoid the effect the variation in quantity could have on the quality cost measure.

Since the base one uses is critical to cost comparisons, it is important that it responds to changes in quantity and is not impacted by extraneous factors. To give an example, consider a situation in which for two time periods, t_1 and t_2, quality costs and the quantity of a service provided by a government remain the same, say, \$200 of quality costs for 150 units of the service provided for each of the two time periods. Assume that direct labor is the base and it costs the government \$5,000 in direct labor at time t_1 and \$7,500 at time t_2. Therefore, based on percentage of direct labor as a measure, quality cost will be 4 percent [(\$200/\$5,000) × 100=4.0] at time t_1, and 2.67 percent [(\$200/\$7,500) × 100=2.67] at time t_2. The difference in quality cost between the two time periods can thus be attributed to some extraneous factors that may have affected the base that did not exist at time t_1. If the extraneous factors did not intervene, the labor costs would have been the same for both time periods.

Other Measures of Base

The use of direct labor as a base raises an interesting question: why not use the units of goods and services as the base instead of direct labor? The problem with this is that it can unnecessarily complicate the process. The complication arises due to the fact that when one evaluates the performance of an organization at the end of an accounting period, one must not only take into consideration the units of goods produced or service provided throughout the year, but also the units that are partially completed. In other words, one must consider both in order to have a true measure of the cost involved: completed as well as partially completed units. This, in turn, makes cost calculations somewhat difficult, as we have seen in the chapter on cost accounting under cost allocation. The calculation becomes even more complicated if product mixes are involved in the production or delivery process. The use of direct labor as a base avoids some of these problems.

Although direct labor is considered useful as a base, there are other types of bases such as cost per equivalent unit, cost as a percent of total revenue, and so on, that can also be used as a quality cost measure. In reality, what is important to remember is that whatever one uses as a base, it must be consistent, be able to produce results that are comparable both over time and across space, and that it is free of biases that can result from variations in cost and quality. One way to ensure that the base one uses is comparable and that it will produce consistent results is to plot them on a two-dimensional plane, with quality costs on the vertical axis and time on the horizontal axis. If the plots produce parallel lines for the bases being compared (i.e., if they move in the same direction), it is an indication that they are comparable and will produce results that will be consistent over time.

Opportunity Costs in Quality Control

In addition to the three categories mentioned above, there is a fourth category of cost that deserves some attention here, especially when dealing with quality costs. It is called the cost of lost income or revenue resulting from poor perception (reputation) in the market of the goods and services an organization provides. For instance, when a government provides a service such as recreation or healthcare that is of much lower quality than what is available in the market, it will affect the perception of the public and, therefore, of the quantity consumed even though the price at which it is offered may be lower than the price of its competitors (assuming that price is not a factor). This type of cost is generally known as opportunity cost, mentioned earlier. Although important as a cost concept, there is a problem with it in that no one quite knows how to accurately measure the amount of revenue lost based on "what if the perception would have been different?" In other words, it is almost impossible to find data based on estimates of "what could have been" that will be reliable enough to draw accurate conclusions on lost revenue.

Several other factors also add to this complication, including some that are directly related to current accounting practices. For instance, conventional accounting practice does not provide enough guidance regarding how to account for lost benefit of added market share (where a government has to compete with the private sector to provide services) or the value of added quantity in moving an organization further along the learning curve. In spite of this difficulty, an organization must have some knowledge of the costs it would incur, even how rudimentary that may be, due to lost revenue from poor perception. One way to deal with the problem is to develop some changes in the current accounting practices consistent with changing conditions in the market, or the economy in general.

A Note on Total Quality Management

A term that frequently appears alongside quality control is total quality management, or TQM. In fact, TQM is the culmination of the quality control movement that started with the work of Walter Shewart in the 1920s. It was subsequently expanded by Deming in the 1940s, Juran in the 1950s (1951), Feigenbaum in the 1950s and 1960s (1961), Crosby in the 1970s (1979), and Teguchi (1962) and Ishikawa (1968) in the 1960s, although their works received more attention in the 1970s, 1980s, and 1990s.

Deming's work is particularly noteworthy because it recognized for the first time that management should be an integral part of quality control, not just the workers who were traditionally held responsible for poor quality, suggesting that about 15 percent or so of the error can be attributed to workers and the rest to the system and process variation. Deming outlined his philosophy on quality in his famous "14 points" that to this day serve as the guideposts for quality improvement. These include, among others, consistency of purpose, frequent inspection, continuous improvement, on-the-job-training, supervision and leadership, interaction within the organization, and recognition for hard work.

Expanding on the works of Shewart and Deming, Juran introduced his famous quality trilogy: quality planning, quality control, and quality improvement. He also emphasized the importance of quality costs. Feigenbaum's work deserves considerable attention because he introduced the concept of total quality control, which became the foundation for TQM. Crosby's approach to quality control was more on rigor and precision, suggesting that the job must be done right the first time and that there is no room for error, calling it a "zero defect" policy. Besides developing the cause-and-effect diagrams we mentioned earlier, Ishikawa is also credited with developing the concept of total quality control. Teguchi's work added a robustness dimension to quality control by suggesting that a product must be able to withstand varying work conditions. Like Juran, his work also emphasized the importance of quality costs.

TQM attempts to integrate the various dimensions into a coherent process within an organization to ensure continuous improvement in the quality of goods and services an organization provides. The end result of TQM is customer satisfaction, which, from the perspective of government, is the public whose opinion on how satisfied they are with the services they receive determines how well a government is performing.

Chapter Summary

Quality control deals with methods, tools, and approaches to diagnose the level of quality, or lack thereof, in a process. Although it is difficult to determine precisely what quality means for many of the goods and services a government provides, it is possible to identify specific areas of service operation where one could use control measures to assure quality. There is a wide range of such measures that one can use for this purpose. The most frequently used measure is a control chart. This chapter has presented several different charts for measuring quality control; in particular, those related to process control. Most of these charts are simple enough that one should not have any difficulty in constructing them. However, two things must be carefully addressed when using the charts: first, the data must be reliable to produce useful results; and second, efforts must be made to identify specific factors that affect quality. In most instances, these factors are preventable, meaning that if properly identified they can be removed from the process before they can cause major damage.

Notes

1 When we are dealing with a dataset in a sample that is normally distributed (i.e., has a bell-shaped distribution), we frequently explain the spread of the data relative to the standard deviation, σ, from the mean, u. We know from our elementary knowledge of inferential statistics that approximately 68 percent of the data in a normal distribution lie within $\pm 1\sigma$ of u, 95 percent within $\pm 2\sigma$ of u, and 99 percent within $\pm 3\sigma$ of u. If the dataset is not normally distributed, Chebychev's inequality provides an alternative explanation of what percentage or fraction of the data will lie within K standard deviations from the mean, given by the expression $1-1/K^2$. Thus, if $K=2$, then we can say that 75 percent of the data will lie within two standard deviations from the mean; that is, $1-1/K^2=1-1/2^2=1-1/4=1-0.25=0.75$; if $K=3$, then 89 percent of the data will lie within three standard deviation from the mean; that is, $1-1/K^2=1-1/3^2=1-1/9=1-0.11=0.89$, and so forth. To give an example, suppose that the performance of students in a government managerial accounting class has a mean of 85 with a standard deviation of 5. Now, expressing the problem in terms of Chebychev's inequality, we can say that 75 percent of the students in the class have scores that are within two standard deviations of the mean. Translated, this means that 75 percent of the students have scores that lie between 75 and 95; that is, $u \pm 2\sigma = 85 \pm 2 \times 5 = 85 \pm 10$.

2 The 3σ control limits for the X-bar chart can be formally expressed as

$$E(\bar{X}) \pm 3SE(\bar{X}) = \mu \pm 3\sigma / \sqrt{n} = \mu' \pm / k\sigma' \quad (1)$$

where $E(\bar{X})$ is the expected value of X-bar (i.e., the mean of X-bar, $\bar{\bar{X}}$), $SE(\bar{X})$ is the standard error of X-bar, $\sigma/\sqrt{n}$ (i.e., standard deviation of a sampling statistic such as the sampling mean, which is the mean of a sampling distribution based on repeated samples), k is a constant representing $3/\sqrt{n}$ that changes with sample size n, and u and σ are population parameters (i.e., population mean and standard deviation).

From the above, and assuming that we already know the population parameters, u' and σ', we can further write the 3σ control limits for the X-bar chart as

$$UCL_{\bar{X}} = \mu' + k\sigma' \quad (2)$$

$$LCL_{\bar{X}} = \mu' - k\sigma' \quad (3)$$

where UCL and LCL are the upper and lower limits of control of X-bar, and the rest of the terms are the same as before.

3 In fact, the ASTM published one of the most authoritative tables in 1951 that provides information on various factors used in quality control analysis with different samples that to this day remains the industry standard.

4 When we do not know the population parameters u and σ, we can use the estimates of u and σ from observed data to determine the control limits for the X-bar chart. That is:

$$\bar{\bar{X}} \pm \left(3\frac{\bar{R}}{d_2}\right)\frac{1}{\sqrt{n}} \quad (4)$$

$$= \bar{\bar{X}}2 \pm \left(\frac{3}{d_2\sqrt{n}}\right)\bar{R}$$

$$= \bar{\bar{X}} \pm A_2\bar{R} \quad (5)$$

where Equation (4) is an estimate of $\hat{u} \pm 3\hat{\sigma}/\sqrt{n}$, with $\bar{\bar{X}}$ representing the estimate of $\hat{\mu}$, $\bar{R}/d_2$ representing the estimate of $\hat{\sigma}$, d_2 is a constant that changes with sample size n, and A_2 is a constant representing $3/d_2\sqrt{n}$. Since d_2 changes with the sample size, A_2 also changes with the sample size.

5 To compute the 3σ control limits for the R chart, we start with the expression

$$\hat{\mu}_R \pm 3\hat{\sigma}_R \quad (6)$$

$$= E(R) \pm 3\hat{\sigma}_R \quad (7)$$

where $E(R)$ is the expected value of R (i.e., the mean of R, $\overline{R}$). Note that $E(R)$ can be estimated from R and σ_R from the relationship

$$\hat{\sigma}_R = kE(R) = k\overline{R} \tag{8}$$

where k is a constant that changes with the sample size.

From the above, we can formally write the upper and lower limits of control for the R chart as

$$UCL_{\overline{R}} = \overline{R} + 3k\overline{R} = (1+3k)\overline{R} = D_4\overline{R} \tag{9}$$

$$LCL_{\overline{R}} = \overline{R} - 3k\overline{R} = (1-3k)\overline{R} = D_3\overline{R} \tag{10}$$

where D_4 and D_3 are constants representing $1+3k$ and $1-3k$, respectively.

6 Since it is a binomial variate, we can formally write the distribution of d as:

$$E(d)=nP \text{ and } Var(d)=nPQ \tag{11}$$

where $E(d)$ is the expected value of d (i.e., number of nonconformities), n is the sample size, P is the proportion of nonconformities, $Var(d)$ is the variance of d, and $Q=1-P$.

From the above, with slight algebraic manipulations, we can write

$$E(p) = E\left(\frac{d}{n}\right) = \frac{1}{n}E(d) = P \tag{12}$$

$$Var(p) = Var\left(\frac{d}{n}\right) = \frac{1}{n^2}Var(d) = \frac{1}{n^2}(nPQ) = \frac{PQ}{n} \tag{13}$$

where $E(p)$ is the expected value of p, and $Var(p)$ is the corresponding variance.

7 If we can assume that we know the population proportion, P, then the control limits for the p chart can be written as:

$$E(p) \pm 3SE(p) = P \pm \sqrt{PQ/n} \tag{14}$$

where $E(p)$ is the expected value of p, $SE(p)$ is the standard error of p, and the rest of the terms are the same as before.

From the above, and assuming that we know the population parameter, P, the 3σ control limits for p chart we can easily write the upper and the lower limits of the p chart, respectively, as:

$$UCL_p = P' + 3\sqrt{PQ/n} \tag{15}$$

$$LCL_p = P' - 3\sqrt{PQ/n} \tag{16}$$

8 If we do not know the population parameter (i.e., the population proportion) we can always estimate it from the observed data such as the number of nonconformities (d) in a sample. That is:

$$\overline{p} = \sum_{j=1}^{m} d_j / \sum_{j=1}^{m} n_j = \sum_{j=1}^{m} n_j P_j / \sum_{j=1}^{m} n_j \tag{17}$$

where $\overline{p}$ is the mean of the fractions of nonconformities, d_j is the number of nonconformities in the jth sample, n_j is the sample of size n_j (for $j=1,2,3\ldots,m$).

It is worth noting that $\overline{p}$ is the unbiased estimator of P, since

$$E(\overline{p}) = \sum_{j=1}^{m} d_j / \sum_{j=1}^{m} n_j = \sum_{j=1}^{m} n_j P_j / \sum_{j=1}^{m} n = P \tag{18}$$

where $E(\overline{p})$ is the expected value of p.

9 The 3σ control limits for c chart, based on the notion that c is a Poisson variate; Equation 13.28 can be written as:

$$UCL_c = E(c) + 3\sqrt{Var(c)} = \lambda + 3\sqrt{\lambda} \quad (19)$$

$$LCL_c = E(c) - 3\sqrt{Var(c)} = \lambda - 3\sqrt{\lambda} \quad (20)$$

where the *UCL* and *LCL* are, respectively, the upper and lower limits of control and the rest of the terms are the same as before.

References

ASTM. (1951). *Manual on quality control of materials*. Philadelphia, PA: American Society for Testing Materials.

Crosby, P. B. (1979). *Quality is free*. New York: New American Library.

Deming, W. E. (1950). *Elementary principles of the statistical control of quality*. Tokyo: Nippon Kagaku Gijutsu Renmei.

Farnum, N. R. (1994). *Modern statistical quality control and improvement.* Belmont, CA: Duxbury Press.

Feigenbaum, A. V. (1961). *Total quality control.* New York: McGraw-Hill.

Grant, E. L., & Leavenworth, R. S. (1996). *Statistical quality control.* New York: McGraw-Hill.

Ishikawa. K. (1968). *Guide to quality control.* Tokyo: JUSE.

Juran, J. M. (1951). *Quality control handbook.* New York: McGraw-Hill.

Kilian, C. S. (1992). *The world of Edwards Deming.* Knoxville, TN: SPC Press.

Montgomery, D. (2005). *Introduction to statistical quality control.* Hoboken. NJ: Wiley & Sons.

Ott, E. R., & Schilling, E. G. (1990). *Process quality control.* Milwaukee, WI: Quality Press.

Tague, N. R. (2004). *Seven basic quality tools: The quality toolbox*. Milwaukee, WI: American Society for Quality.

Teguchi, G. (1962). Studies on mathematical statistics for quality control. Doctoral dissertation, Kushu University, Japan.

12 Beyond Cost and Optimization

This final chapter focuses on several topics that do not belong to cost and optimization in the conventional sense of the term. Nevertheless, they are important in that they allow us to understand that there are factors, behavioral and others, that one must take into consideration to successfully utilize or implement analytical tools such as those discussed here. It does not behoove us to have an idea or a method of solution if it cannot be successfully implemented. Implementation lies at the core of a decision process; it is the means that bridges the gap between theory and practice, between knowledge and the utilization of that knowledge, and between the skill one possesses and making the best use of that skill to solve real-world problems. In its bare essence, implementation means getting a job done (Jones, 1982), but implementation is not an easy task. What makes implementation difficult is the recognition that there are factors, internal as well as external, that an organization needs to consider to successfully carry out the tasks of implementation.

This chapter focuses on several factors that one would find useful during an implementation process: they are partly economic, partly empirical, and partly legal. The chapter also focuses on the role factors such as contingency provision, open communication, and ethical conduct in implementation. To a large measure, the success of an organization depends on the ability, as well as the commitment of those who run the affairs of an organization, to understand and appreciate the role these and other factors play in the implementation process.

Economic Rationale

Implementation costs time, money, and other resources. Even simple and everyday concepts such as buy or lease, managing inventory, and assuring quality in service provision require skill to design, time to analyze, and money to support the process that will eventually utilize them. However, time, skill, and money are scarce resources that must be carefully utilized to realize the goals of an organization. Successful implementation requires that one must fully understand and evaluate the costs and benefits of implementation. In other words, where possible, one must perform formal analyses using methods such as benefit–cost or cost-effectiveness analysis to determine if the benefits of implementation outweigh the costs (Boardman, Greenberg, Vining, & Weimer, 2001) Clearly, for many of the topics discussed throughout the book, including those that have not been discussed, the return must be greater than the cost to justify their use in dealing with a specific problem before they could be formally implemented.

Empirical Relevance

While cost remains an integral part of implementation, it is also a function of the type of knowledge and skill being used to solve a specific problem. For instance, analytical tools that are simple and require minimum information will understandably cost less than those that are complex and require extensive information because they will take less time to design, program, analyze, and finally to implement. At the same time, tools that are simple may not be able to sufficiently capture the complexities of the real world. This does not mean that complex and analytically sophisticated methods and tools are better equipped to deal with real-world problems than those that are simple. What it means is that, regardless of their level of complexity, all methods, tools, and techniques must have certain characteristics that will make them appealing to the decision makers, as well as to those who will implement them. At a minimum, they must be valid, have utility (i.e., serve a useful purpose), and be robust.

Validity means the ability of a method or tool to produce results that better reflect the reality an organization is trying to address. For instance, a method that is unrealistic but elegant will have very little appeal to a decision maker; therefore, it has very little chance of being implemented. Validity also means that it must be empirically relevant in that it must be able to address a problem that is real and can provide practical and effective solutions. Utility, on the other hand, is the value a decision maker places on a method, tool, or technique. To be of any value to a decision maker vis-à-vis to the organization, it must be able to tell the decision makers not only where the problem lies, but also how to fix the problem and the benefit it will produce in the long run when implemented.

Finally, a good method, tool, or technique must be able to deal with problems under changing conditions. In other words, it must be robust to withstand the changing conditions. Decision makers deal with not one, but multiple different problems with varying degrees of complexity every day. They do not have unlimited time or resources to develop new methods, tools, or techniques each time they have to deal with a new problem. Therefore, to be effective, the methods must be flexible to meet the changing needs of an organization. In general, the more robust (i.e., flexible) a method, the better is its ability to deal with problems under changing conditions; with more flexibility, the cost of implementation will also be low.

Legal Considerations

Although cost managers and quantitative analysts such as management scientists are not legal experts, their work can involve legal matters. Consequently, some basic knowledge of legal procedures can be helpful. Perhaps the most important legal procedure is a contract. A contract is a binding commitment between two or more parties involved in an exchange relationship. Most organizations, including government, are required by law to sign a contract agreement before getting into an obligation. Therefore, contracts must be carefully worded to ensure they protect the rights of the parties in the event of any breach of agreement.

Cost is an important element in any contract agreement. Four types of costs are generally associated with legal contracts: fixed cost contracts, open cost contracts, shared cost contracts, and unit cost contracts. In a fixed cost contract, an individual or a firm providing a service, say, to a government, agrees to provide the service at a fixed cost. This includes the actual cost of the service plus any profit the individual or the firm will make on the service. Since government contracts are generally awarded through competitive bidding, the profit is included in the bids. According to this contract, the risk always rests with the bidder, not the government. For instance, if a bid is below the cost of the service, then it is the bidder who must absorb the cost of underbidding. Specification also plays an important role in this type of contract. The bidder, for instance, can be held responsible for failing to provide the service within the specifications. If specifications are not clearly laid out, the responsibility shifts from the bidder to the government (i.e., the contractee).

In an open cost contract, the actual cost of the service is usually not known until it is completed. This can happen in situations in which the service is unique and the government had no prior experience. Since the exact cost cannot be determined beforehand, the service specifications must be spelled out in clear terms to avoid any potential conflict in the future. However, to make sure that the final cost does not go out of proportion, the contract may specify limits for some of the components of cost. Unlike a fixed cost contract, in an open contract the bidder does not have the same responsibility because of the difficulty in determining the final cost in advance. Instead, it is the government who bears a greater share of the risk in the event that there is a cost overrun. Defense contracts are among the best-known examples of this.

In a shared cost contract, on the other hand, the risk is shared by both parties: the bidder and the government. While there are service specifications in this type of contract, the bidder can deviate from the contract provided that no deviation occurs once the job starts, or the service is on the way.

Finally, in a unit cost contract the bidder is paid according to the units or quantities of work done, rather than the job as a whole. The total cost of a contract is determined by multiplying the total units of work by the rate per unit. For instance, if it takes a bidder 300 hours of labor to complete a job and it costs $25 per hour for labor, the total cost to the government of the job will be $7,500 [$\$25 \times 300 = \$7,500$]. Since the bidder is paid on a per unit basis, neither party is held accountable for any possible deviation from contract agreement.

Contingency Provisions

In common parlance, contingencies (also called backup or fall-back options) mean when everything else fails. They are alternatives to deal with uncertainties facing an organization or deviations from a planned goal or objective. Contingencies are necessary because, in spite of our best efforts, things can go wrong. If Murphy's Law teaches us anything, it is that if things can go wrong, they will, but this does not mean that organizations should not be prepared to deal with the consequences of an event that could not have been predicted with certainty; they should. Prudent management requires that preparedness is a much better alternative than doing nothing, especially when potential exists for incurring costs resulting from uncertain events in the future.

The question, then, is: How does one prepare for events that cannot be predicted? There are no hard-and-fast rules for dealing with contingencies. Organizations should develop their own strategies, depending on the resources they have and their ability to command those resources at will when such contingencies become necessary. While this may be true as a general rule, it is important from an operational perspective that organizations develop plans ahead of time to deal with the consequences of unpredictable events. In designing such plans, the fundamental rule is not to consider one but several scenarios that are likely to occur and suggest means for dealing with each one of them.

Scenarios are not wild guesses. They are the results of constructive thoughts, ideas, and exchanges produced through extensive brainstorming by individuals who are recognized as "experts" on problems and issues under consideration. Regardless of who participates in this process, it is important to recognize that scenarios have certain basic characteristics that must be taken into consideration when constructing them. They must be plausible, internally consistent, and they must include all relevant factors. Together, they must serve as the guiding principles for scenario construction (Starling, 1979).

Ideally there are no limits as to the number of scenarios one can construct, but there are costs associated with their implementation. Obviously the more scenarios there are, the more demanding will be the time and resources that will be needed to implement them. No organization has unlimited resources that can deal with all the scenarios that can be realistically constructed. The alternative, therefore, is to restrict them to a few that are plausible or highly likely.

Communication Needs

The primary purpose of government is to serve the needs of the public. Therefore, any decision a government makes has a direct effect on the public and its welfare. To be effective, the decision makers in an organization, particularly in government, must give due consideration to the collective implications of their decisions. A good government must keep the public informed of any decision it makes, whether it is introducing a new idea or implementing a new policy, and how that decision will affect them directly or indirectly. In other words, there must be an open and honest communication between a government and the public it serves.

When a government keeps the public informed of its decisions, it is natural to expect that there will be some public response to those decisions, even if they do not affect them directly. It is the essence of open communication in a civil society. A government communicates with public formally, as well as informally. Formal communication takes place when a government is required either by law, or by precedence to inform the public of its decisions on specific issues that concern them. A good example of formal communication would be a public referendum, where a government must have public support to pass new legislation on a major issue, such as putting a freeze on tax increases, issuing new bonds, or getting rid of an unpopular program. In contrast, informal communication takes place when a government takes initiatives on its own to keep the public informed of its decisions and other activities. Public participation in this case is voluntary.

Communication, therefore, is absolutely necessary to generate support for implementing a new idea, introducing a new system, or dismantling an existing one. The more successful a government is in communicating with the public or the community it serves, the better are its relations with the public and the more effective it will be in receiving their support. Open communication is a *sine qua non* for good government. It is important for an organization to maintain an environment that is conducive to open communication (Wheeler, 1994).

Communication must also be internal, within the organization. The old cliché that all good deeds must start at home also applies to government, which encourages and maintains open lines of communication within its own confines. Good and effective communication, whether it is formal or informal, produces a sense of belonging to those who may otherwise feel left out of the process that affects them, as well as those they serve. The result of such belonging can have profound effects on employee motivation and performance. From the perspective of good government, both are important in realizing common organizational objectives such as reducing waste, promoting work incentives, and improving productivity.

Ethical Conducts

All organizations, public as well as private, operate under a set of ethical codes that frequently exceed legal obligations. Individuals working in an organization are expected to abide by these codes. The first code of ethics for government employees was introduced for local government in 1924 by the International City Management Association (ICMA) as part of a trend that started with the progressive reform movement toward the end of the nineteenth century, to prevent corruption and nepotism in government. In 1958, Congress introduced similar codes of ethics for federal employees, and 20 years later expanded the codes and established the Office of Government Ethics as part of the Ethics in Government Act of 1978 (Henry, 1999). A few years later, the office published the first comprehensive set of codes on ethical standards for public employees.

As noted previously, the American Society for Public Administration (ASPA) also developed a set of standards for practitioners some years ago. If history is any indication, these codes will go through many more changes and revisions in the years to come as the objective conditions administrators have to deal with change.

From an administrative point of view, ethical standards are organized into rules governing an individual employee's relationship with other employees in an organization, as well as with the public. Violation of these rules can not only impair the personal reputation of an individual employee, but also the image and prestige of the organization he or she serves. When an individual code of conduct affects an employee's immediate surroundings but not the entire organization or a majority of its employees, it is called microethics. If, on the other hand, such conduct affects a large number of individuals vis-à-vis the entire organization, it is called macroethics (Gortner, 1981). The latter is infinitely more complex because of the spillover effects it can have on the rest of the organization and the society at large. Examples of both micro- and macroethical problems abound at all three levels of government – local, state, and federal.

An important consequence of ethical misconduct is a decline in organizational productivity. Some organization management scholars believe that productivity is a state of mind that is reflected through ethical excellence in organizational performance (Brown, 1983). If this holds true, then productivity would require a constant infusion of professionalism in organizational management. Since professionalism generally means a commitment to ethical standards, any deviation from those standards will undoubtedly affect productivity. Put differently, there is a direct relationship between ethical behavior and organizational performance. An organization that observes ethical conducts in all its facets is likely to be more productive than an organization that lacks ethical norms and other rules of conduct.

Public employees, in particular public managers, bear an inordinate amount of responsibility in this regard because of the trust the public places in their ability to conduct themselves in spite of occasional breaches of professionalism. Violation of this trust can destroy the foundation that gives an organization the right (i.e., the legitimacy) to function in a civil society.

Future of Cost and Optimization

Organizations are not static or monolithic entities, as they were once perceived. They are constantly changing, forcing the decision makers to change and adopt measures to keep up with the changes as they occur. Increasing demand for goods and services, changing technologies, and rising expectations are placing additional demands on the management to become more responsible, as well as responsive, to those they serve. In government, where changes take place more gradually, one can expect to see the same kind of demand placed regularly on the decision makers to plan ahead, to keep up with changes as they occur. As organizations become complex, the need for increased reliance on cost and optimization will also increase. At the same time, more advanced tools and techniques will emerge to replace the existing ones. Combined with new advancements in computer technology and scientific knowhow, the emerging tools and techniques will provide decision makers with unlimited opportunities to deal with complex problems with greater ease.

Two factors are more than likely to speed up this evolution. One, the future decision makers, bureaucrats, and public administrators will be better trained than their predecessors, thereby making it easy for them to implement change. Their awareness and increasing familiarity with methods, tools, techniques, and modern-day technology will allow them to accept the changes more easily. Two, the availability of high-speed computers with enormous data-processing capabilities (many of which are already available) will make it even more possible to solve complex problems, especially those that are large and unstructured, with relative ease. However, these optimistic scenarios will become feasible only if the organizations concerned are willing to make the necessary changes that will become incumbent upon them as time progresses. As a necessary condition, the organizations must be willing to make sure that internal conditions, especially those that have a direct bearing on those who make decisions and those who carry them out, are suitable to serve the larger goals of the organization.

Chapter Summary

Organizational decision making is concerned with the functions and processes of an organization and how those processes are used to coordinate its resources in order to achieve its goals and objectives with maximum efficiency. Thus, there is a direct relationship between management, efficiency, cost, and optimization. As organizations become increasingly complex, their functions also become complex and their processes more exhaustive. The issues the decision makers are concerned with will also become more diverse. Along with efficiency and productivity, decision makers will be expected to deal with every conceivable aspect of management. Not only that, they will have to be prepared to deal with uncertainties over which they will have very little control.

This concluding chapter has briefly focused on several management issues that are integral to the decision process in an organization insofar as implementation is concerned. They are multidimensional in nature, ranging from simple economic rationale, to contingency provisions, to ethical and legal concerns. Additionally, there are other areas of management that are also important in effective functioning of an organization, particularly in government, such as employee motivation, conflict resolution inside bureaucracy, leadership role, and so forth. Although organizations may not fully utilize many of the cost and optimization tools discussed here, familiarity with them can significantly affect the way in which management makes decisions and the manner in which it tries to implement them.

References

Boardman, A., Greenberg, D., Vining, A., & Weimer, D. (2001). *Cost–benefit analysis: Concepts and practice.* Upper Saddle River, NJ: Prentice Hall.

Brown, D. S. (1983). The managerial ethic and productivity improvement. *Public Administration Review, 36*, 223–250.

Gortner, H. J. (1981). *Administration in the public sector.* New York: Wiley & Sons.

Henry, N. (1999). *Public administration and public affairs.* Upper Saddle River, NJ: Prentice-Hall.

Jones, C. E. (1982). *An introduction to public policy.* New York: Brooks and Cole.

Starling, G. (1979). *The politics and economics of public policy.* Homewood, IL: The Dorsey Press.

Wheeler, K. M. (1994). *Effective communication: A local government guide.* Washington, DC: International City Management Association (ICMA).

Bibliography

Aguila, P. R., Jr., & Petersen, J. E. (1991). Leasing and service contracts. In J. E. Petersen & D. R. Strachota (Eds.), *Local government finance* (pp. 321–338). Chicago, IL: Government Finance Officers Association (GFOA).

Ammons, D. N. (1991). *Administrative analysis for local government: Practical application of selected techniques.* Athens, GA: Carl Vinson Institute of Government, University of Georgia.

Anderson, L. G. (1977). *Benefit–cost analysis: A practical guide.* Lexington, MA: Lexington Books.

Anderson, L. K. (1992). *Managerial accounting.* Cincinnati, OH: Southwestern Publishing Company.

Apostolou, N. G., Brooks, R. C., & Bartley Hildreth, W. (1992). Research and trends in governmental accounting and reporting. *International Journal of Public Administration, 15*(5), 1121–1150.

Bertsekas, D. P. (1995). *Dynamic programming and optimal control.* Belmont, MA: Athena Scientific.

Bingham, R. D. (Ed.). (1991). *Managing local government: Public administration in practice.* Thousand Oaks, CA: Sage.

Brickman, L. (1989). *Mathematical introduction to linear programming and game theory.* New York: Springer-Verlag.

Briers, M. S., & Hirst, M. (1990). The role of budgeting information in performance evaluation. *Accounting, Organization and Society, 15*(4), 373–398.

Brinkerhoff, R., & Dressler, D. (1990). *Productivity measurement: A guide for managers.* Newberry Park, CA: Sage.

Brooks, R. C., & Pariser, D. B. (1996). Local government accounting and the need for audit follow up. In J. Rabin, W. Bartley Hildreth, & G. Miller (Eds.), *Budgeting: Formulation and execution* (pp. 525–531). Athens, GA: Carl Vinson Institute of Government, University of Georgia.

Brown, D. C. (1983). The management ethic and productivity improvement. *Public Productivity Review, 36*, 223–250.

Brown, K. W. (1993). The 10-point test of financial condition: Toward an easy to use assessment tool for smaller cities. *Government Financial Review, 9*(6), 21–26.

Brown, R. E., & Sprohge, H.-D. (1987). Governmental managerial accounting: What and where is it? *Public Budgeting & Finance, 7*(3), 35–46.

Brown, R. E., Myring, M., & Gard, C. G. (1999). Activity-based costing in government: Possibilities and pitfalls. *Public Budgeting & Finance, 19*, 3–21.

Busch, D. M. (1991). *The new critical path method: The state-of-the-art modeling and time-reserve management.* Chicago, IL: Probus Publishing Company.

Carr, D. K., & Litterman, I. (1993). *Excellence in government: Total quality management in the 1990s.* Arlington, VA: Coopers and Lybrant.

Chan, J. C. (1981). Standards and issues in governmental accounting and financial reporting. *Public Budgeting and Finance, 1*, 55–66.

Chappels, T. M. (1999). *Financially focused quality.* Boca Raton, FL: St. Lucie Press.

Christopher, W. F., & Thor, C. G. (1995). *The 16-point strategy for productivity and total quality.* Portland, OR: Productivity Press.

Coleman, A. M. (1995). *Game theory and its applications to the social and biological science.* Boston, MA: Butterworth-Heinemann.

Coltman, M. M. (1987). *Hospitality management accounting.* New York: Van Nostrand Reinhold.

Cooper, R. (1997). *Target costing and value engineering.* Portland, OR: Productivity Press.

Cullins, J. G. (1998). *Public finance and public choice.* Oxford: Oxford University Press.

Denardo, E. V. (1982). *Dynamic programming models and applications.* Englewood-Cliffs, NJ: Prentice-Hall.

Di Benedetto, C. A. (1986). *Game theory in marketing management: Issues and applications.* Cambridge, MA: Marketing Science Institute.

Dorst, R. B. (1986). *Introduction to linear programming: Applications, & extension.* New York: Marcel-Dekker.

Droms, W. O. (1986). *Finance and accounting for non-financial managers.* Reading, MA: Addison-Wesley.

Fallou, C. (1971). *Value analysis to improve productivity.* New York: Wiley Interscience.

Finkler, S. A., Ward, D. M., & Baker, J. J. (2007). *Essentials of cost accounting for health care organizations*. Boston, MA: Jones and Bartlett Publishers.

Forrest, E. (1996). *Activity-based management: A comprehensive implementation guide.* New York: McGraw-Hill.

Fountain, J. R., Jr. (1990). Future changes in financial reporting: A primer on GASB. *Intergovernmental Perspective, Fall*, 35–37.

Fowler, T. C. (1990). *Value analysis and design.* New York: Van Nostrand Reinhold.

Freeman, R. J., Shoulders, C., & Allison, G. (1988). *Governmental and non-profit accounting: Theory and practice.* Englewood-Cliffs, NJ: Prentice-Hall.

Friedman, M. (1996). Calculating compensation costs. In J. Rabin, W. Bartley Hildreth, & G. Miller (Eds.), *Budgeting: Formulation and execution* (pp. 265–275). Athens, GA: Carl Vinson Institute of Government, University of Georgia.

Garner, C. W. (1991). *Accounting and budgeting in public and non-profit organization.* San Francisco, CA: Jossey Bass Publishing.

Geiger, D. R. (2010). *Cost management and control in government: Fighting the cost war through leadership driven management.* New York: Business Expert Press, LLC.

Gibbous, R. (1992). *Game theory for applied economists.* Princeton, NJ: Princeton University Press.

Glick, P. E. (1990). *A public manager's guide to governmental accounting and financial reporting.* Chicago, IL: Government Finance Officers Association (GFOA).

Gore, A. J. (1993). *Creating a government that works better and costs less.* Washington, DC: US Government Printing Office.

Greenberg, L. (1973). *A practical guide to productivity management.* Washington, DC: Bureau of National Affairs.

Hackbart, M. H., & Ramsey, J. R. (1993). Public cash management: Issues and Practices. In T. D. Lynch & L. Martin (Eds.), *Handbook of comparative public budgeting and financial management* (pp. 293–312). New York: Marcel-Dekker.

Halachmi, A. (Ed.) (1999). *Performance & quality measurement in government: Issues and experiences.* Burke, VA: Chateline Press.

Hammel, F. C. (with P. G. Goulet) (1988). *Breakeven analysis: A decision-making tool.* Washington, DC: US Small Business Administration, Office of Business Development.

Harrison, A. (1997). *A survival guide to critical path analysis.* Boston, MA: Butterworth-Heinemann.

Hart, D. J. (1990). How activity accounting works in government. *Management Accounting, September*, 36–40.

Harvey, N. (2007). Use of heuristics: Insights from forecasting research. *Thinking and Reasoning, 13*(1), 5–24.

Havens, H. (1983). Integrating evaluation and budgeting. *Public Budgeting and Finance, 3*(2), 102–113.

Hay, L. E., & Engstorm, J. H. (1993). *Essentials of accounting for governmental and not-for-profit organizations.* Homewood, IL: Irwin.

Herbert, L., Killough, L. N., & Steiss, A. W. (1984). *Governmental accounting and control.* Monterey, CA: Brooks/Cole Publishing Company.

Herzlinger, R. E., & Nitterhouse, D. (1994). *Financial accounting and management control for non-profit organizations.* Cincinnati, OH: Southwestern Publishing Company.

Hey, J. D. (1991). *Experiments in economics.* Cambridge, MA: Blackwell.

Hood, C. (1995). The new public management in the 1980s: Variations on a theme. *Accounting, Organizations and Society, 20*, 93–109.

Horngren, C. T. (1996). *Introduction to management accounting.* Upper Saddle River, NJ: Prentice-Hall.

Ives, M. (1992). 25 years of state and local government financial reporting: An accounting standards perspective. *Government Accounting Journal, Fall*, 1–5.

Ives, M., Patton, T. K., & Patton, S. R. (2013). *Introduction to governmental and not-for-profit accounting*. New York: Pearson.

Jackson, A., & Lapsley, I. (2003). The diffusion of accounting practices in the new "managerial" public sector. *The International Journal of Public Sector Management, 16*, 359–372.

Kasozi, E. (2013). *Cost accounting management*. Munich: GRIN Verlag.

Kelley, J. T. (1984). *Costing government services: Guide for decision making.* Washington, DC: Government Finance Officers Association (GFOA).

Kelly, A. (2003). *Decision making using game theory: An introduction for managers.* Cambridge: Cambridge University Press.

Khan, A. (1997). Capital rationing, priority setting, and budget decisions: An analytical guide for public managers. In R. T. Golembiewski & J. Rabin (Eds.), *Public budgeting and finance* (pp. 963–974). New York: Marcel-Dekker.

Lapsley, I., & Mitchell, F. (Eds.) (1996). *Accounting and performance measurement.* London: Paul Chapman Publishing Limited.

Lee, D. R. (1987). On the pricing of public goods. *Southern Economic Journal, 49*(1), 99–105.

Lee, R. C. (1991). Life-cycle costing. In D. N. Ammons (Ed.), *Administrative analysis of local government: Practical applications of selected techniques* (pp. 100–105). Athens, GA. Carl Vinson Institute of Government, University of Georgia.

Levin, H. M. (1983). *Cost-effectiveness: A primer.* Beverly Hills, CA: Sage.

Li, S. (1996). *Analyzing efficiency and managerial performance: Using sensitivity scores of DEA models.* New York: Garland Publishers.

Mazmanian, D. A., & Sabatier, P. A. (1981). *Effective policy implementation.* Lexington, KY: Lexington Books.

McKenna, C. K. (1980). *Quantitative methods for public decision making.* New York: McGraw-Hill.

McMillan, J. (1992). *Games, strategies, and managers.* Oxford: Oxford University Press.

Meade, D. M. (2006). A manageable system of economic condition analysis for governments. In H. A. Frank (Ed.), *Public financial management* (pp. 383–419). New York: Taylor & Francis.

Mikesell, J. C. (2014). *Fiscal administration: Analysis and application for the public sector.* Belmont, CA: Wadsworth Publishing Company.

Miller, A. D. G. (2009). *Managerial cost accounting in federal government: Providing useful information for decision making*. Washington, DC: AGA CPAG.

Miller, G. J. (1993). Cost–benefit analysis. In M. Holzer (Ed.), *Public productivity handbook* (pp. 253–279). New York: Marcel-Dekker.

Mishan, E. J. (1982). *Cost–benefit analysis.* London: Allen and Unwin.

Moder, J. J. (1983). *Project management with CPM, Pert, and precedence diagramming.* New York: Van Nostrand Reinhold.

Montgomery, D. C. (1996). *Introduction to statistical quality control.* New York: Wiley & Sons.

Morris, P. (1994). *Introduction to game theory.* New York: Springer-Verlag.

Mudge, A. E. (1971). *Value engineering: A systematic approach.* New York: McGraw-Hill.

Mundel, M. A. (1985). *Motion and time study: Improving productivity.* Englewood Cliffs, NJ: Prentice-Hall.

Mushkin, S. (Ed.) (1972). *Public prices for public profits.* Washington, DC: Urban Institute Press.

Nahmias, S. (1993). *Production and operations analysis.* Homewood, IL: Irwin.

Nash, S. G. (1996). *Linear and non-linear programming.* New York: McGraw-Hill.

Oakland, J. S., Turner, M., & Oakland, R. (2014). *Total quality management and operations excellence: Text with cases*. New York: Routledge.

Oliver, L. (2000). *The cost management toolbox: A manager's guide to controlling costs and improving profits*. New York: American Management Association.

Pearce, D. W. (1983). *Cost–benefit analysis.* New York: St. Martin's Press.

Pierce, L. W., & Rust, K. I. (1991). Government enterprises. In J. E. Peterson & D. R. Strachota (Eds.), *Local government finance* (pp. 393–416). Chicago, IL: Government Finance Officers Association (GFOA).

Points, R. (1990). Recent developments in accounting and financial management in the United States. In A. Premchnd (Ed.), *Government financial management* (pp. 334–336). Washington, DC: International Monetary Fund.

Poister, T. H., & Streib, G. (1994). Municipal management tools from 1976 to 1992: An interview and update. *Public productivity review, 18*(2), 115–126.

Prichard, R. E. (1980). *The lease/buy decision.* New York: AMACOM.

Rayburn, L. G. (1996). *Cost accounting: Using a cost management approach.* Chicago, IL: Irwin.

Reed, B. J. (1997). *Public finance administration.* Thousand Oaks, CA: Sage.
Riahi-Belkaoui, A. (1986). *The learning curve: A management accounting tool.* Westport, CT: Quorum Books.
Richardson, K. (1998). The effect of public vs. private decision environment on the use of the net present value investment criteria. *Journal of Public Budgeting, Accounting, and Financial Management, 10*(1), 21–52.
Ringquest, J. L. (1992). *Multiobjective optimization: Behavioral and computational consideration.* Boston, MA: Kluwer Academic Publishers.
Rivenbark, W. C. (2005). A historical overview of cost accounting in local government. *State and Local Government Review, 37*(3), 217–227.
Robinson, M. (1998). Capital charges and capital expenditure decisions in core government. *Journal of Public Budgeting, Accounting, and Financial Management, 10*(3), 354–374.
Rubinstein, A. (1998). *Modeling bounded rationality.* Cambridge, MA: MIT Press.
Ryan, T. P. (1989). *Statistical methods for quality improvement.* New York: Wiley & Sons.
Shafritz, J. M., Russell, E. W., & Borick, C. P. (2013). *Introduction to public administration.* New York: Pearson.
Sierksma, G. (1996). *Linear and integer programming: Theory and practice.* New York: Marcel-Dekker.
Silver, E. A. (1998). *Inventory management and production planning and scheduling.* New York: Wiley & Sons.
Simpoon, R. (1997). *Managing finance and information.* Washington, DC: Pitman.
Smith, D. K. (1991). *Dynamic programming: A practical application.* New York: Ellis Horwood.
Smith, J. (1989). *Learning curve for cost control.* Norcross, GA: Industrial Engineering and Managing Press.
Solenberger, H. M. (1995). *Managerial accounting.* Cincinnati, OH: Southwestern (College) Publishing Company.
Sorensen, J. E. et al. (1996). Managerial accounting. In J. Rabin, W. Bartley Hildreth, & G. Miller (Eds.) *Budgeting: Formulation and execution.* Athens, GA: Carl Vinson Institute of Government, University of Georgia.
Starling, G. (1982). *Managing the public sector.* Homewood, IL: Dorsey Press.
Steiss, A. W. (1982). *Management control in government.* Lexington, MA: Lexington Books.
Sternberg, R. J., & Sternberg, K. (2012). *Cognitive psychology*. Belmont, CA: Wadsworth – Cengage Learning.
Stewert, J. D. (1984). The role of information in public accountability. In T. Hopwood & C.R Tomkins (Eds.), *Issues in public accounting* (pp. 14–25). Oxford: Phillip Allan.
Stout, R, Jr. (1980). Management, control, and decision. In R. Stout, Jr. (Ed.), *Management or control? The organizational challenge* (pp. 98–124). Bloomington, IN: Indiana University Press.
Stramatis, D. H. (1997). *TQM engineering handbook.* New York: Marcel-Dekker.
Swift, J. A. (1998). *Principles of total quality.* Boca Raton, FL: St. Lucie Press.
Swiss, J. E. (1991). *Public integration systems.* Englewood Cliffs, NJ: Prentice-Hall.
Tankersley, W. B., & Grizzle, G. E. (1994). Control options for the public management: An analytical model for designing appropriate control strategies. *Public Productivity Review, 8*(I), 1–18.
Tenner, A. R., & Detoro, I. J. (1992). *Total quality management.* Reading, MA: Addison Wesley Publishing Company.

Thai, K. V. (1997). Govt'l auditing. In R. T. Golembiewski & J. Rabin (Eds.), *Public budgeting and finance* (pp. 585–608). New York: Marcel-Dekker.

Thompson, F. (1997). Matching responsibilities with tactics: Administrative controls and modern government. In R. T. Golembiewski & J. Rabin (Eds.), *Public budgeting and finance* (pp. 551–574). New York: Marcel-Dekker.

Tolley, K. (1995). *Evaluating the cost-effectiveness of counseling in health care.* New York: Routledge.

Truett, L. J. (1995). *Managerial economics: Analysis, problems, and cases.* Cincinnati, OH: Southwestern Publishing Company.

Vajda, S. (1992). *Mathematical games and how to play them.* New York: Ellis Harwood.

Vazakidis, A., Karagiannis, I., & Tsialta, A. (2010). Activity-based costing in public sector. *Journal of Social Sciences, 6*(3), 376–382.

Walker, R. C. (1999). *Introduction to mathematical programming.* New York: Pearson.

Weygandt, J. J., Kimmel, P. D., & Kieso, D. E. (2015). *Managerial accounting: Tools for business decision making.* Hoboken, NJ: Wiley Global Education.

Woods, M. D. (1994). *Total quality accounting.* New York: Wiley & Sons.

Ziebeli, M. T. (1991). *Management control systems in non-profit organizations.* San Diego, CA: Harcourt Brace Jovanovich.

Zimmerman, J. L. (1997). *Accounting for decision making and control.* Chicago, IL: Irwin.

Zionts, S. (1974). *Linear and integer programming.* Englewood-Cliffs, NJ: Prentice-Hall.

Index